Ap
Advance

TRANSPORTATION ENGINEERING

Proceedings of the Second International Conference

Sponsored by the
Urban Transportation Division
and the
Highway Division
of the
American Society of Civil Engineers

and the
Federal Highway Administration
Minnesota Department of Transportation
The Center for Transportation Studies,
  Department of Civil and Mineral Engineering,
  University of Minnesota

in cooperation with
American Association of State Highway and Transportation
  Officials (AASHTO)
Council of University Transportation Center
German Roads and Transportation Association
Institute of Transportation Engineers
International Road Federation
International Symposium on Automative Technology and Automation
Operations Research Society of America
Permanent International Association of Road Congresses
Transportation Research Board

Minneapolis, Minnesota
August 18-21, 1991

Edited by Yorgos J. Stephanedes and Kumares C. Sinha

Published by the
American Society of Civil Engineers
345 East 47th Street
New York, New York 10017-2398

ABSTRACT

This proceedings, Applications of Advanced Technologies in Transportation Engineering, consists of extended abstracts of papers presented at the Second International Conference held in Minneapolis, Minnesota, August 18-21, 1991. It is concerned with the challenges and opportunities faced by the transportation engineering profession as it prepares to enter the 21st century. While these challenges cover a wide spectrum including increasing congestion, continuing problems with safety and environmental degradation, and acute budget constraints, technical innovations offer opportunities to address these problems. Major advances in the fields of automation, information systems, and telecommunications have the potential of opening new horizons in transportation engineering by developing new products and procedures while making substantial improvements in cost savings and productivity. The papers focus on such topics as: 1) automatic vehicle monitoring, 2) advanced traveler information systems, 3) image processing and vehicle detection, 4) real-time traffic control, 5) advanced rail systems, and 6) automated incident detection. In addition, these papers present specific applications of these different technologies and national and state programs with particularly important results, including SHRP innovation projects.

Library of Congress Cataloging-in-Publication Data

Applications of advanced technologies in transportation engineering: proceedings of the second international conference, Minneapolis, Minnesota, August 18-21, 1991/edited by Yorgos J. Stephanedes and Kumares C. Sinha.
    p.   cm.
    "Sponsored by the Urban Transportation Division and the Highway Division of the American Society of Civil Engineers . . . [et al.]"
    Includes bibliographical references and index.
    ISBN 0-87262-818-3
    1. Transportation engineering—Congresses. I. Stephanedes, Yorgos J. II. Sinha, Kumares C. (Kumares Chandra) III. American Society of Civil Engineers.
TA1005.A67   1991
629.04—dc20   91-24882
                                                                       CIP

The Society is not responsible for any statements made or opinions expressed in its publications.

Authorization to photocopy material for internal or personal use under circumstances not falling within the fair use provisions of the Copyright Act is granted by ASCE to libraries and other users registered with the Copyright Clearance Center (CCC) Transactional Reporting Service, provided that the base fee of $1.00 per article plus $.15 per page is paid directly to CCC, 27 Congress Street, Salem, MA 01970. The identification for ASCE Books is 0-87262/91. $1 + .15. Requests for special permission or bulk copying should be addressed to Reprints/Permissions Department.

Copyright © 1991 by the American Society of Civil Engineers,
All Rights Reserved.
Library of Congress Catalog Card No: 91-24882
ISBN 0-87262-818-3
Manufactured in the United States of America.

PREFACE

The challenges and opportunities faced by the transportation engineering profession, as it prepares to enter the 21st century, are unique. These challenges cover a wide spectrum including increasing congestion, continuing problems with safety and environmental degradation, and ever more acute budget constraints. However, there are also opportunities offered by the timely application of technical innovations to address these challenges. Major advances in the fields of automation, information systems and telecommunications have the potential of opening new horizons in transportation engineering by developing new products and procedures while making substantial improvements in cost savings and productivity. Recognition of the importance of innovation and new technologies for the advancement of transportation motivated the organization of this conference. The success of the first conference held in San Diego, California in February, 1989 and the need to follow up on the rapid changes in the field, both in theory and in practice, led to the Second International Conference on Applications of Advanced Technologies in Transportation Engineering.

Since the first conference, the U.S. Department of Transportation and a number of state transportation agencies have begun to focus increasingly on what society can gain from the advances in real-time management and information tools, and in controlling the movement of vehicles for increased speed under improved safety. In March 1991 a number of transportation professionals converged in Washington, D.C. to establish the Intelligent Vehicle Highway Society (IVHS) of America that would address the application of these advancements. Individual state programs, such as the California PATH program and the Minnesota Guidestar program, were established to achieve similar objectives at the state level. In Europe, DRIVE is ready to enter its second phase, and Japan is proceeding with a similar coordinated effort. The present conference provides a forum to exchange the experiences of these programs so far. Over 150 transportation professionals are participating in the conference, with contributions of both theoretical and practical nature. While the conference is limited to surface transportation applications, it focuses on a number of major areas:

- Automatic Vehicle Monitoring, Detection and Location
- Advanced Traffic Management Systems
- Advanced Traveler Information Systems
- Automated Vehicle and the Highway
- Image Processing
- Communications and Standards
- Human Factors and User Interfaces
- Policy Issues
- Dynamic Guidance and Real-Time Traffic Control
- Advanced Rail Systems
- Automatic Incident Detection
- Robotics
- Neural Network Applications

The conference covers the above topics in the form of regular and invited sessions, panel discussions and tutorials. In particular, the inclusion of tutorials continues the direction started by the first conference, where tutorials succeeded in allowing a large number of professionals to become acquainted with a field not well covered by traditional academic curricula.

While the conference sessions have focused individually on the topics listed, several sessions touch on more than one area as these are interrelated. For instance, the sessions on image processing cover applications in detection and highway maintenance. Similarly, the sessions on incident detection cover both hardware and software. Other sessions cover topics such as improvements in theoretical models or safety applications that cut across areas. In addition, the conference covers specific applications and national and state programs with particularly important results, including SHRP innovative projects.

This volume has been assembled so that participants can have a ready reference to the discussions at the conference. The papers appearing here are extended abstracts summarizing the work and ideas presented at the conference. While abstracts were peer reviewed, the papers appearing in this volume have not been reviewed by the Conference Committee and have been reproduced from camera ready copy submitted by the authors. However, all authors and presenters are invited to submit complete papers for possible publication in forthcoming special issues of the *ASCE Journal of Transportation Engineering*. Discussions of papers or topics considered by the conference can also be submitted to the *Journal*.

Numerous individuals and groups contributed to the success of the conference. The conference has been organized as part of the activities of the ASCE Urban Transportation Division's Committee on Advanced Technology and the ASCE Highway Division Committee on Planning and Economics. The financial sponsors of the conference are ASCE, the Federal Highway Administration, the Minnesota Department of Transportation, and the Center for Transportation Studies of the Department of Civil & Mineral Engineering of the University of Minnesota. Other participating organizations and members of the Conference Committee are listed in the Acknowledgements. We especially appreciate the support provided by Gene Ofstead of the Minnesota Department of Transportation. Special thanks are due to the Conference Steering Committee, Conference Planning Committee, and the Local Conference Committee.

Yorgos J. Stephanedes, PE
Minneapolis, Minnesota

Kumares C. Sinha
West Lafayette, Indiana

May, 1991

CONTENTS

M1. AUTOMATIC VEHICLE MONITORING

Measuring Section-Related Traffic Data by Correlation Methods—A New Approach to Traffic Monitoring and Control
   E. Pfannerstill 1
Inductive Loop Detection/Testing with DSP Techniques
   Boris Donskoy 6
Automatic Vehicle Classification System
   Wen-Min Pan, Geok K. Kuah, Shi-Lin Su and An-Sheng Wang 11
Analysis of Automatic Vehicle Identification Technology and Its Potential Application on Florida's Turnpike
   Edward A. Mierzejewski, Michael C. Pietrzyk and William L. Ball 16

M2. ADVANCED INFORMATION SYSTEMS I

Analysis of Roadside Equipment and Central Facilities for Innovative Driver Information and Traffic Control Systems
   M. Cremer 21
Traffic Reporter: A Real-Time Commuter Information System
   Mark P. Haselkorn, W. Barfield, J. Spyridakis, B. Goble and M. Garner 26
SINOD—A Location Reference Tool for Road Information Systems
   Antonio Lemond de Macedo 31

M3. TRANSPORTATION MODELS I

Continuum Modeling of Traffic Dynamics
   Panos G. Michalopoulos, Ping Yi and Dimitrios E. Beskos 36
Application of Fuzzy Set Theory to the Analysis of Transportation Problems
   Shinya Kikuchi 41
Operation of Freeway Simulation Tools on I-35W
   James L. Wright, James Aswegan, Ping Yi and Gene Hicks 46
Heavy Traffic Control—Measurement Techniques and Control Strategies
   Reinhart D. Kuhne 51

M6. IMAGING APPLICATIONS FOR HIGHWAY MAINTENANCE I

Assessment of the State-of-the-Art of Robotics Applications in Highway Construction and Maintenance
   Tong Zhou and Thomas West 56
Applications of Robotics and Automation in Highway Maintenance Operations
   Bahram Ravani and Thomas West 61
Perception and Control for Automated Pavement Crack Sealing
   Chris Hendrickson, S. McNeil, D. Bullock, C. Haas, D. Peters, D. Grove, K. Kenneally and S. Wichman 66

Highway Pavement Surfaces Reconstruction by Moire Interferometry
Sidney A. Guralnick, Eric S. Suen and Christian Smith 71
Design Considerations for Automated Pavement Crack Sealing Machinery
Steven A. Velinsky and Kenneth R. Kirschke 76

M7. REAL-TIME URBAN TRAFFIC CONTROL I

Comparison of Traffic Signal Systems in Australia and North America
K.J. Fehon 81
Implementation of a Centralized Traffic Signalization Control System with
Enhanced Capabilities
A. Ceder and A. Shmilovits .... *
Fully-Distributed Control of Signal Networks
George List, Siew Leong and Saaju Paulose 86
Cars: A Demand-Responsive Traffic Control System
J. Barcelo, R. Grau, P. Egea and S. Benedito 91

M9. VEHICLE DETECTION AND IMAGE PROCESSING I

Testing and Feasibility of VIPS for Traffic Detection
Alypios E. Chatziioanou, Stephen L.M. Hockaday, Carl A. McCarley
and Edward C. Sullivan 96
Incident Detection Through Video Image Processing
Panos G. Michalopoulos 101
Traffic Queue Detection Using Image Processing
A. Rourke and M.G.H. Bell 106
Report of the DEVLONICS Video Based Traffic Detector System (DRS-CCATS/DTMSA)
Frans Lemaire and Marc Coussement 111
A Commercial Machine Vision System for Traffic Analysis
Sal D'Agostino .... *

M10. TRAFFIC MANAGEMENT I

ATMS—What Can It Accomplish
Alberto J. Santiago, Hobih Chen and Ammar Kanaan .... *
Advanced Technology Traffic Redirection and Access Control System
Dominick J. Gatto .... *
An Application of Advanced Technologies for Freeway Traffic Management—An Indiana
Case Study
Michael J. Cassidy, Kumares C. Sinha 116
Evaluation of Reliability of Road Network for Better Performance, Advanced
Management and Future Network Design
Hiroshi Wakabayashi and Yasunori Iida 121

M12. PANEL ON POLICY ISSUES

Public Agency Issues in the Implementation of IVHS
S. Edwin Rowe 126
Institutional Considerations in the Evolutionary Development of IVHS
John J. Fearnsides .... *
National IVHS Legislation: What It Can and Can't Do
Thomas A. Horan .... *

---

*Manuscript not available at time of printing.

Technological Determinism vs. Consumer Preferences: The IVHS
Implementation Dilemma
  D. Sperling .... *

T2. VEHICLE DETECTION AND IMAGE PROCESSING II

Stereo Vision Onboard a Vehicle for Obstacle Detection
  Jean-Luc Bruyelle and Jack-Gérard Postaire 131
Automatic Object Identification in Road Traffic
  Martin Lipičnik, Danijel Rebolj and Tomaž Tollazzi 136
Stereo Vision with Scan-Line Cameras for Intrusion Detection on L.R.T. Tracks
  Luc Duvieubourg, J.-P. Deparis and J.-G Postaire 142

T3. APPLICATIONS OF OPERATIONS RESEARCH METHODS

The Heavy Vehicle Electronic License Plate Program and Crescent
Demonstration Project
  C. Michael Walton 147
Using Real-Time Location Information for Hazardous Materials Shipments
  Mark A. Turnquist 152
An Analytical Framework for Minimizing Freeway Incident Response Time
  Kostas G. Zografos and Teti Nathanail 157
Cost Savings from AVMC Systems in Trucking
  Susan F. Hallowell, Edward K. Morlok and Lazar N. Spasovic 162
A Fleet Management System for Road Transportation
  P.R. Schrijver and H.G. Sol 167

T4. TUTORIAL ON ADVANCED TRAVELER INFORMATION SYSTEMS

Advanced Traveler Information Systems *The Vision Beyond*
  H. Milton Heywood 468
TravTek System in Orlando as a Case Study
  James Rillings .... *

T5. AUTOMATIC GUIDANCE AND TRAFFIC CONTROL

Optimal Ramp-Metering Control for Freeway Corridors
  Yorgos J. Stephanedes and Kai-Kuo Chang 172
Application of Control Theory to Regulate Traffic Flow on Periurban Ringways
  Neila Bhouri and Markos Papageorgiou 177
A Dynamic Modelling Framework of Real-Time Guidance Systems in General
Urban Traffic Networks
  R. Jayakrishnan and Hani S. Mahmassani 182
Routing Based on Anticipated Travel Times
  L. Rilett and M. Van Aerde 183

T6. ADVANCED INFORMATION SYSTEMS II

Geographic Database for IVHS Management
  T.A. Yang, S. Shekhar and P.A. Hancock 188
New Tasks for New Technique in Car Parking
  Ian C. Hilton .... *

---

*Manuscript not available at time of printing.

Information Technology for Urban Transport Modelling
Boris L. Shmulyian, Andrei V. Fedotov, Leonid M. Heifits, Shamil S. Imelbayev and Galina M. Kirsanova .... *

T7. ADVANCED RAIL SYSTEMS I

Maglev Technology: A Look at Guideway Construction and Maintenance Concerns
R. Scott Phelan and Joseph M. Sussman 193
Two Approaches for Characterizing Images of Rail Surface Flaws
Roemer Alfelor, Thomas Short, Behnam Motazed and Sue McNeil 198
Development of a Support System for Inspection and Maintenance of Railway Concrete Bridges
H. Osada, S. Okuaki, A. Matsuoka, Y. Shinoe, S. Tottori and K. Yamazumi .... *
Computer Simulation and Disposition in Railway
Leon Kos and Werner Kretschmer 203
Rail Replacement Planning: A Knowledge System Approach
Rabi G. Mishalani and Carl D. Martland 208

T8. SHRP-IDEA

Rapid Air Void Measurement in Fresh Concrete Using Fiber Optics Technology
Farhad Ansari .... *
Non Contact Laser Inspection of Pavement Structures
Ron L. Smith .... *
Portable Corrosion Rate Monitor Using Ultra-Low Frequency
Digby McDonald .... *
Automated Process Controller for Asphalt Using NMR Technology
Robert Pearson .... *
Smart Structure Technology for NDE of Concrete and Asphalt
Robert E. Shannon .... *
Intelligent Weather Prediction Systems for Localized Snow Hazard Prediction
Elmar R. Reiter .... *

T9. AUTOMATED VEHICLE AND HIGHWAY I

Potential Freeway Capacity Effects of Automatic Vehicle Control Systems
Steven E. Shladover 213
A Preliminary Systems-Level Evaluation of Automated Urban Freeways
Robert A. Johnston and Dorriah L. Page 218
Institutional Barriers to IVHS Introduction
Michael L. Patten and John M. Mason, Jr. 223

T10. IMAGING APPLICATIONS FOR HIGHWAY MAINTENANCE II

CADD—Database Application for Facility Inspections
Axel J. Pollak and Rosalind Pierce-Spring 228
Automated Analysis of Pavement Distress Data
H.N. Koutsopoulos, Rabi G. Mishalani and Allen B. Downey 233
Priority Rating of Highway Maintenance Needs Using Neural Network Models
T.F. Fwa and W.T. Chan .... *
Pavement Image Processing Using Neural Networks
Mohamed S. Kaseko and Stephen G. Ritchie 238

---

*Manuscript not available at time of printing.

T11. REAL-TIME URBAN TRAFFIC CONTROL II

Design of HCM Signal Timings Using Signal Expert
   J.A. Stewart and M. Van Aerde 243
Design and Evaluation of Multi-Band Progression Schemes
   Nathan H. Gartner, Susan F. Assmann, Fernando Lasaga and Dennis L. Hou 248
An Intelligentual Real-Time Traffic Control System Suited to Chinese Cities
   He Guoguang, Lu Baichuan and Liu Bao 252
An Application of Expert Systems to Traffic Signal Control
   S. Manzur Elahi, A. Essam Radwan and K. Michael Goul 258
Bus Pre-Emption: A Real Time Control Strategy for Privatized Transit Operation
   Snehamay Khasnabis and Bharat B. Chaudhry 263

T14. COMMUNICATIONS AND STANDARDS

Advanced Software Design and Standards for Traffic Control
   Darcy Bullock and Chris Hendrickson 268
Error Rate Measurements of RDS-FM Transmissions—Application to an RDS-Beacon
   M. Heddebaut, M. Berbineau and M. Szelag 273
Fiber Optic Communication Systems for Freeway Traffic Management
   D. Bowen Tritter 278

T15. TRAFFIC MANAGEMENT II

Traffic Signals and At-Grade LRT
   K.J. Fehon and W.A. Tighe 283
Traffic Control System on the Hanshin Expressway
   Tsuyoshi Yoshino, Takeshi Matsuo and Toshiharu Hasegawa 288
Computer Aided Engineering Analysis for Transportation Links & Management Options
   A.K. Gupta and P.V.S.S. Ravi Prasad 293
Simulation of Transport Management Systems
   A.P. Artynov, G.A. Kondratjev and A.I. Vasilchenko .... *

T16. ADVANCED SAFETY APPLICATIONS

Safety of IVHS: Methods for Determination of Accident Levels for AVCS-1 Devices
   A. Hitchcock 297
Design for IVHS Safety: Technology Assessment
   Wei-Bin Zhang 302
Taming the Silicon Steed—Perceptions of Risk
   John C. Keller and Paul P. Jovanis 307
Variable Message Signs as Part of a Roadway Communications System
   Jonathan Upchurch .... *
Exploring Headsup Displays for Driver Workload Management in Intelligent Vehicle Highway Systems
   M. Coyle, S. Meir, S. Shekhar, A. Yang, J. Caird, P. Hancock and S. Johnson 312

T17. PANEL ON DETECTION TECHNOLOGY

Productization and Deployment of the AUTOSCOPE Video Detection System
   William Russell .... *
Recent Developments in Vehicle Detection Technology
   George Palm .... *

---

*Manuscript not available at time of printing.

Infrared Detector Developments
Michael A.G. Clark and Andrew Hodge 317
The California Inductive Loop Radio Demonstration Project
Sam Taff, Walt Winter, Clint Staley and Ronald Nodder 322

W1. DYNAMIC ROUTE GUIDANCE

The Contribution of Dynamic Route Guidance and Information Systems to More Efficient Traffic Management
Jurg M. Sparmann 327
Individual Traffic Information and Route Guidance from RDS-TMC until Beacon Communication
Wolf Zechnall 332
Dynamic Route Guidance—The "DRIVE" Project Car-Goes
J.D. Turner 337
The Integrated Dynamic Traffic Management Project: Operation ULIISSE of Lyon
M. Lourd .... *
WEGWIJS-AMSTERDAM—A dual mode routeguidance plan
J.J. Klijnhout 343
Cooperative Traffic Management—A Solution to Improve Traffic Flow by Introducing Dynamic Route Guidance and Drivers Information
Heinz Sodeikat .... *

W2. VEHICLE DETECTION AND IMAGE PROCESSING III

Video Image Processing for Toll Operation Evaluation
B. Daviet, J.M. Morin, J.M. Blosseville and V. Motyka 353
The Use of MEIS Imagery for Transportation Engineering Applications
James G. Linders .... *
An Adaptive Video Image Analysis System for Transport Applications
K.W. Dickinson and C.L. Wan .... *

W3. ADVANCED TECHNOLOGIES IN STATE IVHS PROGRAMS I

A Review of European Developments in Route Guidance and Navigation Systems
Ian Catling 358
Assessment of IVHS Research Efforts in Japan and Future Directions
Masahiko Katakura and Mitsuru Saito 363
Car Wars—The DOTs Strike Back
Jerry L. Hautamaki and Katharine S. O'Hara 368

W5. AUTOMATIC INCIDENT DETECTION ALGORITHMS

A Catastrophe Theory Approach to Freeway Incident Detection
Lisa Aultman-Hall, Fred L. Hall, Yong Shi and Bradley Lyall 373
A Low Pass Filter for Incident Detection
Yorgos J. Stephanedes and Athanasios P. Chassiakos 378
Automatic Incident Detection Using Image Processing Techniques: A Specific System Used in INVAID
J.M. Blosseville, V. Motyka, N. Djemame and F. Lenior 383
A Distributed Real Time Knowledge-Based System Using Video Image Processing for Junctions Automatic Incident Detection
S. Sellam, A. Boulmakoul and J.C. Pierrelee 388

---

*Manuscript not available at time of printing.

W6. TRANSPORTATION MODELS II

A Real-Time Traffic Diversion Model: A Conceptual Approach
  A.G. Hobeika and Y. Zhang 393
A New Approach for Real-Time Prediction of Traffic Demand-Diversion in
Freeway Corridors
  Eil Kwon 398
Dynamic Stochastic Equilibrium Model in Multiple Vehicle Type
Transportation Networks
  Shogo Kawakami and Zhimin Xu 403
Dynamic Estimation of Freeway Demand Patterns and a Stochastic Programming
Approach to Freeway Ramp Metering
  Gary A. Davis 408
Effectiveness of Real Time Information Strategies in Situations of
Non-Recurrent Congestion
  Srinivas Peeta, Hani S. Mahmassani, Richard Rothery and Robert Herman 409

W7. ADVANCED RAIL SYSTEMS II

Contact-free IC Cards for New Railway Ticket Systems
  Shigeo Miki, Noboru Nagai, Hiroshi Matsubara, Hideaki Yoshino and Koichi Goto ... 414
Real-Time Technology for Operating Control of Railway Transport
  V.I. Gritsenko, V.A. Bogemskij, A.P. Lapa and S.L. Trukhin .... *
Fault-Tolerant Computing Architecture for MF 88 and MP 89 Rolling Stock
  H. Bordenave 419
Experiments with Formal Specifications on MAGGALY
  P. Dauchy and P. Ozello 423
The Safety Microprocessor-Based Architectures used in Dedicated Guided-Way Land
Transportation Means
  E. El. Koursi and A. Stuparu 428

W8. AUTOMATED VEHICLE AND HIGHWAY II

Simulation of a Vehicle Platoon Control System for Automatic Highway Using the
Fuzzy Control Concept
  S.J. Liu and A.A. Frank 433
Vehicle Lateral Guidance Using a DSP Based Vision System
  Mahlon Heller, Robert Trahms and Sompol Chatusripitak 438
Coupled Vehicles: A Safe and Efficient Approach to Roadway Automation
  Farhad Bolourchi and Paul P. Craig .... *
Observation and Modeling of Mobile Environment
  Hannu Hakala, Jari Kaikkonen and Pentti Mattila 443

W9. IMAGING APPLICATIONS FOR HIGHWAY MAINTENANCE III

Measuring Highway Inventory Features Using Stereoscopic Imaging System
  Hosin Lee, Michael A. Weissman and Joseph P. Powell 448
Flexible Pavement Distress Evaluation Using Image Analysis
  Lan Li, Paul Chan and Robert L. Lytton 473
Subsurface Pavement Structure Inventory Using Ground Penetrating Radar and a
Bore Hole Camera
  M. Inagaki, H. Tada, A. Kasahara, H. Tomita and T. McGregor 453

---

*Manuscript not available at time of printing.

W10. ADVANCED TECHNOLOGIES IN STATE IVHS PROGRAM II

Minnesota Department of Transportation Experience in the Application of Advanced Traffic Management Systems
    Richard A. Stehr 458
Academic Research Efforts in the Guidestar IVHS Program
    Robert C. Johns 463
The Michigan Metropolitan Transportation Center and Operational Field Test Program for IVHS
    Robert E. Maki .... *

Subject Index 479

Author Index 483

---

*Manuscript not available at time of printing.

Measuring section-related traffic data by correlation methods - a new approach to traffic monitoring and control

Elmar Pfannerstill[1]

Abstract

Traffic monitoring and control systems require reliable realtime data on traffic flow for an efficient performance. A new 'class' of data, the so-called section-related data such as journey speed and traffic density can be obtained automatically by application of pattern recoginition principles and correlation methods to traffic engineering. The so-called MAVE-system incorporates this approach: it analyses traffic condition on the basis of section-related and local data, delivers an 'electronic road map' for authorities and drivers and derives strategies for traffic control.

Introduction

Due to economic and ecological restraints there is an increasing demand for traffic monitoring and control systems in order to operate road traffic within the existing road network more efficiently.

All such systems for the improvement of traffic flow, including automatic incident detection, route guidance systems, traffic information broadcasting, etc. require reliable and real-time information on the actual traffic condition in each section of a road network.

Measuring section-related traffic data

Present traffic data acquisiton systems are usually based on the collection of local data, such as passing vehicles' speed or volume, which are picked up at certain points along the road. Using these local data it is only possible to estimate the traffic conditions in the road section between the measuring points.

---

[1] Dr.-Ing. Elmar Pfannerstill, ave Verkehrs- und Informationstechnik GmbH, Jülicher Str. 336, D-5100 Aachen, Germany

A high density of such measuring points is usually required for sensitive and fast-reacting traffic monitoring systems (e.g. automatic incident detection and tail-back warning).

A more reliable basis for the above-mentioned objectives can be obtained by measuring section-related data such as journey time and traffic density, thus allowing a deeper insight into the actual traffic conditions (Fig. 1). For this purpose it is necessary to identify vehicles entering and leaving the observed road section, i.e. they have to be 're-identified' at the end of each section.

The MAVE-system is capable of performing this: it uses vehicle patterns which are not unique for each individual vehicle, but are significant to such an extent that groups of vehicles may be identified at the end of a road section in order to calculate the above-mentioned section-related data. This is achieved with a relatively low number of data acquisition points in comparision with common systems based on local data.

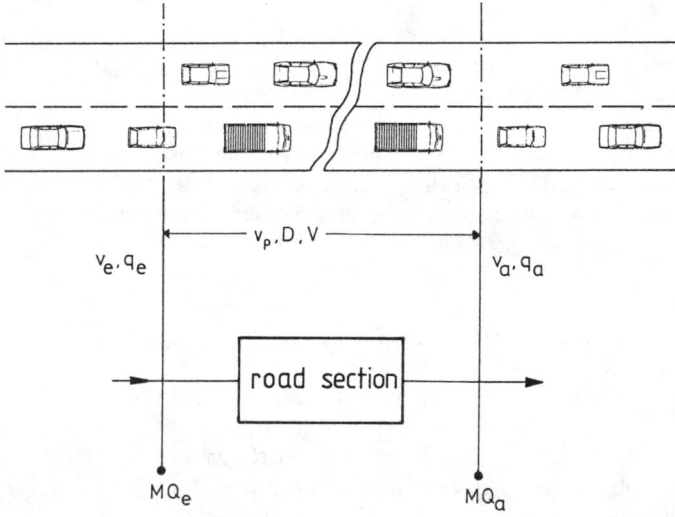

Fig. 1: Road section with traffic flow described by:
   local data: v_e: speed / entry q_e: volume / entry
                 v_a: speed / exit q_a: volume / exit

   section-related data: v_p : journey-speed
                           D : traffic density
                           V : dissipation (overtaking)

Vehicle recognition and correlation

Vehicle-characteristic patterns are obtained by the use of the inductive loop detector (Other vehicle detectors may be possible, if they deliver sufficient vehicle-characteristic information). The impedance of a loop of several windings of wire, which are buried in the road surface is changed by passing vehicles. A detector transforms these changes into a voltage signal u(t), which is characteristic of the undercarriage of each car (Figure 2).

The detector signatures u(t) are normalized in order to eliminate non-vehicle characteristic influences (e.g. actual speed or detector sensitivity). The samples of such a normalized, digitized vehicle signature are treated as components of a pattern vector $\vec{x}$.

Information compression methods are used to extract a few vehicle characteristic features from this pattern vector $\vec{x}$ and to transform it into pattern vector $\vec{x}_T$. Pattern vectors $\vec{x}_T$ representing vehicles from the entry of the road section are compared with vectors $\vec{x}_T'$, representing vehicles from its exit. Vehicles are regarded as identical, if a certain level of similarity is reached /3/, /4/.
Within a platoon of the length L the number of 'identical' vehicles is counted. This leads to a correlation function $\varphi(m)$ /1/, which is smoothed by an adaptive lowpass filter function in order to limit the effects of dissipation by overtaking, recognition errors, etc. Finally, a smoothed correlation function $\phi(m,t)$ is obtained /5/.

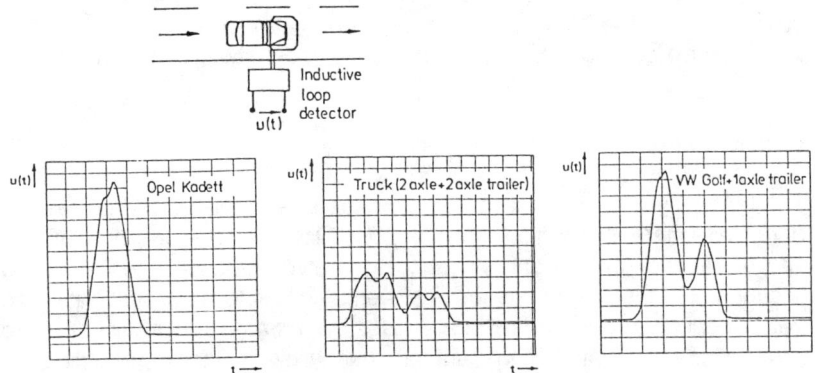

Fig. 2: Vehicle patterns from the inductive loop detector

At a certain time t a platoon of L vehicles is taken out of a vehicle sequence at the beginning of a road section and the correlation function $\varphi(m)$ is calculated by successively comparing it with the vehicle sequence at the end of the specific section. This is done at successive time intervals Δt, and thus finally a sequence of correlation functions $\varphi(m,t)$ (as shown in Figure 3) is obtained. The maximum of each correlation function $\varphi(m,t)$ indicates the 're-identification' of each platoon L. From the instant of the re-identification the journey time T for the platoon can be obtained and hence its mean speed v_p for passing the road section can be determined. The actual traffic density D can easily be calculated from the number of vehicles that have entered the section after the platoon which has just been re-identified. These data can be differentiated for trucks/passenger cars or for different lanes, which leads to a very accurate picture of the actual traffic condition /2/.

Results

The above-mentioned principles are incorporated in the MAVE system (Modular system for traffic data acquisiton) which is successfully used on various motorway sections of Germany and abroad.

The example of Figure 3 shows vehicle correlation functions and the obtained journey speed and traffic densities measured on a 2.85 km section of motorway A4 (Cologne-Aachen) on a Friday afternoon.
In this way the traffic data for each road section is collected and transmitted to the monitoring centre. Here it is used for classification of the actual traffic condition ('level of service') within each road section and, finally, within the whole network.
Based on this information variable message signs or other traffic control strategies may be activated, such as the distribution of "electronic road maps" to police and other authorities or to radio stations and drivers.

Conclusion

The MAVE system which is based on pattern recognition and correlation principles offers for the first time the possibility of gathering automatically section-related traffic data describing traffic situations **within** road sections. It is therefore superior to systems collecting only local data. The traffic monitoring and control center uses this as a reliable basis for traffic information and traffic management, thus allowing increased efficiency and safety of road networks.

References

/1/ Böhnke, P.: "Beitrag zu einer Systemtheorie von Objektfunktionen." Dissertation RWTH Aachen, 1980. RMI-Verlag, Aachen, ISBN 3-921924-02-2.

/2/ Böhnke, P.; Pfannerstill, E.: 'A System for the Automatic Surveillance of Traffic Situations.' ITE Journal, Vol. 56 (1986), Washington.

/3/ Pfannerstill, E.: "Ein Verfahren zur Merkmalsextraktion aus Linienmustern zur Zuordnung von Fahrzeugkollektiven." Dissertation RWTH Aachen, 1983. RMI-Verlag, Aachen, ISBN 3-921924-04-9.

/4/ Pfannerstill, E.: 'A Pattern Recognition System for the Re-identification of Motor Vehicles.' Proceedings 7th International Conference on Pattern Recognition, Montreal, July 30-August 2, 1984, IEEE Computer Society, Los Angeles, ISBN 0-8186-0545-6.

/5/ Ziegler, R.: "Die Strecken-Systemfunktion zur Beschreibung des Verkehrsflusses auf Schnellstraßen". Diss. T.H. Aachen, 1985.

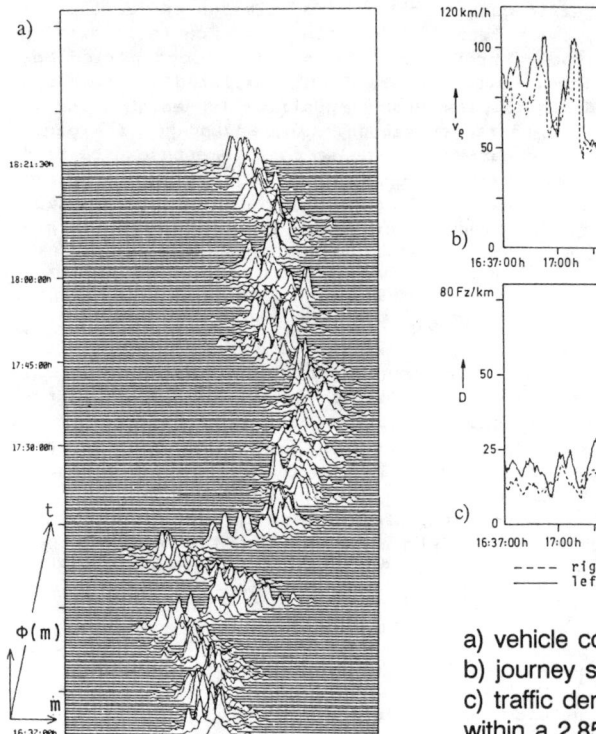

a) vehicle correlation functions
b) journey speed
c) traffic density
within a 2.85 km road section

Fig. 3: Traffic condition on motorway A4 Cologne-Aachen on a Friday afternoon

INDUCTIVE LOOP DETECTION/TESTING WITH DSP TECHNIQUES

Boris Donskoy, Mark Price, DVP, Inc. [1]

Abstract

Vehicle traffic control is recognized as an effective way to relieve mounting traffic congestion in large urban areas across the country. The most common method of vehicle detection is in sensing a change in inductance through the use of wire loops buried beneath the road surface. DVP, Inc. has just completed the development of an advanced digital instrument specifically designed for IL detection, testing and characterization. The advantage of a digital instrument is that a series of precise measurements can be made with a single data acquisition and processing system. Digital technology enables implementation of clearer display formats, storage and hardcopy of measurement data, remote control interfaces, etc.

Background

IL sensing is a simple but effective technology that has been used in roadways for vehicle surveillance and traffic control systems since the late 1960's. Inductive loops used in the transportation industry are coils of wires that are embedded in the roadway to detect vehicles passing above. Usually, the coils are the part of an oscillator circuit in the inductive loop detector contained in the traffic control unit. The size of the loop, the number of turns, and the length of the lead-in combine with the detector characteristics to determine the oscillator frequency (typically in the range from 20 to 200 kHz). The IL energizes the loop and provides oscillator gain. The loop network consists of three major components:

       o The buried inductive wire loop
       o The lead-in cable
       o The detector amplifier's input capacitor

Mr. B. Donskoy is a Member of Technical Staff; Mr. Price is the President of DVP, Inc.; 2401 Research Blvd. Suite 200 Rockville, MD 20850. (301-670-9282).

The network's inductance (L) is primarily found in the loop and the lead-in cable (which can contribute approximately 200 μH for long runs). The capacitance of the network is primarily found in the IL input capacitor and in the lead-in cable (when long runs are required). The alternating current flowing through the loop generates an electromagnetic field around the loop. Any conductor entering the field (such as the frame and sheet metal of a car) will absorb energy from the electromagnetic field thereby decreasing the inductance of the loop and raising the resonant frequency of the network.

Modern loop systems are dependable and remain trouble-free after years of service. Nevertheless, over a period of time, failures of inductive loops can be caused by a variety of problems. The loop windings may become shorted from worn or stressed insulation or may break or short to ground. Connectors to the loop may corrode, or the loop wire itself may deteriorate or break. The lead-in wires may become damaged through corrosion. Any number of failures may eventually occur through the hostile environment, road modifications, or deterioration over time. These failures, though relatively infrequent, will eventually cause gradual degradation of the loop's performance and its eventual failure. To properly service a faulty loop and perform initial installation now requires several pieces of test equipment.

Instrumentation is an area that sometimes lags behind many other technological advances. Conventional analog measurements usually require calibration and adjustment to produce accurate and consistent results and usually are targeted at a narrow set of specific parameters and do not allow change of testing frequencies. The latter is important because at frequencies above 20 kHz the effective loop inductance becomes more and more frequency dependent. For this reason it is important to measure the loop parameters (L, Q, r) at the loop's operating frequency (which requires measuring the loop frequency when connected to the detector electronics, then disconnecting the loop to directly measure the inductance at the measured frequency).

DVP's instrument was designed to be a fully automated, hand-held, IL detector/tester. Our instrument performs IL measurements at any frequency within 5 -88 kHz range, allows direct measurement of loop sensitivity, and with proper cabling, it performs instantaneous in/out substitution of existing detector units. This feature enables effective troubleshooting of IL/detector systems in real life conditions. This paper presents details on the theory of operation, design, trade-offs, applications, and use of this instrument.

Theory of Operation

The DVP IL tester uses digital signal processing to make a broad range of measurements such as inductance, internal resistance and Q-factor of inductive loops. All-digital sinewave generation and analysis techniques ensure a high degree of accuracy, stability, and overall capability of the instrument. The advantage of using digital techniques are flexibility in measurement, consistency in results, and direct programmability of higher mathematical functions.

Inductance, resistance and quality factor of inductive loops are the primary parameters that are important in the operation of the IL detectors. For most cases, an inductive loop can be modeled with a simple impedance Z = r + jWL

r = series (dc) resistance of the loop (Ω)
L = Inductance of the loop (Henries)
W = 2*π*f
f = frequency

The Q of the inductor is a measure of its quality, and is defined by: Q = W*L/r

Any inductor produces a phase shift between the sinewave voltage driving it and the current flowing through it. The amount of phase shift is proportional to the "L" and "r" parameters of the coil and the frequency of the driving source. A practical simplification of this idea can be achieved by inserting a known impedance (or resistance) in series with the loop (Fig. 1).

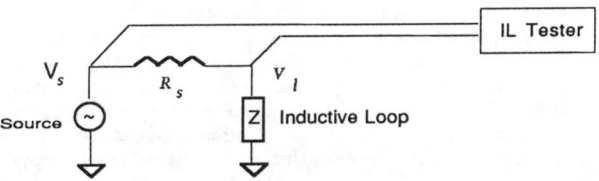

Fig. 1   IL Tester Measurement Circuit

To measure the current passing through the loop, the voltage across a known source resistance (Rs) is measured. The instantaneous current is:

I(t) = Vr(t)/Rs = (Vs(t)-Vl(t))/Rs   [1]

where   Vr(t) = voltage drop across resistor Rs
           Vs(t) = source voltage
           Vl(t) = voltage across load (DUT)
           Rs   = known source resistance

By computing the magnitude and phase of Vs(t) and Vl(t) for a sinusoidal voltage, the complex impedance of the IL can be computed by:

Z = [Rs*(cos(θ)-Vl/Vs)/(Vs/Vl+Vl/Vs-2*cos(θ))]
    + j [Rs*sin(θ)/(Vs/Vl + Vl/Vs - 2*cos(θ)]   [2]

where θ = phase angle difference between Vs(t) and Vl(t)
          Vl = magnitude of Vl(t)
          Vs = magnitude of Vs(t)

Assuming the IL is purely resistive, the phase angle θ is zero and Equation 2 reduces to:

$$Z = Rs*Vl/(Vs - Vl) \quad\quad\quad [3]$$

Now we have to measure and compare two voltage vectors; a voltage vector applied to the loop, and a voltage vector of the internal generator source. The magnitude ratio of the vectors and a phase angle between them are used to calculate "r" and "L" of the loop and from them the Q of the loop.

$$L = (Rs/W)*\sin(\theta)/(Vs/Vl+Vl/Vs-2*\cos(\theta)) \quad\quad [4]$$

$$r = Rs*(\cos(\theta)-Vl/Vs)/(Vs/Vl+Vl/Vs-2*\cos(\theta)) \quad [5]$$

Implementation

The IL test instrument combines a precision vector (phase and magnitude) meter with a frequency synthesizer. To take measurements in a non-energized loop, an interactively programmable frequency is generated by a high-speed digital signal processor (TMS320C25) and converted to analog form by a D/A converter. This signal is injected into the IL (see Fig. 2).

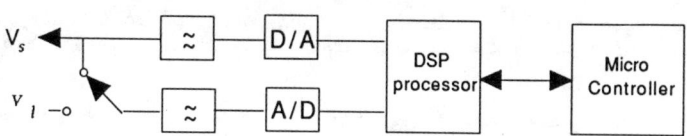

Fig. 2   IL Tester Block Diagram.

The source and load voltages are filtered, sampled, and converted to the digital form by an A/D converter. The DSP microprocessor performs all time critical signal processing. Post-processing is done on a general-purpose microprocessor (V40), which also controls the LCD display and keyboard. All components are CMOS, allowing for low-powered battery operation over a broad temperature range. Using the microcontroller, sophisticated functions such as context-sensitive help, intelligent results interpretation, and tailored displays are possible. A small power converter/charging unit permits operation over an extended voltage range from a variety of input sources.

To accommodate the broad range of L, and r values, the test instrument can autorange the source impedance Rs. The unique feature of this instrument is that it can measure frequency and voltage of the multichannel (pulsed) IL detectors by quick "locking" on the detector signal during "on" periods—something that no existing IL tester can do.

Instrument's Performance and Use

The DVP, Inc. test instrument can be used to quickly diagnose faulty inductive loop/detector (IL) systems, qualify initial loop installations, and help with loop preventive maintenance. It is a high reliability, hand-held device that operates either from an external DC source or from an internal rechargeable battery (up to 10 hours of continuous operation).

The following parameters can be measured by the IL tester:

1. Inductance measurement (0.05 to 15 mH)
2. Loop resistance (0.5 - 5000 ohm)
3. Quality factor (1 -75)
4. Capacitance (2 -500 nF)
5. Loop sensitivity (deltaL/L) (0.015% min) - change of L with and without vehicle present.
6. IL detector voltage (0.2 - 2.0 V)
7. IL detector frequency (5 -88 kHz)

Several test configurations are supported:

1. Off-line loop testing: The inductive loop is disconnected from the detector; the IL Tester is then connected to the loop and energized with the IL Tester's generator. Tests 1 to 5 can then be performed.

2. Off-line detector testing: As above, the IL is disconnected and the IL Tester is connected to the loop detector. Tests 6 and 7 are supported.

3. On-line testing: The IL Tester is inserted between the loop and the detector. First, tests 6 and 7 are performed to measure detector frequency and voltage. Second, the IL Tester is set to generate the signal with the measured parameters. Finally, the loop detector is switched off the loop and IL Tester is switched in. Tests 1-5 are then performed. This mode of operation requires special cabling between the loop, detector, and the IL Tester, but once in place can be done solely from the instrument's front panel. DVP's IL Tester/Detector is currently being field tested in Washington, D.C. and Delaware DOT. Work on the instrument was sponsored by the FHWA through the SBIR program.

The authors would like to express gratitude to Mr. M. Mills from FHWA for his advice and support during this project.

Automatic Vehicle Classification System

by Wen-Min Pan, Geok K. Kuah, Shi-Lin. Su, and An-Sheng Wang

Abstract

China has recently completed a number of freeways, all of which were financed in part by the Ministry of Communications and local governments and in part by loans from the World Bank, with debt serviced by toll revenues. While freeway congestion is not a problem in China because of low traffic demand, the Chinese Toll Authority is faced with two major problems associated with toll collection. They are (a) appropriate vehicle classification and (b) toll revenue management, in particular the prevention of fraud in toll collections. This paper presents an automatic vehicle classification (AVC) system to assist manual toll collection for Guangzhou-Fusheng and Xian-Lingtong freeway. The system determines vehicle types and appropriate toll rates which are then collected manually.

Introduction

The use of automatic vehicle identification (AVI) technology for toll collection is gaining wide acceptance in the West as well as in some Eastern Countries such as Singapore and Hong Kong because of its potential to facilitate toll collection; thereby, reducing toll traffic congestion (Davis et. al, 1989 and Gravelle and Walker, 1990). However, the benefit of the AVI technology in non-stop toll collection can only be fully realized if the system is able to correlate its data with data from other traffic monitoring systems, such as automatic vehicle classification or weigh-in-motion (Gravelle and Walker, 1990), e.g., it is useless to weigh a truck unless its type is first known.

Numerous studies reported the technologies used in developing automatic vehicle classification equipment. Pursula and Kosonen (1989) discussed the use of analog signals produced by loop detectors in vehicle classification. Vehicle classification is performed according to the characteristics (the length, height, and form) of the analog signals generated by the passing vehicles.

Wen M. Pan, Visiting Professor,Center for Transportation Studies & Research, New Jersey Institute of Technology, Newark, NJ07102. (201)596-3355.

Geok K. Kuah, PhD, PE, Principal Transportation Engineer, DeLeuw, Cather & Company, 110 William Street, New York, NY 10038. (212) 266-8527

Shi-Lin Su, Lecturer, and An-Sheng Wang, Engineer, Department of Traffic Control, Xian Institute of Highway, China.

Miyasako et. al (1989) developed an ultrasonic vehicle classifier with which vehicles are classified into categories by comparing their profile, as measured by an overhead ultrasonic transducer, as they pass under it, with a library of standardized profile data. Pursula's design required that vehicles travel at uniform speed; an operational condition not found in most manual toll collection facilities. Miyasako's design cannot achieve the level of accuracy needed in vehicle classification for toll collection. Due to the shortcomings of the above designs, this paper used infra-red sensors and a microprocessor to develop an automatic vehicle classification system for the toll freeways in China.

Criteria in Vehicle Classification

A survey of vehicle types in several Chinese cities including Beijing, Xian and Guangzhou, showed that there are at least 741 different types of vehicles operating on the Chinese highways. The multitude of vehicle types is due mainly to the numerous vehicle import sources. There is not a widely accepted standard in vehicle classification for the Chinese vehicles.

Most toll road assess toll structure based on the passenger/freight carrying capacity of a vehicle. Table 1 shows the capacity-based scheme used by two recently completed freeways and a scheme adopted by the Ministry of Public Security. Like many other freeways, the toll official of this study adopted a capacity-based classified scheme similar to those used on other freeways.

Vehicle Category	Vehicle Type	Ministry of Public Security	Freeway Sheng Yang-Dalian	Freeway Xian-Shangyang	Criteria Used (Guangzhou-Fusheng)
Small	Passenger	# Pass < 19	# Pass ≤ 19	# Pass ≤ 19	# Pass ≤ 19
Small	Truck	GVW ≤ 4.5 (1)	FW ≤ 2.5	FW ≤ 2	FW ≤ 2.5
Medium	Passenger	N/A	20 ≤ # Pass ≤ 39	20 ≤ # Pass ≤ 49	20 ≤ # Pass ≤ 49
Medium	Truck	N/A	2.5 ≤ FW ≤ 7	2 ≤ FW ≤ 5	2.5 ≤ FW ≤ 7
Large	Passenger	# Pass ≥ 20	# Pass ≥ 40	# Pass ≥ 50	# Pass ≥ 50
Large	Truck	GVW > 4.5 (1)	7 < FW ≤ 15	5 < FW ≤ 20	7 < FW ≤ 20
Special	Truck	N/A	FW > 15	FW > 20	FW > 20

(1) GVW (Gross Vehicle Weight) includes freight weight & vehicle weight in tons
FW = Freight Weight in tons

Table 1. Existing Criteria for Vehicle Classification
(Based on Vehicle Carrying Capacity)

The capacity-based scheme is not suitable in automatic vehicle classification, however, because the number of passengers and vehicle weight are difficult to measure automatically. To achieve automatic classification, parameters based on vehicle dimensions are used. The selected parameters are: a) vehicle height, b) number of axles, c) axle spacing, and d) diameter/chord length of wheel.

Figure 1 illustrates the logic of vehicle classification based on the dimension-based scheme. As shown, the allocation of a vehicle from one class to another is dependent on whether or not the threshold values of the criteria are exceeded.

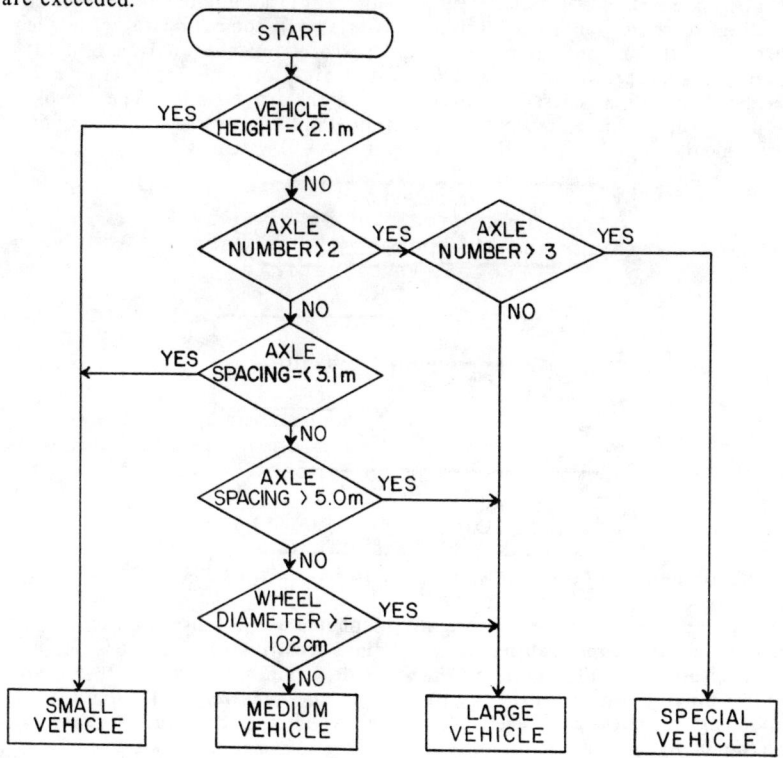

FIG. 1 FLOW CHART-DECISION LOGIC OF AUTOMATIC VEHICLE CLASSIFICATION

To evaluate the accuracy of the dimension-based scheme relative to the capacity-based, the 741 different vehicle types surveyed were classified using both schemes, and the classification results were compared. The analysis results are shown in Table 2.

Actual Vehicle Class	Total Vehicles Tested	Detected Veh Class				# of Mis-Class
		Sm	Md	Lg	Spec	
Small	338	334	4	0	0	4
Medium	163	7	154	2	0	9
Large	186	0	2	182	2	4
Special	54	0	0	1	53	1
Total	741					18

Table 2. Mis-classification Matrix

Automatic Vehicle Classification System

The AVC system can perform four functions, it can (a) sense vehicle presence and count the actual number of vehicles in the traffic stream by separating the vehicles, (b) classify the vehicles into a number of categories according to a set of predetermined criteria, (c) communicate with the central computer to process toll data and perform revenue analysis and (d) display the appropriate toll rate. It consists mainly of a microprocessor and several sets of infra-red sensors located along the travel lane. The sensors are used for vehicle separation/detection, and for vehicle dimension measurement. Figure 2 shows the layout of the AVC system.

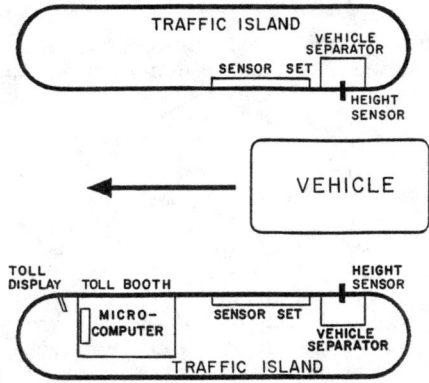

FIG. 2. Layout of the Automatic Vehicle Classification System

Each sensor used in the AVC system is a photoelectric sensor consisting of two parts: an infra-red transmitter and a light sensitive receiver. By placing the transmitter and receiver side-by-side, opposite each other, an uninterrupted beam from the transmitter to the receiver will form. The output signal from the receiver changes when the continuous beam is interrupted by the passage of a vehicle. Once the measurements are taken, vehicles are classified based on these measurements. An Intel MCS-51 series single-chip microprocessor 8031 is used for processing data and performing logic decisions related to vehicle classification.

The interface between the separator/other sensors and the microprocessor is through GAL (Generic Array Logic) which pre-processes the detected signals. After a vehicle is classified, information related to its class is communicated through serial interface ports, RS-232C or RS-442C, to a microcomputer located in the toll booths. The computer performs the necessary evaluation to determine the appropriate toll rate and subsequently displays the rate to the toll collector and the driver.

Field Tests

The device was tested under a wide variety of operational and adverse weather conditions and was implemented on the Guangzhou-Fushang Freeway during the months of June and July 1990. The results of the most recent test on the Xian-Lingtong Freeway in February, 1991 (Table 3), were very satisfactory. The percent of mis-classification is under 1.0%.

Test Date	Vehicle Category	Number of Vehicles			% of Mis-classifications
		Tested	MisCounted	MisClassified	
Feb., 1991	Small	170	0	1	
	Medium	63	0	1	
	Large	17	0	0	
	Special	2	0	0	
	Total	252	1	2	0.8%

Table 3. Field Operation Tests of AVCS

Summary and Conclusions

This paper presents an automatic vehicle classification system used in conjunction with manual toll collection methods on the Chinese toll freeways. The use of the automatic vehicle classification system resolved the problems associated with toll management on Chinese toll freeway, namely, (a) the multitude of vehicle classes and (b) toll collection fraud.

The conventional criteria used by the toll authority in vehicle classification, such as number of passengers and vehicle weight, are able to be measured automatically. For automatic vehicle classification, other criteria based on vehicle dimension were used. The criteria selected for the AVC system are (a) vehicle height, (b) number of axles, (c) axle spacing, and d) diameter/chord length of wheel. High accuracy in vehicle classification was achieved with the system because logical decision rules are used to evaluate threshold values of the selected criteria instead of exact vehicle dimensions of each vehicle. Several field tests performed on Guangzhou-Fushang Freeway gave very satisfactory results.

Appendix

1. Davies, P., N. Ayland and C. Hill, "Automatic Vehicle Identification for Non-Stop Toll Collection - The Virginia Experience," Proceedings of The 2nd International Conference on Road Traffic Monitoring, IEE, U.K., 1989, 133-136.

2. Gravelle, K.P. and C. Walker, "Technical Requirements for Effective Application of Automatic Vehicle Identification (AVI) Technology to Highway Systems," Compendium of Technical Papers of The ITE 60th Annual Meeting, 1990, 24-32.

3. Miyasako, T. et al, "The Ultrasonic Vehicle Profile Classifiers," Proceedings of The 2nd International Conference on Road Traffic Monitoring, IEE, U.K., 1989, 110-113.

4. Pursula, M. and I. Kosonen, "Microprocessor and PC-Based Vehicle Classification Equipments Using Induction Loops," Proceedings of The 2nd]International Conference on Road Traffic Monitoring, IEE, U.K., 1989, 24-28.

ANALYSIS OF AUTOMATIC VEHICLE IDENTIFICATION
TECHNOLOGY
AND ITS POTENTIAL APPLICATION ON FLORIDA'S TURNPIKE

Edward A. Mierzejewski[1], Michael C. Pietrzyk[2]
and William L. Ball[3]

The Florida Department of Transportation contracted with the Center for Urban Transportation Research (CUTR) at the University of South Florida to conduct an analysis of automatic vehicle identification (AVI) and its potential application on Florida's Turnpike. The project included three phases: (1) a review of the state of the art of AVI technology, (2) survey research relating to the attitudes and characteristics of current Turnpike patrons, and (3) an evaluation of the specific application of AVI technology to Florida's Turnpike.

AVI TECHNOLOGY

Automatic vehicle identification in the toll collection industry uniquely identifies vehicles as they pass specific points without requiring any action by the driver or an observer. As illustrated in Figure 1, AVI systems consist of three basic elements: a vehicle-mounted transponder/tag, an adjacent reading device, and a master computer system for data processing.

The AVI process is relatively simple: information that identifies a vehicle is encoded onto the transponder/tag. As the vehicle passes through the facility, the transponder/tag is activated to transmit the coded data to a roadside reading device. At this point, the data are checked for validity before being transmitted to the master computer for processing and storage. The entire transaction process is complete in a fraction of a second.

---
1. Deputy Director for Engineering, 2. Senior Research Associate, 3. Research Associate; Center for Urban Transportation Research, University of South Florida, Tampa, Florida 33620-5350.

AUTOMATIC VEHICLE IDENTIFICATION

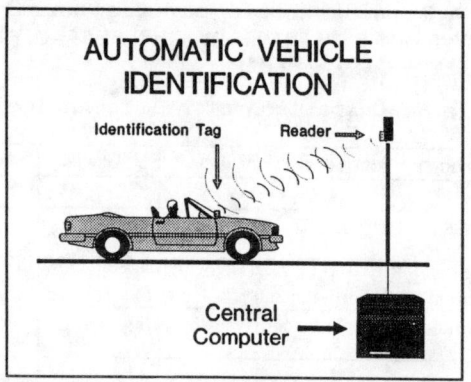

Figure 1. Typical AVI Configuration

Four major AVI technologies were considered:

Optical/infrared (bar code) technology uses a laser scanner to scan a bar code decal sticker that uniquely identifies the vehicle.

Inductive loop technology uses an antenna imbedded in the pavement to communicate with a transponder mounted on the underside of the vehicle.

Radio frequency (RF)/microwave technology employs microwave frequencies and a vehicle-mounted transponder, which contains a small internal receiving antenna, and an internal transmitter.

Surface acoustical wave (SAW) technology uses a low power radio frequency signal from the AVI reader which is captured by the transponder antenna, energizing a lithium crystal, which sets up a surface acoustical wave.

AVI IMPLEMENTATION ISSUES

CUTR examined a wide range of implementation issues which impact the feasibility of AVI.

Technology - Based on information provided by vendors, visits to various AVI facilities, and meetings with representatives of current AVI installations, relative comparisons were made. Seven major issues were identified and each technology was given a relative score of high, medium, or low for each of the issues.

Payment System Issues - Issues related to payment systems include toll structures, prepayment vs. post-payment, and methods of payment. Advocates of a premium toll structure contend that AVI users should pay an additional charge for the convenience offered by the system; others argue that discount toll rates should be offered in order to maximize participation. Finally, many contend that AVI users should pay the normal toll rate since the convenience offered by the system will provide enough incentive to participate.

Table 1. Comparison of AVI Technologies

ISSUES/TECHNOLOGIES	RF/Microwave	SAW	Inductive Loop	Bar Code
Reliability	high	medium	high	low
Resistance to Duplication (security)	medium	high	medium	low
Potential for Multiple Reads (speed vs. reliability)	high	high	low	low
Resistance to Interference (lane-to-lane)	low	low	high	high
Tolerance to Environment	high	high	medium	low
Simplicity of Tag (cost)	low	medium	low	high
Health Safety	high	high	high	high

The issue of pre-payment vs. post-payment was also evaluated. It was determined that a post-payment system should not receive further consideration since the burden of collecting delinquent accounts would be an additional responsibility for the Turnpike.

Methods of payment that should be provided under a pre-payment system include cash, check, credit card, and electronic funds transfer.

Legal Issues - Two major issues were identified and reviewed: compliance with the Turnpike's trust indenture if discount tolls were used, and the legality of photographic enforcement of violations. Because current Florida Statutes require that toll violations be enforced on-site, the main deterrence technique has been the use of gates on toll lanes. To realize the maximum benefit of AVI, gates should be eliminated, necessitating the use of active enforcement measures. Photographic enforcement is a viable alternative, which will require legislative changes.

Ownership/Finance Options - There are three major ownership arrangements available for AVI implementation: the agency owning and operating the system, a vendor owning and operating the system, and various lease agreements.

Traffic Operations - Traffic control measures (pavement markings, signing, channelization) will need to be revised to safely and efficiently accommodate the use of AVI. It is particularly important that a safe method for the movement of toll collectors across AVI lanes be identified.

Computer system requirements - AVI can be added as an upgrade to a state-of-the-art, microprocessor-based toll plaza. The Turnpike is currently acquiring a state-of-the-art computer system which will be able to accommodate AVI.

Dedicated vs. Mixed-Use Lanes - AVI can operate in either dedicated lanes or in mixed-use lanes. Dedicated lanes can be within a conventional toll plaza configuration that would allow AVI users to pass through at speeds of 10 to 15 miles per hour or as newly-constructed lanes separated from the conventional plaza configuration to accommodate speeds up to 55 miles per hour. Alternatively, a mixed-use lane is one that accepts AVI users as well as conventional patrons who pay a collector or use an automatic coin machine.

Capacity by lane type - Typical capacities were estimated for each lane type, including manual (350 vph), automatic (500 vph), mixed AVI (700 vph), dedicated AVI (1,200 vph), and express AVI (1,800 vph).

FLORIDA TURNPIKE PATRON SURVEYS

Three survey efforts were conducted to help assess the potential of AVI on Florida's Turnpike: personal interviews, a mail-back survey, and focus groups.

Personal Interviews - Over 7,000 personal interviews were conducted to determine user characteristics, vehicle type, occupancy, residency, trip purpose, and frequency of Turnpike use. Key findings were that 60 percent were five or more day per week users, and that 80 percent were using the Turnpike for work or business trips.

Mail-Back Survey - Of the 10,000 mail-back surveys distributed, over 2,000 were returned. The mail-back survey established other patron characteristics and perceptions toward the concept of AVI, with 90 percent of respondents indicating they would use AVI if discount tolls were offered and 68 percent indicating they would use AVI if the toll structure remained the same.

Focus Groups - Two focus groups were assembled, one representing commercial users and one representing private commuters. Both groups reacted positively to AVI. The most important advantages to commercial users were the ability to track vehicles and improve accounting procedures. The commuter representatives also reacted positively but were more sensitive to the costs factors.

COST-EFFECTIVENESS ANALYSIS

An analysis was conducted to determine the cost-effectiveness of AVI at Tamiami Plaza in Dade County. The analysis compared the cost of construction programs with and without AVI to the "no-build" alternative, through the year 2015. Because each improvement alternative was designed to maintain a queue length criterion of 300 feet, when compared to the no-build alternative, the estimated road user benefits were virtually the same in each alternative.

To handle year 2015 demand levels with a conventional plaza configuration will require 26 toll lanes. By comparison, lane requirements can be reduced to 22 lanes, 18 lanes, and 14 lanes, with AVI use of 10 percent, 30 percent, and 50 percent, respectively. Economic indicators are summarized below:

Table 2. Economic Feasibility Indicators

	(1) Benefits (present value)	(2) Costs (present value)	(3) Benefit/Cost Ratio (1)/(2)	(4) Net Present Value (1) - (2)
Current FDOT Plan	$50,789,781	$30,789,055	1.65	$20,000,725
10% AVI Participation	$50,877,415	$25,116,791	2.03	$25,760,623
30% AVI Participation	$50,946,719	$22,249,835	2.29	$28,696,883
50% AVI Participation	$51,016,022	$16,595,674	3.07	$34,420,348

Even with a minimal rate of participation, the AVI alternative is superior to the non-AVI alternative. In the absence of AVI, more lanes are required to maintain the same level of service.

CONCLUSIONS AND RECOMMENDATIONS

Several recommendations were made as a result of the findings of the feasibility study:
1. Florida's Turnpike should implement AVI.
2. Both passenger cars and commercial vehicles should be accommodated.
3. Traffic control practices should be revised as necessary to safely accommodate AVI.
4. Legislative changes should be sought to permit photographic enforcement of toll violators, thus permitting the removal of toll gates.
5. A variety of pre-payment options should be offered: cash, check, credit card, and electronic funds transfer.
6. Normal toll rates should be charge to AVI users, premium rates should not be utilized.
7. The transponder should be provided to patrons at a modest monthly payment of $2 to $3. If an optional scanner system is selected, a one-time fixed cost of $1 to $3 would be appropriate.

Analysis of Roadside Equipment and Central Facilities for Innovative Driver Information and Traffic Control Systems

M. Cremer

Abstract: *Following a top down analysis of a selected set of innovative RTI functions, we identify in this paper the requirements and characteristics of the roadside equipment and central installations which are necessary to realize and support the functions. In this context synergetic effects are highlighted by showing that many subfunctions as well as most physical units support more than one function. In addition the requirements for data transmission, reaction time and data processing capabilities are roughly specified for the different units.*

INTRODUCTION

Supported by the rapid development in microelectronic technology and communication engineering there has been new initiatives in recent years to improve the quality of future vehicular traffic substantially ([1], [2]). These activities are aimed at giving to the driver better information about the conditions he will face on his trip and possible alternatives. For this purposes a reliable data and information basis is necessary together with efficient data channels which transmit detailed information by selective dissemination strategies. A comprehensive information about the current traffic state provides the necessary basis for more sophisticated and more efficient traffic responsive control strategies for intersections, corridors and road networks. In this paper we analyse the most relevant functions among the proposed innovations putting special emphasis on the roadside and central installations which are required for an efficient performance of these functions. The set of functions to be considered here is

- incident and obstacle warning (F1), - medium range pre-information (F2)
- route guidance (F3), - network control (F4), - trip planning (F5).

Other functions which are included in some of the major research programs, like vehicle monitoring, distance warning, cooperative driving etc. are con-

---

Technical University of Hamburg-Harburg, Lohbrügger Kirchstr. 65, 2050 Hamburg 80, Germany

fined to a local neighbourhood or to the inside of a vehicle and require less support by the roadside infrastructure. They are therefore omitted here.

FUNCTIONS DESCRIPTION

In table 1 a rough scheme of the decomposition of the functions and the required equipment is listed together with the supporting algorithmic tools. The following description of the functions shows how they are performed in an appropriate way and how they are interacting.

F1 Incident and obstacle warning. First an obstacle or an incident has to be identified and classified by a camera system with a fast image processor (obstacle), by a crash sensor (accident) or by an alarm button pushed by the driver of an involved or a passing vehicle. Next a warning message is sent to oncoming vehicles and possibly to a near beacon (short range microwave) and additionally to a radio relay station. The information is passed to the concerned local control station, the place is localized and, depending on severity, an emergency service is called. A warning signal is sent repeatedly via radio (RDS, GSM) to the drivers which use the disturbed road. Then a message qualifying the reduced capacity is sent to the upper level controllers which update their traffic predictions and take into account the incident for route guidance and network control decisions.

F2 Medium range pre-information. This function comprises driver information about static road conditions (dangerous slope, road construction, stop sign ahead, bridges, tunnels, splitting of route etc.), weather conditions (fog, black ice, heavy rain, wet surface, gusty winds) and traffic conditions (incidents, congestion and temporary traffic regulations). Critical traffic conditions and warnings should be transmitted via short range microwaves from car to car with high priority. Additionally, any such information is to be stored, if possible, at an upstream beacon to be conveyed to passing vehicles. Static information can be transmitted to the vehicle by beacons or via radio which may be carried out part by part well ahead of the relevant road section. The message has to be stored within the car and released in time to the driver on display or by artificial voice. Any critical situation, traffic or weather related, should be passed to the controller of the next higher level to be considered in route guidance or network control. Accident or congestion warning should be supported by model based state estimation on local controller level to have an accurate picture about the length of congestion at present and in near future.

F3 Route guidance. Static route guidance is based on navigation (dead reckoning, GPS), a digital map and a route selection routine. It can be largely concentrated within invehicle elements or be partly supported by roadside components (beacons) like in the Berlin LISB system ([3]). The digital map should be updated at special beacons (e.g. at gasoline stations) involving a considerable amount of data exchange. *Dynamic* route guidance takes into account transient traffic states and temporary weather conditions

TRAFFIC CONTROL SYSTEMS

Table 1 Analysis of system functions

functions	subfunctions	roadside equipment	central equipment	data processing tools
F1 Incident and obstacle warning	detection; localization and classification; warning of driver; warning message dissemination	beacons with receiver + transmitter (two way)	radio relay station (GSM) with two-way communication; control stations (on var. levels); emergency services	image processing; warning dissemination strategies; conventional inc. det. algorithms
F2 medium range pre-inform.	static information; data/inform. collection; short range comm.; communication with central stations	beacons (two-way); traffic signs with markers; sensors for weather conditions; traffic flow detectors	radio relay stations; local control stations	medium range state estimation; plausibility checks; elementary data en- and decoding
F3 Route guidance	digital map; navigation; traffic state monitor.; traffic prediction; optimal route recommendation	beacons (two-way); sensors for weather conditions; traffic flow detectors	radio relay stations; local control stations; regional control stat.; suprareg. control stat.; data banks; diverse services; satellites	navigation routines; static and dyn. route selection; state estimation; O-D estimation; traffic prediction; route optimization
F4 Network control	network charac.; traffic state monitor.; traffic prediction; control actions; weather conditions observation	traffic flow detectors; beacons (two-way); sensors for weather; variable mess. signs; control signals	radio relay stations; local, regioanl, suprareg. control stations; data banks; meteorological services	traffic state and O-D estimation; traffic prediction; trip param. calculation; control strategies
F5 Trip Planning	digital map; traffic state monitor.; medium (long) term prediction; weather cond. obs.; informations generat.; public transport	traffic flow detectors; weather rel. sensors; beacons (data coll.)	radio relay stations; local, regional, suprareg. control station; data banks; meteor. services; data channels for home terminals; public transp. control	state (O-D) estimation; traffic prediction (medium to long range); trip param. calculation; route selection; transp. mode selection; parking facilities monit.

and therefore involves assistance from roadside and central facilities. For this traffic state monitoring has to be performed in the local controllers based on local sensor data and trip data delivered by single vehicles at beacons or via radio. Moreover, for longer trips traffic evolution has to be predicted to guarantee that a recommendation is justified for the duration of a trip. Depending on the length of a trip, these - preferably model based - predictions are performed on the appropriate level making use of the expected traffic demand values and O-D characteristics which are stored in data bases. Road capacities have to be determined with respect to traffic and weather conditions. On this basis route selection has to be performed - at least in a rough manner - within central control stations. Encoded recommendations are transmitted via beacons or radio involving a moderate amount of data transmission.

F 4 Network control. From the viewpoint of road administration authorities, traffic flow within a network has to be controlled to avoid congestion and accidents. For this a hierarchical control system seems to be most appropriate where local problems are solved locally whenever this is possible and regional problems are solved by area wide measures. Control measures may be conventional ramp metering, traffic diversion via variable message signs, speed control and - more innovatively - single vehicle route guidance. A complex system of sensors and receivers provides the necessary data basis which is transformed into a complete picture of the current traffic state by model based estimation techniques. Within this hierarchical system higher level controllers define set points for lower level control. Thus, route guidance may be influenced by network control strategies.

F 5 Trip planning. For medium and long distance trips future information systems will provide recommendations with regard to the best starting time, the best route and the best transportation mode (including public transport). The required information basis is partly the same as for route guidance and network control but includes long distance predictions over a longer time horizon. This will be performed in higher level control centers (regional, supraregional, national) which use the current traffic state as estimated by local stations with estimates for the traffic demand supplied by data banks and other sources (e.g. the inquiries of the trip planning system itself) to calculate the expected traffic evolution on a macroscopic basis. Then a selection routine determines the recommendations which fit best to the specifications of the driver. This function may include booking of parking spaces and public transport (trains, ferries) as well as the particular requirements of specific driver groups (fleet management) which often are defined as separate functions.

SPECIFICATION OF TECHNICAL UNITS

From the function analysis as outlined above requirements and specifications could be concluded for roadside and central system units. Because of lack of space, here the main points are listed in table 2. From this functional coordination an overall architecture has been designed ([2]) including the partial architectures of the units and flow diagrams that describe the decision making and data processing flow within the units.

Table 2 Technical units and their tasks

techn. units	RTI functions	data processing requirements	information	update rate
traffic sensors	F2 F3 F4 F5	counting classification averaging simple checks	traffic counts vehicle classes mean speed	≈ 1 sec
weather sensors	F2 F3 F4	surveillance of thresholds plausibility checks	fog, wind black ice rain, snow wet surface	≈ 5 min
beacons	F1 F2 F3 (F4)	data handling checking routines message generation + selection	warnings medium range pre-information route recommend. trip data	0,2 to 5 min
local control stations	(F1) F2 F3 F4 (F5)	section state (O-D) estimation short range pred. local control data transm. contr.	section state (current + future) control signals (VMS + invehicle) trip parameters	1 min
radio relay stations	F1 F2 F3	accident localiz. message generation + selection data transm. contr.	warnings est. trip param. recommendations trip data	0,2 to 5 min
higher level control stations	(F3) F4 F5	long range pred. trip param. calcul. network control route selection data bank managmt.	area state control signals set points recommendations alternat. transp	10 min

REFERENCES

[1] PROMETHEUS: Functions and how to achieve Prometheus objectives. - Internal Rep., Stuttgart, July '89

[2] Willis, D.K.: Intelligent Vehicle Highway Systems - A Summery of Activities Underway, Worldwide. - Proc. of the AATT Conference, San Diego, California, Feb. '89.

[3] Hoffmann, G.: LISB - An Individual Route Guidance and Information System in Berlin.- Proc. IFAC/IFIP/IFORS Conf. CCCT, Paris, Sept.'89

[4] Cremer, M.: An integrated architecture for future freeway information and control systems. - Eng. Found. Conf. on Traffic Management, Palm Coast, Florida, April '91

Traffic Reporter: a Real-Time Commuter Information System

Mark Haselkorn,[1] Woodrow Barfield,[2] Jan Spyridakis,[3] Brian Goble,[4] Margaret Garner[5]

Abstract

Traffic Reporter is a PC-based graphical, interactive, traveler information system being developed to meet the informational needs of Seattle-area commuters. The goal of *Traffic Reporter* is to influence commuter behavior and decision making by delivering up-to-the-minute, useful traffic information. *Traffic Reporter's* design and development are based on a thorough study and analysis of the behavior and traffic needs of Seattle-area commuters.

Traffic Reporter receives traffic data from freeway detectors and converts these data to information that can be explored both for general freeway conditions and specific trip information. That information can then be delivered to (1) commuters, (2) traffic reporters, and (3) traffic managers, planners, and engineers.

This paper describes the current version of *Traffic Reporter* and focuses on how it provides graphically-based commuting information to motorists. It also briefly describes the design and development of *Traffic Reporter* to date, evaluation of it in its current state, and plans for its expansion and enhancement.

Introduction

Traditionally, traffic problems in the Seattle area have been solved by either expanding existing roads or building new roads. But, there are significant costs and considerations in such solutions. Thus, several years ago, a UW interdisciplinary research team began to look for a way to maximize the

---

[1]Professor and Chair, Department of Technical Communication.
[2]Associate Professor, Department of Industrial Engineering.
[3]Assistant Professor, Department of Technical Communication.
[4]Research Consultant/Software Design, Department of Technical Communication.
[5]Writer/Editor, Department of Technical Communication.
College of Engineering, University of Washington, Seattle, Washington 98195.

efficiency with which commuters use existing transportation facilities. After thoroughly investigating and analyzing alternative approaches, this team chose a real-time, PC-based, graphical, interactive, traveler information system. The traveler information system being developed for the Seattle area is called *Traffic Reporter*.

This paper focuses on how *Traffic Reporter* currently works. We also briefly describe the design and development of the current prototype as well as plans for future expansion and enhancement of *Traffic Reporter*.

Designing and Developing *Traffic Reporter*

Traffic Reporter's ultimate goal is to improve traffic flow by influencing commuter behavior and decisions concerning alternative routes, departure times, and transportation modes. Our strategy to achieve this goal was to isolate the particular behavior we wanted to modify and then focus on those drivers most likely to alter that behavior. While it would seem ideal to impact as many commuters as possible, this is neither feasible nor desirable. Significant improvements in traffic flow can be achieved by influencing the behavior of a relatively small percentage of commuters, while influencing the behavior of a high percentage of commuters would just move the problem to another part of the transportation system.

Influencing such commuter behavior and decision making can only be achieved by first understanding the behavior and information needs of commuters--the first stage in designing and developing *Traffic Reporter*. Therefore, we have seen the development of *Traffic Reporter* primarily as a communication problem and secondarily as a technology problem.

To understand the behavior and information needs of Seattle-area commuters, we first needed to find out more about them and their current use of motorist information. To obtain this information, we conducted two extensive on-road surveys of commuters traveling on Seattle's major freeways. The UW research team administered questionnaires, with 72 variables, to 15,000 on-road commuters, of which 6,000 (40%) questionnaires were returned (creating the most extensive database of its kind).

From the results of these on-road surveys, four types of commuters were identified: *pre-trip changers*, commuters willing to change departure time, route, or travel mode before departure (16%); *route and time changers*, commuters willing to change departure time and route (40%); *route changers*, commuters willing to change route (21%); and *non-changers*, commuters unwilling to change departure time, route, or travel mode (23%). From these groups, we identified the commuters most likely to change some aspect of their commuting behavior. See Barfield *et al.* (1991) and Spyridakis *et al.* (1991) for further details on these surveys.

Based on our understanding of Seattle-area commuter behavior, traffic information needs, and delivery preferences, we designed and developed a working prototype of *Traffic Reporter*. The screen design and functions of this system are described in the next section.

Using *Traffic Reporter*

Traffic Reporter's most important function is to process traffic data collected from the freeway into up-to-the-minute information that can be delivered directly to commuters. The process begins with data collected as vehicles travel over detectors embedded in the pavement of freeway lanes. Detectors are spread along four major interstate and state routes at approximately half-mile intervals. A microprocessor gathers the detector data for one second and sends these data to a central mainframe.

The next stage takes place at the Traffic Systems Management Center in Seattle. There, a mainframe receives the one-second microprocessor data and produces one-minute summaries of these data. These summaries are then transmitted to a personal computer. Once a minute, *Traffic Reporter* converts these data to estimated travel speeds and times. These estimates are then displayed on *Traffic Reporter's* graphical screen.

Data Display. *Traffic Reporter's* main screen displays a graphical representation of a map of a northern corridor of I-5. This corridor begins in downtown Seattle and runs north for about 15 miles. (See Figure 1.) This freeway map shows all north and south entry and exit ramps on that freeway corridor. The screen also includes a legend showing four color-coded speed ranges (green for 50+ mph, yellow for 35-49 mph, purple for 20-34 mph, and red for 0-19 mph). The area between each freeway sensor station is colored according to one of these four speed ranges.

A status bar at the bottom of the screen shows the current time and date as well as the time the displayed information was gathered. This feature informs the user of the recency of the information. A fifth color (blue) indicates a malfunctioning station; this feature was specifically designed for the transportation engineers at the Traffic Systems Management Center. Users of *Traffic Reporter* interact with the system by using a mouse.

Access and Display of Traffic Data. While *Traffic Reporter's* main screen presents useful traffic information, *Traffic Reporter* has other screens that also contain useful features. For example, let's say a commuter wants more specific traffic or trip information. Instead of viewing a speed range, he or she may want to view the actual average mph at a particular location. To view average mph, the commuter points the cursor on a particular section of I-5 and clicks the mouse button. The mouse button then turns into a magnifying glass and zooms in on that section. The magnified version of I-5 is displayed in a dialog box to the left of the freeway map, showing mean speed rates at specific stations instead of broad speed ranges.

COMMUTER INFORMATION SYSTEM

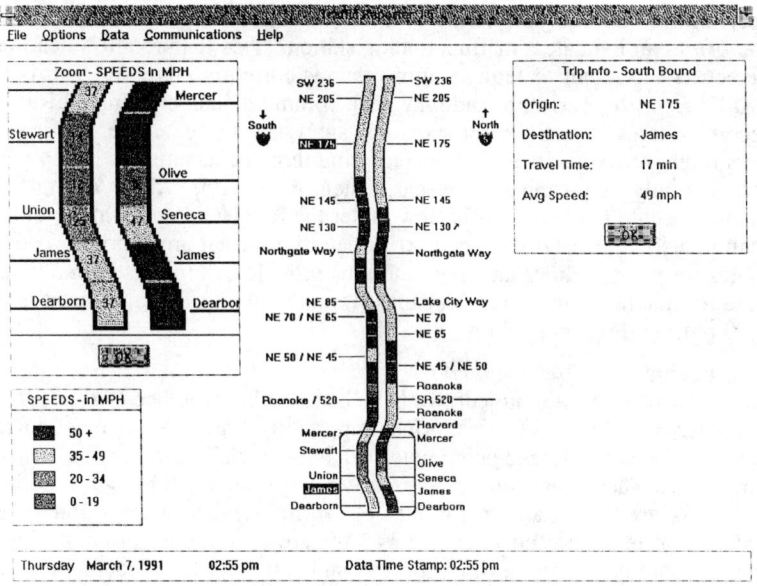

Figure 1. *Traffic Reporter's* graphical screen. Main screen in center shows broad speed ranges. Zoom box in upper left screen shows mean speed rates at specific stations. Box in upper right screen shows estimated speed and travel time between two ramps.

Or, a commuter may want to know the average speed and travel time between two freeway ramps. In that case, he or she clicks the mouse button on any two ramps. A dialog box to the right of the freeway map shows estimated speed and travel time between those two ramps.

Typically, a commuter may want to select the "best" exit or entry ramp for a given trip. To select an exit ramp, a commuter double-clicks the right mouse button on the origin ramp. A table appears showing estimated speeds and travel times to all possible exit ramps currently displayed. To select an entry ramp, the commuter double-clicks the left mouse button on the destination ramp. The same table appears, this time showing estimated speeds and travel times to the destination ramp from all possible entry ramps currently displayed.

In addition to providing traffic information to commuters, *Traffic Reporter* provides information that is of specific interest to traffic managers and engineers. First, *Traffic Reporter* can display up-to-the-minute data on volume (number of vehicles passing over each detector) and occupancy (percentage of time vehicles are over each detector). Second, *Traffic Reporter* can indicate a malfunctioning station. This feature allows earlier detection and repair of faulty stations than is currently possible. Third, *Traffic Reporter* can store and play back commute data, creating a historical record of freeway activity for statistical analyses. For example, traffic planners might redesign an HOV lane requiring three occupants per car so that it requires only two occupants per car. Then, with *Traffic Reporter*, traffic planners could compare traffic flow under the two HOV conditions. This feature also allows *Traffic Reporter* to determine what are "normal" commutes for various times and days and thus provide commuters with a broader base for making commuting decisions (for example, an average speed for 5:00 p.m. on Friday is 18 mph).

Evaluating *Traffic Reporter*

Throughout the project, we have been asking whether *Traffic Reporter* is useful or effective. Most importantly, we have attempted to determine if it will influence commuter behavior by delivering real-time information. These questions are currently being answered through usability tests. We are testing all three groups of *Traffic Reporter* users: commuters, traffic reporters, and traffic engineers. We are conducting formal usability tests on commuters and traffic reporters and soliciting less structured input from traffic engineers. Preliminary results of usability tests show that the graphical interface is intuitive, easy to use, and effective in communicating traffic information.

Planning Ahead

In the next phase of *Traffic Reporter's* development, we will accomplish the following: (1) expand *Traffic Reporter* to cover other Seattle-area freeways, as well as separate out express and HOV lanes from regular lanes; (2) enhance *Traffic Reporter* by improving the conversion and accuracy of sensor data and by adding features, such as a prediction mode; (3) deliver *Traffic Reporter* directly to commuters via TV and radio traffic reporters, as well as develop a touch-screen version of *Traffic Reporter* for delivery via kiosks; (4) explore integrating *Traffic Reporter* with other methods of delivery, such as dedicated cable TV, PCs at homes and places of business, and in-car delivery, (5) use *Traffic Reporter* to record data for monitoring traffic patterns and establishing norms of freeway performance; and (6) continue to evaluate *Traffic Reporter's* usefulness and effectiveness.

References

Barfield, W., Conquest, L., Haselkorn, M., and Spyridakis, J. 1991. "Integrating Commuter Information Needs in the Design of a Motorist Information System," Transportation Research, 25A, 2/3, 71-78.

Spyridakis, J., Barfield, W., Conquest, L., Haselkorn, M., and Isakson, C. 1991. "Surveying Commuter Behavior: Designing Motorist Information Systems," Transportation Research, 25A, 1, 17-30.

SINOD - A Location Reference Tool for
Road Information Systems

António Lemonde de Macedo[1]

Abstract

This paper focuses on the setting up of a location reference system - named SINOD - envisaged as a basic requirement for on going actions and for expected developments in the field of road information systems in Portugal.

The general characteristics of the system, some underlying conceptual aspects of its implementation and conclusions drawn from its practical application are presented.

Introduction

At the Transportation Networks Department of the National Laboratory for Civil Engineering (LNEC), in Lisbon, studies are under way in order to assist in the development of road information systems in Portugal.

In fact, solutions are needed for a set of problems concerning not only improvements on the existing systems, but also for the implementation of advanced integrated road and transportation information systems. Furthermore the recent Portuguese membership of EEC has increased its participation in several European programs (e.g. DRIVE) urging also for adequate responses in that field.

In this context, the need for a suitable road data location reference system became apparent, owing to known limitations of "traditional" reference methods, and to their inability to satisfy present requirements that arise from technological advance in road data acquisition and transmission systems.

The development of such reference system took place at LNEC, as a thesis work (Macedo,1988), which was presented for its author's qualification as Research Officer.

----
[1] Research Officer, Head of Traffic and Safety Division, Transportation Networks Department, LNEC, Av. Brasil 101,1799 Lisboa Codex, Portugal.

The present paper is based on that work and on its following developments. It starts by mentioning the main conceptual aspects that were taken into account. The resulting system (SINOD), whose characteristics are summarized, was tested and evaluated through its implementation in a chosen sample of the road network (about 1000 km),in a district, with very satisfactory results. As a consequence, SINOD is now being expanded to the Lisbon Region, covering three districts, by means of a cooperative action between the Portuguese Highways Administration (JAE) and LNEC, as the first step for its implementation over the entire network of national roads. Some aspects of this practical experience are also reported.

Conceptual Background

Following a general systems approach, the definitions of some relevant concepts took place prior to the setting up of the reference system.

The concept of "information system" within a road organization was defined, in its broader sense, as the set of all the information situations and relations among its components, thus not restricted only to computerized systems for data storage and retrieval.

The concept of "road occurrence" is a basic element in the conceptual background of the work. It was adopted to designate any physical characteristic (e.g. a pavement), or any event (e.g. an accident) related to the road, and capable of generating information.

The fact that a physical location is associated with many of those "occurences" constitutes the main link to the so-called "Geographic Information Systems" whose components attributes are usually classified as "locational", "non-locational" and "time".

In this context, road reference operations can be defined as the acquisition of data items related to the "locational" attributes of road occurrences. They assume the existence of a whole system that turns those operations possible - the reference system - which may include one or more reference methods and their referentials. In Fig. 1 a general representation of this conceptual structure is shown.

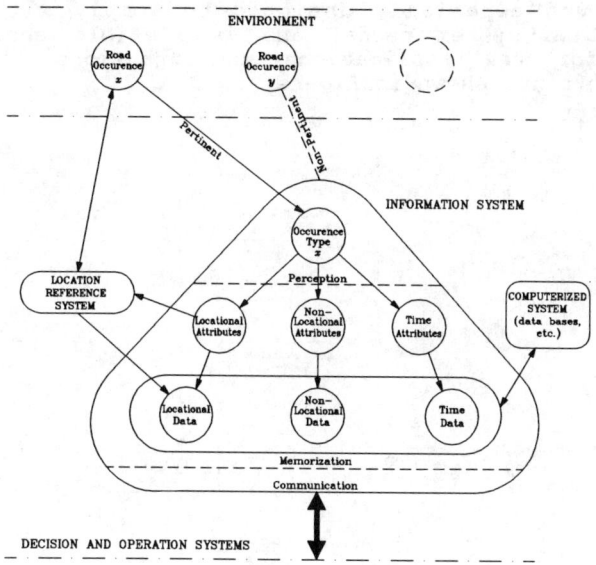

Fig. 1 - Conceptual framework

System's Description

The main components which interrelate within the SINOD Reference System are:
- a road network description mode;
- a location reference method;
- a reference data sub-system.

The description mode represents the road network as a "graph" through an adequate selection of a number of nodes (reference points) and links, according to clearly defined criteria. The identification code for each node relates it to a geographic area rather than to a specific road, which is one of the conditions for the desired spatial stability of the system. The links are associated to the carriageways (not exactly to the roads) of the network; this condition being required by some of the capabilities of the system.

The reference method assumes the network as described above. It consists of a well defined and coherent set of procedures (related to road devices and/or documents produced within the system) which were conceived for the physical location of the different types of "occurences" that can be expected.

Finally, the reference data sub-system is built up over a relational structure of the information, to be gathered for each reference point and link. This subsystem was especially designed to comply with road information system needs, and is flexible enough to allow for new applications and expansions. Its main components are shown in figure 2.

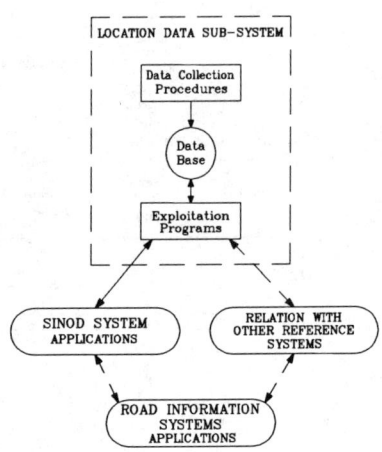

Fig. 2 - Basic Components of the data sub-system

Implementation Aspects

The implementation of the SINOD Reference System is under way along the Portuguese principal interurban road network (about 10000 km). It has started and is being concluded in the Lisbon Region (about 3000 km) where nearly 2000 reference points have been selected, coded, and their pertinent data collected and stored.This work involves a large number of interrelated activities, in field and in office, which have been programmed for a one year term in this first phase.

It must be pointed out that, despite its previous studies and field tests, the continuing practice with the system's implementation process has proven to be a very important source of learning. It has not only contributed to various improvements and adjustments in the system's components as shown in fig.3 (Macedo and Arsénio, 1991), but has also introduced marginal benefits, such as:
- the development of equipments and techniques (e.g. related to distance measurements along the roads);
- the detection of undefined situations on the existing network, and the need for their quick and coherent resolution.

Other important issues have been the compelling effect of this work on other necessary developments in the field of road information systems, and its articulation with other related on going projects (e.g. a pavement management system being implemented for the same network, or a more general road and transport data access network - RAIAR project - under way for the public administration).

Despite inevitable drawbacks inherent in every system, we are convinced that the SINOD System can fulfill its objectives, and constitute a comprehensive, stable, effective and efficient tool for the data location reference needs of our road organizations.

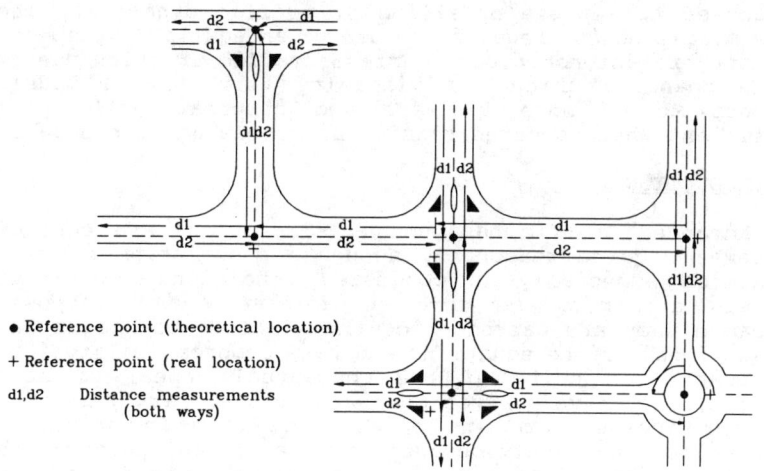

● Reference point (theoretical location)
+ Reference point (real location)
d1,d2 Distance measurements (both ways)

Fig. 3 - Reference points location and field measurements process.

Appendix

MACEDO, A.L. (1988). Sistemas de Informação Rodoviária - Estudo de um Sistema de Referência para o caso de Portugal.Tese, Proc. 93/13/9330, LNEC, Lisbon.

MACEDO, A.L.; ARSÉNIO, E.M.(1991). Implantação do Sistema SINOD na Rede Rodoviária da Região de Lisboa, Relatório nº2, Proc. 93/1/9865, LNEC, Lisbon.

CONTINUUM MODELLING OF TRAFFIC DYNAMICS

Panos G. Michalopoulos[1], Ping Yi[1],
Dimitrios E. Beskos[2], and
Anastasios S. Lyrintzis[3]

ABSTRACT

Two new efficient higher-order continuum models are proposed to improve modelling of traffic dynamics. The new models do not require the use of an equilibrium speed-density relationship, and address traffic friction due to lane changing through a viscosity term. Both models appear to be more accurate and computationally more efficient when compared to similar models and field data.

INTRODUCTION

Advanced traffic management, simulation, and control schemes require reasonably accurate description of flow dynamics especially at congested conditions. Continuum models appear to be more suitable for such purposes because they are macroscopic, they include both time and space in the state equations and take compressibility into account. In spite of the conceptual appeal of such models, they have not yet been widely used partly because of our inexperience in implementing them in practical situations and partly because of some needed improvements in their formulation. This paper introduces and implements two new formulations. The effectiveness of the implementation is evaluated through qualitative and quantitative tests. Computational efficiency is also demonstrated in comparison to other models.

BACKGROUND

1. Simple Continuum Models
A simple continuum model includes the continuity equation

---

[1]Dept. of Civil & Mineral Eng., Univ. of Minnesota, Mpls, MN 55455
[2]Dept. of Civil Eng., Univ. of Patras, GR-26110, Patras, Greece.
[3]Dept. of Aerospace Eng. & Mechanics, Univ. of Minnesota, Mpls, MN 55455

$$\partial q/\partial x + \partial k/\partial t = g \qquad \text{(Model 1)}$$

where $q=q(x,t)$ is the traffic flow (volume) in veh/h, $k=k(x,t)$ is the traffic density in veh/mile, g is a generation term, x is the distance coordinate along the direction of travel and t denotes time. In the simple continuum concept it is assumed that

$$q = ku_e$$

where u_e is the mean traffic speed in miles/hr; its steady-state value is determined through an empirical or theoretical relationship such as:

$$u = u_f[1-(k/k_j)^\alpha]^\beta,$$

where u_f is the free flow speed, k_j is the jam density and α and β are positive constants. For instance, for $\alpha=\beta=1$ one obtains the Greenshields (1934) equation of state. Because it is essentially steady-state in nature and does not take into account the effect of the acceleration, the simple continuum model does not faithfully describe traffic flow dynamics.

2. Higher-Order Continuum Models

A higher-order continuum model includes, in addition to the continuity Eq. (1), a momentum equation that involves acceleration.

Payne (1979) was among the first to introduce a higher-order continuum traffic model that includes a momentum equation in addition to the continuity equation. This equation has the form

$$\frac{du}{dt} = \frac{\partial u}{\partial t} + u\frac{\partial u}{\partial x} = -\frac{1}{T}[u-u_e(k)] - \frac{v}{k}\frac{\partial k}{\partial x} \qquad \text{(Model 2)}$$

where the first term on the right hand side represents relaxation to equilibrium, the second represents anticipation, T is the relaxation time, v is the anticipation parameter, and $u_e(k)$ is the steady-state mean speed-density relation discussed in connection with the simple continuum models. Rathi et al (1987) and Ross (1987) reported that the relaxation of flow speed to the steady-state speed is too slow and that the model may not capture the real traffic dynamics under abrupt geometric and traffic volume changes within a short period of time.

Papageorgiou (1989) modified Model 2 by including a couple of additional parameters, ζ and κ; when discretized it takes the form

$$u_j^{n+1}=u_j^n+\Delta t\zeta[-u_j^n\frac{(u_j^n - u_{j-1}^n)}{\Delta x_j} - \frac{1}{T}(u_j^n-u_e^n(k_j))+v\frac{(k_{j+1}^n - k_j^n)}{(k_j^n+\kappa)\Delta x_j}]$$

(Model 3)

Furthermore, another parameter is introduced in the calculation of traffic volumes to account for possible discontinuities. This model represents an improvement over Model 2. However, the physical meaning of the added parameters is not clearly explained, and the implementation of this model requires additional amount of effort to estimate those parameters.

Other models (Phillips 1978, Kühne 1989, Ross 1987) have similar characteristics to Models 2 and 3. However, these models either have not been practically implemented or have limited use in practical applications.

TWO SIMPLIFIED FORMULATIONS OF THE MOMENTUM EQUATION

Because of the limitations of existing continuum models, we introduce two new formulations of momentum equation. They represent a considerable simplification because they do not employ a steady-state u_e-k relationship. The first formulation is heuristic and is based on existing higher-order modelling. An improvement is sought by the introduction of a traffic friction term to address the lane changing effects. The complete formulation was given as:

$$\frac{du}{dt} = \frac{1}{T}(u_f - u) - G\frac{\partial u}{\partial t} - \alpha\, k^{\beta-1}\frac{\partial k}{\partial x} \quad \text{(Model 4)}$$

The first term on the right side of Eq.(4) represents relaxation as in all existing higher-order existing models. However, this model does not employ the steady-state speed u_e but the free flow speed u_f, and the contribution of this term is effective only when traffic is relaxing from a lower speed to a higher one. The second term on the right side of Eq.(4) represents traffic friction and is used to take into account lane changing.

The second formulation of the momentum equation (Model 5) was developed by treating traffic flow as a viscous, compressible mass. This equation was derived from the fluid flow momentum equation by substituting fluid flow pressure with traffic flow density. This Model does not require the use of an equilibrium speed-density relationship. In addition, it includes a viscosity term which serves to handle traffic friction.

We have discretized Model 4 with the Lax scheme and upwind scheme with flux vector splitting and Model 5 with simple Euler's method. Through stability analysis, these parameter values are chosen at this time for Model 4:
t_o=50 sec   r=0.8 α=500-600 mph^2 β=0.01

MODEL TESTING

A qualitative test was performed with a bottleneck case to check the process of queue formation and dissipation. A quantitative test on congested flows was then performed using data collected from an urban test site on Interstate Highway 35W in Minnesota. Based on the deviations of simulation results from the field data, the mean absolute error (MAE) and mean square error (MSE) were computed. Computational efficiency was also compared by using the computer execution time of Model 1 as the basic unit. Similar results were obtained when other space and time step sizes were used and in the following only when $\Delta x=200$ ft, $\Delta t=1$ sec is presented unless specified otherwise. The u_e-k curve for implementing existing models was estimated from the aforementioned data. Test results for the two cases are summarized in figure 1 and table 1, respectively.

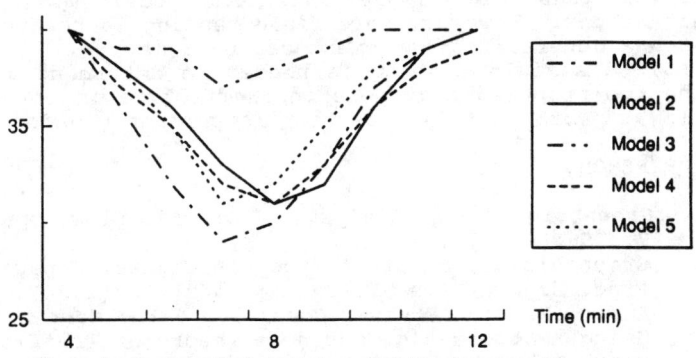

Figure 1 Queue end trajectories in qualitative testing case

Table 1 Error Indices and Computational Efficiency

Models(4)	1	2	3	4 (upwind)	4 (Lax)	5
MAE (1) (% diff)	82 (15.26)	59 (8.10)	46 (7.84)	23 (4.46)	34 (6.43)	40 (7.23)
MSE (2) (std dev)	5871 (90.02)	4662 (74.80)	3725 (66.85)	808 (31.15)	1470 (42.00)	3583 (59.43)
Comp Index(3) (Time Improv)	1.00	1.07 (-7%)	1.10 (-10%)	0.86 (+14%)	0.97 (+3%)	0.83 (+17%)

(1) MAE, veh/5 minutes;
(2) MSE, veh^2/5 minutes; std. deviation, veh/5 minutes;
(3) Computer execution time index, compared with Model 1 and based on IBM-PC 386-25Mhz machine.
(4) for Model 1, $\Delta x=100$ ft, $\Delta t=1$ sec.

CONCLUSIONS

Based on limited testing, the following conclusions can be drawn:

1) Qualitatively our models (3 and 4) describe traffic dynamics in a reasonable manner and the produced trajectories follow that of the simple continuum model solved by the method of characteristics. (Michalopoulos et al 1980).
2) When tested with congested flow data, our models produce lower errors (4.46-7.23%) compared with other models (7.84-15.26%). Model 4 (upwind version) was the most accurate, and Model 5 was the fastest (24-27% faster than Models 2,3 and 17% than Model 1).
3) Our models do not require the use of a u_e-k relationship. This feature saves not only a significant amount of effort in acquiring such a relationship, but also computation time as it can be seen from table 1. Therefore, our models are more practical in field applications.
4) All the higher-order models seem to be significantly better than the simple continuum model in congested situations. However, more field testing is needed before such a generalized statement can be made.
5) More field testing is needed to determine which of the two proposed models is more effective in interrupted flows. Work in this direction is currently under way.

REFERENCES

1. Greenberg, H., An Analysis of Traffic Flow, Oper. Res. 7, 79-85 (1959).
2. Greenshields, B. D., A Study in Highway Capacity, Proc. Highway Res. Board, 14, 448-477 (1934).
3. Kühne, R. D., Freeway Control and Incident Detection Using a Stochastic Continuum Theory of Traffic Flow, Proc. 1st Int. Conf. on Applied Advanced Technology in Transportation Engineering, San Diego, CA, 287-292, ASCE, New York, 1989.
4. Michalopoulos, P. G., G. Stephanopoulos and V. B. Pisharody, Modelling of Traffic flow at Singalized Links, Transpn. Sci., 14, 9-41 (1980).
5. Payne, H. J., FREFLO: A Macroscopic Simulation Model of Freeway Traffic, Transpn. Res. Rec., 772, 68-75 (1979).
6. Papageorgiou, M., J. M. Blosseville and H. Hadj-Salem, Macroscopic Modelling of Traffic Flow on The Boulevard Peripherique in Paris, Transpn. Res. B, 23B, 29-47 (1989).
7. Phillips, W. F., Kinetic Model for Traffic Flow, Pub. No. DOT/RSPD/DPB/50-77-17, U.S. Dept. of Transportation, Washington, D. C. 1978.
8. Rathi, A. K., E. B. Lieberman and M. Yedlin Enhanced FREFLO Program: Simulation of congested Environments, Transpn. Res. Rec. 1112, 1987, pp61-71.
9. Ross, P., Traffic Dynamics, Proc. 67th Annual Meeting of Transp. Res. Board, November 1987.

Application of Fuzzy Set Theory to
the Analysis of Transportation Problems

Shinya Kikuchi, ASCE Member[1]

Abstract

Potential of applying fuzzy set theory is discussed. Fuzziness in the nature of transportation engineering problems is identified and typical fuzzy theory techniques are presented. Application areas are listed.

1. Introduction

Many subjects in transportation planning and engineering are often characterized by subjective, ill-defined, ambiguous, and vague. Evaluation of alternatives, traveler's route choice, driver's perception and reaction are typical examples of problems involving subjective judgement. Traditionally such problems have been dealt with in the framework of mathematics based on binary logic. This logic, which has contributed enormously to the progress of science for the past centuries, cannot deal effectively with the decision maker's feeling of ambiguity, uncertainty and vagueness. Fuzzy set theory is based on a set theory which allows for the vague boundary of a set, and thus, it enables the analysis of problems involving ambiguity and uncertainty. Although it has already been a quarter of a century since Zadeh introduced fuzzy set theory, the theory has seen increasing level of acceptance in engineering in the last decade.

It appears that some transportation problems can be dealt with using fuzzy set theory. In fact, a sizable number of attempts have already been made to apply fuzzy set theory to transportation planning and engineering. This paper discusses the application of fuzzy set theory to transportation problems for its potential, limitations, and examples.

2. Fuzziness in Transportation Problems

The nature of many transportation planning and engineering problems is associated inherently with uncertainty. Some of the reasons are:
- An important function of transportation planning process is to predict the consequences of decisions, including that of the status que. The decision cannot be experimented in laboratory environment, rather the

---

[1]. Associate Professor, Civil Engineering Department, University of Delaware, Newark, Delaware 19716

consequences are found only in the future.

- Decisions are often made based on imprecise information. Accuracies are often inconsistent among different data sets used. Data pertaining to human factor, such as perception and feeling, add ambiguity to the information.

- The objective of the decision maker is vague so that the target value of the objective may not be clear. Further, the plan intends to achieve many objectives without clear priorities among them.

- Reasoning and inference for decisions are based on fuzzy logic, in which the implication and the propositional statements are ambiguous due to the fuzziness of the meaning of words and the interpretation. A typical example is the definition of the level of service in Highway Capacity Manual in which many qualifying statements exist.

These characteristics of the problems make fuzzy set theory a suitable approach to some of transportation problems. Traditionally, the unknown has been treated as a random variable of probability theory; however, it may not be just random but rather may be fuzzy variables, and mathematics of fuzzy set theory can be used for the analysis.

For example, estimated travel time of a traveler has both the probabilistic and subjective uncertainty. Delays due to traffic signals, accidents and congestion may be caused by random events. The driver, however, has some controls as to the speed of the vehicle, and choice of route. Furthermore, the estimate may be made conservatively or risky depending on the circumstances and the judgment of the decision maker. Thus, the estimated travel time can be treated as a fuzzy number.

3. Fuzzy Algorithms Potentially Applicable to Transportation Problems

Several algorithms of fuzzy set theory can be potentially applicable dealing with the uncertainty encountered in transportation problems. They are:
Fuzzy data handling techniques
Fuzzy optimization
Fuzzy inference and control
Fuzzy system
Fuzzy measures
Each is explained briefly here.

<u>Fuzzy Data Handling Methods.</u> To conduct analysis based on fuzzy numbers, fuzzy arithmetic operations, fuzzy regression and fuzzy clustering techniques are available. The extension

principle allows computation of equations involving fuzzy numbers. The ordinary regression equation assumes that the relationship between the independent and dependent variables is crisp, and that any scattering of data points are considered to be the random error of the data. In other words, an underlying deterministic relationship is assumed to exist. The fuzzy regression, on the other hand, assumes that the relationship itself is fuzzy, and thus, introduces fuzzy numbers for the coefficients and also considers data points are fuzzy numbers. The scattering of the points are assumed not as the result of the data error but rather inherent to the phenomena.

An extension of the fuzzy data handling includes the fuzzy GMDH(Group Method for Data Handling) by which non-linear relationship between parameters are determined. Fuzzy clustering analysis is also based on the assumption that the data points cannot be classified between different groups in a crisp manner, and that it designates a fuzzy set for each group and assign a membership grade to each data point for its degree of belonging to individual clusters.

Fuzzy Optimization. The ordinary optimization process minimizes or maximizes the objective function given a set of constraints. In the fuzzy optimization, both the objective and the constraints are considered vague and to form fuzzy sets. It is to find the solution which maximizes the least satisfaction of both the objective or the constraint sets. Thus, the objective and the constraints function the same purpose of defining the solution set. This formulation allows multi-objective programming and also accepts prioritizing the constraints.

Fuzzy optimization technique is suited for problems in which both the objective and constraints are not clearly defined; for example, when the objective is to control cost "around " a certain value to meet the budget level, or when the constraints are not known exactly and certain allowance is available. The fuzzy optimization approach can be used for the strategic level of planning. The trade-off between the costs of obtaining precise information to find the optimum versus obtaining less accurate data with a lower cost and find a less optimum solution must be evaluated.

The concept of fuzzy optimization has been applied to the traditional optimization algorithms, such as fuzzy linear programming and fuzzy dynamic programming.

Fuzzy Logic and Inference. Fuzzy logic provides foundations for "approximate reasoning". It represents the majority of applications of fuzzy set theory to the so-called fuzzy based products today. Fuzzy logic deals with not only the true or false of propositions but also the indeterminate or unknown condition. This applies in the case of future events or when the rule contains a vague statement, for example, "If the speed is high, then keep a long headway".

The essence of fuzzy logic is the inference method based on modus ponens. Given that, if x=A then y=B, then if x=A' then y =C'. The proposition can be vague so that it can be written in fuzzy relationship. When x=A' is not exactly A, still an inference as to the condition of y can be made. The consequence can then be inferred with fuzzy membership grade.

Furthermore, if applied in the expert system, several rules can be fired at the same time. This allows the combination of different rules to derive the consequence. This combination of fuzzy rules seems to represent the human decision process well.

Fuzzy System. A complex system in which input, output and states are known fuzzy is a fuzzy system. The fuzzy system can be used for inferring the effect of a certain fuzzy input in a multi-stage process, or for analyzing the system performance. For example, from a given vague state, the cause of the state can be inferred using the fuzzy relationship between the cause and effect. Among the applications of the fuzzy system are for diagnosing the cause of an accident, inferring a consequence of a decision in a complex system involving chain reactions.

Fuzzy Measures. The fuzzy measure is a measure which indicates the likelihood of occurrence of an event. Such event could be either crisp or fuzzy. For many years, probability has been the only measure of uncertainty and the mathematical tool for dealing with randomness. Unlike probability, fuzzy measure is not restricted by the additive constraint. Several fuzzy measures are proposed including the possibility and necessity measures. These measures are based on the idea that the membership function of a fuzzy set represents the possibility distribution, and thus, the how much the set belongs to another set (either fuzzy or crisp) can be measured by the way which the two sets intersect. The fuzzy measure can be used for measuring vague objects by a vague scale, predicting vague outcome, estimating weights for evaluation, perception of distance and time. Combining with the technique of fuzzy integral, it can be used to rank alternatives. It can also be used for determining the relationship between two fuzzy numbers.

4. Example Applications of Fuzzy Algorithms to Transportation Problems

Application of fuzzy set theory may encompass many types of transportation problems. Some of promising applications are listed under the same titles as above:

Fuzzy data handling methods
- Classification and organization of data, delineation of zone boundaries.
- Travel demand forecasting, in particular, regression analysis of trip generation models.
- Cost estimate and economic analysis.

Fuzzy optimization methods
- Fuzzy linear programming for the strategic level planning of resource allocation, scheduling, and network analysis.
- Fuzzy dynamic programming for network analysis.
- Multi-objective programming.

Fuzzy Logic and Inference
- Traffic assignment model.
- Traffic control and signal timing.
- Traffic flow analysis.

Fuzzy System
- Site and traffic impact analysis.
- Land use modeling.
- Traffic accident analysis and evaluation of accident mitigation measures.

Fuzzy Measures
- Evaluation of driver perception of safe separation of vehicles and stopping distance.
- Evaluation of alternative plans involving subjective judgement.
- Evaluation of vehicle design, passenger comfort, safety.

5. Conclusion

The essence of fuzzy set theory lies in:

1. Its ability to analyze natural and human related phenomena using multi-valued logic. Measures of feeling and ambiguity can be expressed by a continuous variables. This approach is suited for analyzing problems involving perception and subjective judgement.

2. It can be used within the framework of the traditional mathematics. Therefore, many of the traditional OR techniques can be modeled with fuzzy set theory.

3. It does not explain the reason why the system is fuzzy rather it merely describes the phenomena.

Terms such as fuzziness, vagueness, ambiguity and nebulous have not been considered "desirable" in the traditional analytical approach. Definitive crisp statements and logic have been preferred for reasoning. This approach, however, has not been sufficient to allow for the analysis and control and reasoning involving human judgement. Fuzzy set theory should offer an alternative to complement the traditional mathematical approach in transportation engineering and planning.

OPERATION OF FREEWAY SIMULATION TOOLS ON I-35W

James L. Wright[1], P.E., James Aswegan[1],
Ping Yi[1], Gene Hicks[1]

INTRODUCTION AND SUMMARY

Throughout the past 15 years, staff of the Minnesota Department of Transportation (MN/DOT) have explored and experimented with several traffic simulation models. Efforts began in the middle 1970's when FREQ6 (Jovanis 1978) was tested on I-35W. Unfortunately, after several tests the effort failed and the work was stopped for several years. In the middle 1980's, the Dept. of Civil Engineering of the University of Minnesota, in conjunction with MN/DOT, began development and testing of a continuum flow model called KRONOS (Michalopoulos 1986). Early testing of this model left users frustrated because of long computer run times and apparent erroneous output. Other models, including TRAFLO (Payne 1978) from the United States Department of Transportation (US/DOT), were also explored, but eventually forgotten because of the large effort required to collect data, test output, and because of lack of the skills required to operate the models. In 1988, MN/DOT management renewed its interest in freeway analysis through simulation. The University of Minnesota was also highly motivated to form collaborations with MN/DOT. This motivation came from the University's interest in participating in the Intelligent Vehicle Highway System program and also from the Center for Transportation Studies which is striving to tie MN/DOT and the Civil Engineering Department closer together.

The effort began with a review of the traffic simulation models available. FREQ10 and KRONOS6.2 were identified as being most likely to offer success because they were user friendly and had continuous support from the developers, while TRAFLO was not used due to complexity in coding the input data and lack of graphic output. A test site was chosen on I-35W on a 8.5 miles

---

[1]Office of Traffic Management, Minnesota Dept. of Transp., Metropolitan District, St. Paul, MN 55155

section in south Minneapolis. Some preliminary runs were
then made with test data from I-35W. Reviewing the
results, MN/DOT identified several deficiencies:
1) Lack of understanding of traffic flow theory and
the modeling aspects.
2) Unable to effectively operate the models.
3) Inadequate traffic data to validate and determine
the accuracy of the models.
4) Model output was very difficult to interpret
because of 1) and 2).

With these deficiencies identified, a more rigorous approach was proposed. A project development plan (figure 1) was established and the material following will discuss our experience in three major areas of the plan, data collection, model validation, and model output.

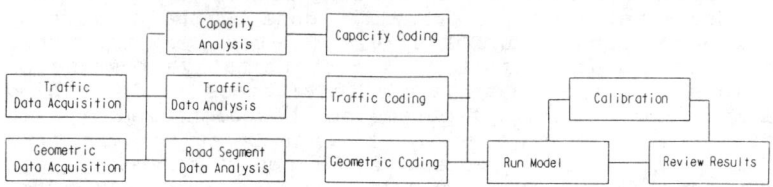

Figure 1   Project Development Plan

DATA COLLECTION

Traffic data is needed at the beginning and end segments of the test site as well as at the entrance and exit ramps to provide the model with traffic demand input. Data at other intermediate segments are also needed to compare with simulation results. Because the loop detectors on I-35W can only provide traffic volumes and occupancies and in general speed information is not available, a comprehensive data collecting effort was made in conjunction with the University of Minnesota. A time correlated data set of traffic volumes, speeds, occupancies, and vehicle classifications was acquired to cover three morning peak periods and three afternoon peak periods. The total data collection effort lasted three weeks in November, 1989 and required a total of 70 people. Figure 2 shows a sample layout plan of the data collection.

MODEL VALIDATION

Model validation involved qualitative model assessment and quantitative model testing. Qualitative model assessment examines model response to various roadway and traffic situations including:

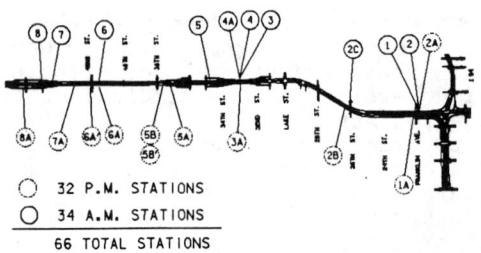

○ 32 P.M. STATIONS
○ 34 A.M. STATIONS
66 TOTAL STATIONS

Figure 2   Data Collecting Stations

1) Entrance ramps with high merging volumes
2) Exit ramps with traffic spillback
3) Congestion propagation at bottlenecks
4) Capacity reductions at weaving areas

The assessment work was primarily done by the university model developers. MN/DOT simulation team, on the other hand, focused its effort on the quantitative testing of the models to evaluate their effectiveness in describing real traffic. A set of parameters has been used by each model to provide degrees of freedom in calibrating the model as it replicates real field observations. These parameters include one or multiple flow-density curves, and a capacity measure at each roadway segment. The flow-density curves used in both FREQ10 and KRONOS6.2 were estimated from the data collected. In capacity estimations, the initial approach was to follow the Highway Capacity Manual. Further analysis led to using the loop detectors and choosing the highest flow recorded. Because capacity is sensitive to traffic mix, weather, flow conditions, driver attention and incidents, the initial capacity selection was only used as an operating maximum. Capacity was eventually used to "fine tune" the model operation so that model output replicates field data as much as possible. Comparisons of traffic volumes as well as speeds to the field data were made and the final results of calibration showed that:
• 90% of volumes were within 9% error
• 90% of speeds were within 17% error

In order to visualize the comparisons, three dimensional contour graphs were created. A speed contour map was displayed in figure 3 where it can be clearly seen that model output compares quite well with data.

MODEL OUTPUT

In order to examine whether the calibrated models can be used for a generalized situation, a prediction run was made by using data obtained in 9 days in November, 1990. Test results were shown in figure 4 where approximately

85% of the predicted volumes are within 10% of error. A 90% confidence interval band based on the field data was established (figure 5) to visualize the "fit" where the solid line represents model output.

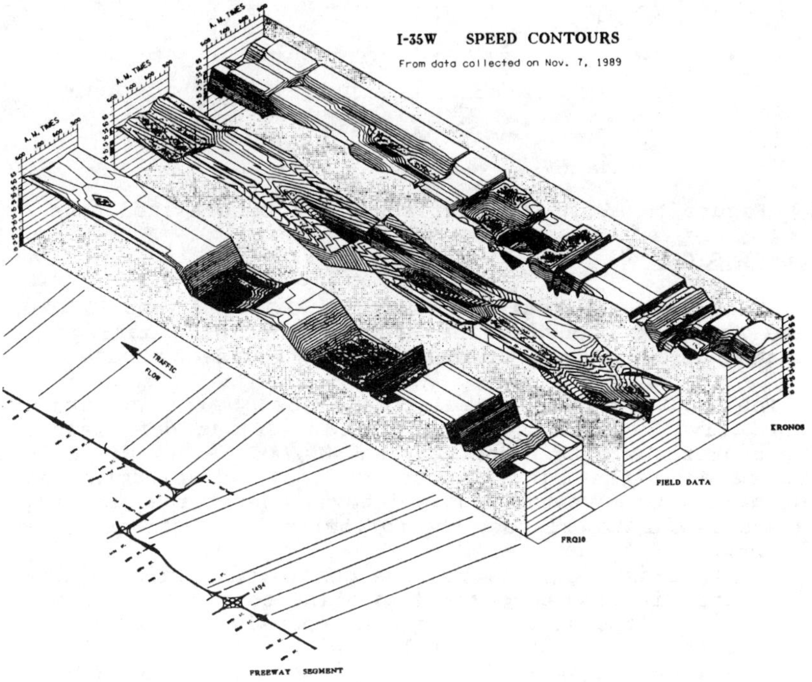

Figure 3 Speed Contours

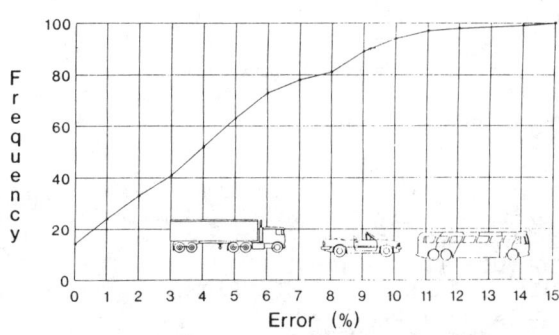

Figure 4 Cumulative Frequency of Flow Errors

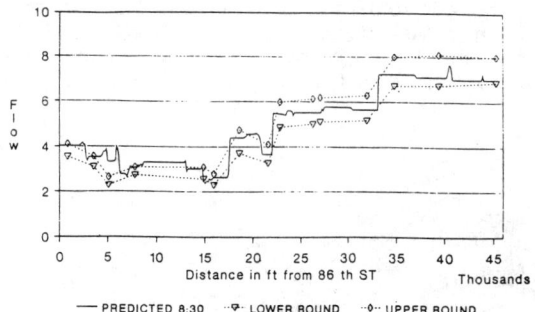

Figure 5   Volume Comparisons in a Generalized Case

CONCLUSIONS

The employed simulation models, KRONOS6.2 and FREQ10, are accurate enough to be used for decision makings in planning, operation, maintenance, and incident management.

The MN/DOT traffic simulation effort began in earnest with five deficiencies identified. These deficiencies were overcome within one year and MN/DOT is now preparing to make further investments in model development and preparing to move the models into an operational setting. Based on our experience, the following conclusions can be drawn:
- It is important to provide continuous training in traffic flow theory and modeling to simulation team.
- It is imperative that a comprehensive set of field data be available to validate a traffic flow model
- It takes a small internal team of motivated and skilled individuals coupled with computer resources and a university development team to produce long term results.

ACKNOWLEDGEMENT

The authors would like to thank David Berg, Kimberly Berggren, and Hyun A. Moon for their enormous contributions to this project.

REFERENCES
1. Jovanis, Paul P., Waiki Yip, and Adolf May. FREQ6PE-A freeway priority Entry Control Simulation Model. Institute of Transp. Studies, Univ. of California, Berkeley, 1978 (UCB-ITS-RR-78-9).
2. Michalopoulos, P. G., and R. Plum, KRONOS-4, Final Report and User's Manual, 1986.
3. Payne, H. J., FREFLO: A Macroscopic Simulation Model of Freeway Traffic: Version 1-User's Guide, ESSOR Report, ES-R-78-1, 1978.

Heavy Traffic Control
Measurement Techniques and Control Strategies
Reinhart D. Kühne[1]

Abstract

Advanced technologies for traffic detection for freeway control systems are described. mm-wave radar detectors are used with a correlation of vehicle reflexion patterns to acquire section related traffic variables like density and travel time. With density as crucial parameter control strategies can be designed which are appropriate especially for heavy traffic. Traffic is regulated within line control and network control systems. Line control uses traffic dependent speed limit, lane attachment, passing prohibition and general warning. Network control influences traffic by alternative route guidances. The optimization of the control systems is done by an expert system.

Introduction

Traffic flow on freeways is described with macroscopic variables. These variables are local variables which refer to traffic conditions in the local vicinity of the measurement site like

traffic volume q [veh/h]
mean speed   v [km/h, averaging interval e.g. 5 min]

and segment related variables which refer to traffic conditions of a whole stretch like

traffic density ρ [veh/km]
travel time t_{travel} [h]

Besides the elementary relation $q = \rho v$ an empirical volume density relation exists $q = Q(\rho)$ (fundamental diagram which devides traffic flow into stable (free traffic) and unstable regimes (congested traffic). The border corresponds to the maximum capacity. In this static description local traffic variables and section related variables can be converted. In a dynamic description where density and mean speed show temporal and spatial variations this conversion is no longer possible. Non-homogeneous traffic flow with temporal and spatial variations is characteristic for dense traffic. Heavy traffic control therefore needs a direct measurement technique for section related

[1] Daimler-Benz AG, Research Center Ulm, Wilhelm-Runge-Strasse 11, W-7900 Ulm, West Germany

variables. These variables regard the traffic dynamics between two measurement sites and take into consideration the spreading of disturbances which are omnipresent in heavy traffic.

Measurement technique for section related variables

As a non-local measurement technique a correlation method is presented which correlates the vehicles series at neighboured measurement sites. When vehicles are re-identified, travel time and traffic density can easily be deduced from time lag and vehicle number lag respectively

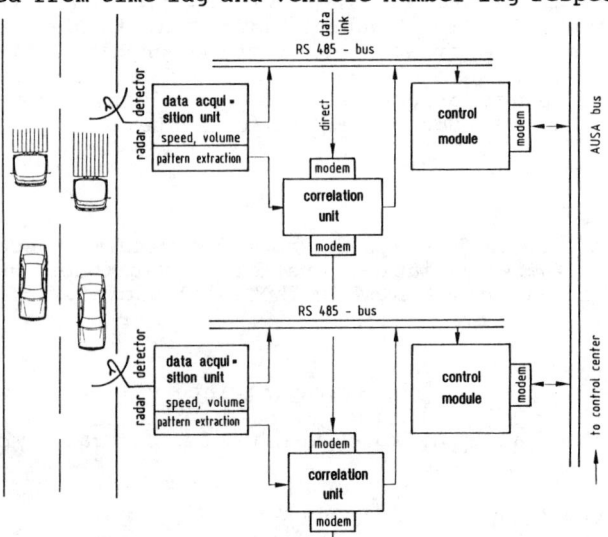

Fig. 1: Measurement set-up for detection of local and section related variables

Figure 1 shows the set-up for section related traffic variable detection combined with standard local variable detection. As passive reidentification criterion the radar reflection patterns of a cw Doppler traffic radar transmitter receiver unit are used. The Doppler radar uses mm-waves of 61 GHz transmission frequency. The Doppler shift of the reflected signal is directly proportioned to the vehicle speed; the envelope of the signal gives a characteristic finger print of the vehicle contours which can be analysed by pattern recognition techniques. Two examples of reflection patterns are represented in Figure 2 in normalized and reduced form. Originally 32 000 values are scanned with a scanning rate

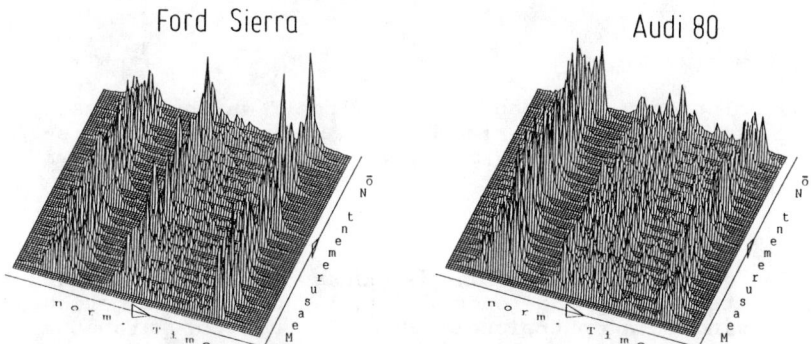

Figure 2: Normalized radar reflection patterns from a 61 GHz CW Doppler traffic radar transmitter receiver unit.

of 16 kHz. By chirpzet Fourier transformation and cutting out the Doppler frequency in a backtransformed and normalized form 200 values are sufficient for presenting the whole reflection pattern information. Application of pattern recognition techniques with covariance matrix formation can reduce the signature information further. Finally a data exchange rate for pattern features transmission less than 1200 bit/s is reached.

Heavy traffic flow as turbulent fluid motion

For efficient traffic control not only a measurement technique for detecting heavy traffic data but also a traffic flow model is needed which is appropriate describing dense traffic. A continuum description of traffic flow uses the equation of continuity

$$\rho_t + (\rho v)_x = 0$$

and an acceleration equation with relaxation to the equilibrium speed-density relation $V(\rho)$ and an anticipation of traffic conditions downstream together with a small viscosity

$$v_t + v v_x = \frac{1}{\tau}(V(\rho)-v) - c_o^2 \frac{\rho_x}{\rho} + \upsilon_o v_{xx}$$

Characteristic is the convection nonlinearity $v v_x$ which expresses acceleration due to velocity gradients. It describes saturation and overproportional amplification of small disturbances.

In the limit of breaking and spreading profiles ($\tau \to 0$) the solution with an arbitrary initial function $v(x,t=0)$

= f(x) reads

v = f(x-vt)

The solution v is dependent on v itself which expresses a feedback loop. The feedback meachnism can be discretized by stroboscopic mapping and simplified to a quadratic recursion

$$v_{n+1} = r\, v_n\, (1-v_n)$$

This quadratic recursion is known as logistic map (Schuster, 1988) and leads to regular or chaotic motion depending on a control parameter. The control parameter can be derived from the slope of the speed density relation (Kühne,1987). The competition of the nonlinearity effects saturation and over proportional amplification produces critical fluctuation near threshold. These critical fluctuations serve as early warning criterion for traffic break down. In Figure 3 two examples of iterates of the logistic map starting from v_o = 0.7 are given, both from the inter-mittent region and both 1 % apart from the critical control parameter.

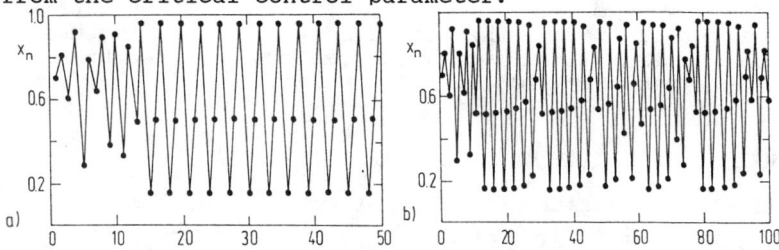

Figure 3: Iterates of the logistic map starting from v_o = 0.7 a) regular; b) chaotic (After Hirsch et al. 1982).

The irregularities and the strong spreading of local measurement data in the dense traffic regime are thus explained by means of the chaos theory similar to the turbulence description of fluid motion. The advantage of this interpretation is a time series analysis which determines the time scale for systematic predictions and the time scale for which only statistical predictions are possible.

Freeway control systems and control strategy

As a most modern example in Europe of a heavy traffic control system the line control and alternative route guidance system Bavaria North Würzburg-Nürnberg may serve. On 16 sign bridges with variable traffic signs (compare figure 4) over a length of 24 km traffic dependent speed limits, lane attachment, passing prohibition for trucks, and general warnings are displayed on variable traffic signs. The algorithm for switching the different signs is based on the above derived early warning criteria detecting the erratic character of traffic flow just before traffic break down by means of the standard deviation of the speed distribution. The traffic control system consists also of an alternative route guidance system (figure 4) which diverses traffic at 3 decision points depending on a traffic prognosis on the main route and on the alternative route. The installed overload prevention program solves the above posted equations for

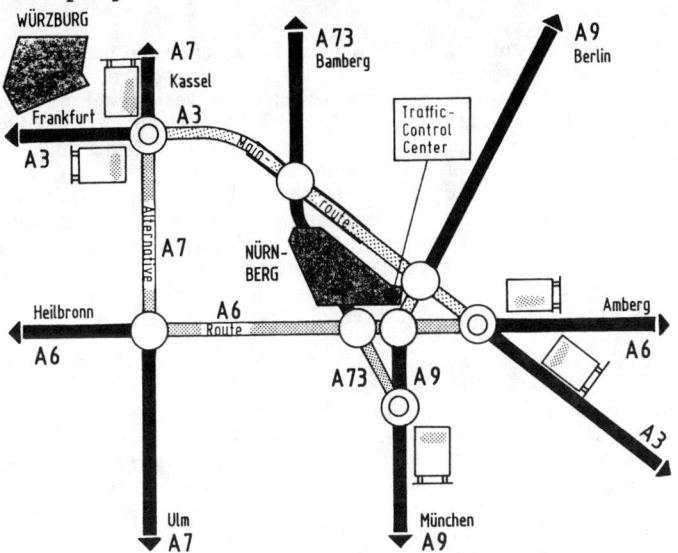

Figure 4: Heavy traffic control system Bavaria North with line control and alternative route guidance.

optimization of the threshold values in the control logic an expert system is under consideration. Measurable values like actual speed and standard devition of speed distribution as well as assessment variables like public reactions and accident rate are input variables for the expert system for fixing the thresholds and cycle times.

References

Hirsch, J.E. et al.(1982) Theory of Intermittency
       Phys. Rev. 25A, pp. 519-533
Kühne, R.D. (1987) in: Gartner, N; Wilson, N (edts.)
       Transp. and Traffic Theory Elsevier, pp. 119-138
Schuster, H.G. (1988) Deterministic Chaos, VCH Publishers

Assessment of the State-of-the-Art of Robotics Applications
in Highway Construction and Maintenance

Tong Zhou and Thomas West

Automation and robotization in highway construction and maintenance are currently under active development in the United States and other countries around the world. Although road construction and maintenance have been well-mechanized over the past decades, it is urgent and feasible to adapt new robotic technology to develop a series of highway construction and maintenance robots that increase the productivity and reduce costs, provide maximum safety to workers and the traveling public, and improve the service quality. As highway robots will be subject to a semi- or unconstructed, harsh and dynamic environment, robustness in design and manufacture is of paramount concern. A thorough literature search has been conducted and this paper presents the status of research and development in this area and highlights the possible applications of robotics in highway construction and maintenance. This paper is a condensed version of a full report.[Zhou]

1. Introduction

While automation typically used in manufacturing appears to be notably befitting the relatively simple and periodic work that highway construction and maintenance normally demands, road construction and maintenance have many distinct features from those of manufacturing industries. For instance, road work is performed over a large area with workers moving from one location to another. In addition, handling large amounts of material and the harsh work environment make road construction an unattractive area for robots. The expense to the California Department of Transportation (Caltrans) for highway maintenance alone reached 1.43 billion dollars in 1989. During the past decade, annual maintenance costs have greatly increased, aggravating the maintenance engineer's difficult task of coping with rising traffic volume and public demands for increased service. While applying robots to roads is not easy to conceive and is a more challenging task to scientists and engineers, it is an inevitable trend as has been seen in other industries.

---

1   Associate Professor, Department of Mechanical Engineering, California State University, Sacramento, CA 95819.
2   Electrical Engineer, California Department of Transportation, 5900 Folsom Boulevard, sacramento, CA 95819.

2. The State-of-the-Art

Although there are no commercial applications for robots in highway construction and maintenance as yet, there are encouraging results from a number of experimental studies and surveys. Road construction and maintenance tasks have a significant potential for gradual automation due to the repetitiveness and magnitude of the activities. Relatively flat and straight, complete with navigational signing, the highway may be considered as a semi-structured environment. Highway systems can be mapped and stored in a computer and the road signing could be used to locate and position robots. Although some initial efforts may tend toward machine automation rather than robotics as defined by the Robotics Industry Association, the ultimate goal is maximum utilization of robotics technology to automate highway operations. The core of much research has been in the development and testing of prototypes of robot systems that could be forerunners of commercially available machines.

2.1 Developments in Construction automation

Skibiewski and Hendrickson have previously reported the following three examples of automation for road construction and maintenance. For completeness, examples previously reported are briefly mentioned here.[Skibiewski]

Miller Formless Systems Co. of France has developed four automated slipform machines for sidewalk curb and gutter construction. All the slipforming machines have the ability of operating in a playback mode while following a preset and pre-cleaned path of work.

A robotics excavator prototype has been developed at Carnegie Mellon University. The excavator uses a sensor-built surface model to plan its digging action and interprets sonar data to determine the presence and location of buried obstacles. Appropriate trajectories are generated and executed by the backhoe and six-degree-of-freedom manipulator, providing nine degrees-of-freedom for tool positioning and location.

Spectra-Physics of Dayton, Ohio, has developed a microcomputer-controlled, laser guided grading machine. Laser light receptors mounted on the equipment measure the height of the blade relative to an established laser plane.

The remotely operated John Deere excavator, used for excavating and handling hazardous materials, has one color camera mounted to the right-front fender and another mounted on the roof of the cab to give the operator a clear view of the work area. The operator can reposition the roof camera to view gages and warning lights.

Takenaka construction company in Japan has developed a system which uses robots to arrange the steel reinforcing bars, and to distribute, compact, level and finish the concrete. A computer simulation system calculates the proper path and flow rates for the concrete distribution robot. The concrete leveling robot is a small bulldozer which uses laser leveling instrumentation to control the height of the blade and an automatic navigation system to control the path. The surface finishing robot also has a tracked mobility system with automatic navigation, and uses eight trowels which completely rotate around the vehicle so as to smooth the surface of the concrete both in front and back of the tracked vehicle as it moves. The latest version of this robot finishes concrete at the rate of 300 m^2 per hour, the equivalent of three human workers.

A general purpose positioning system has been developed in France which consists of a rotating laser transmitter installed at a fixed point on the work site. The mobile robot is equipped with suitable sensors providing geometrical information which is processed in real time by a computer carried on the mobile robot.

A pavement cutting robot system has been constructed in Japan on a trial basis to automate pavement cutting when a road is to be cut and recovered. The accuracy required for pavement cutting is approximately +/_ 20 mm. In addition, more than 70% of the slurry was recovered, the noise level averaged about 2dB lower than conventional machines, and the operator could operate the machine from a safe position because most of the cutting work, except for aspects of placing the guide tape, is performed by remote control or by automatic operation.

Other concrete cutting robots such as the AWH Cutting System (Taisei of Japan) and the Abrasive Jet Cutting Robot (Kajima of Japan) which are remotely controlled and have tactile sensors, are currently being used at work sites.[Kobayashi]

2.2 Developments in maintenance automation

The Japanese companies Kandenko and Tokyo Electric Power have jointly developed a new robot for repairing the reinforced concrete structures of underground cable conduits. The robot does this by applying a super high-pressure water technique to scrape off concrete that has deteriorated by exposure to salt. The water jet technique enables the work to be done in a clean environment with little noise or dust. The robot will automatically scrape off the deteriorated concrete at a speed of 2 m^3/hour. There is no danger that cable, piping or any other fitting will be damaged in the process since the robot will work to a prescribed depth to a high degree of accuracy. It is sufficiently versatile to be able to clean off the rust from reinforcing steel at the same time.[Rudall]

A rapid runway repair (RRR) equipment development project is under way at the University of Florida and the U.S. Air force Tyndal Base. The autonomous performance of rubble removal, crack filling, and nondestructive testing, among other functions, is being designed. An important benefit to the Air Force for implementing such a system will be the removal of humans from a life threatening work environment in combat situations.[Skibiewski]

A program for maintenance automation entitled, Automated Highway Maintenance Technology (AHMT) is being conducted jointly between The California Department of Transportation (Caltrans) and the University of California at Davis (UC Davis) and is currently two years into development of semi-automated or automated highway maintenance equipment.

The first and second phases of a project to automatically place highway raised pavement markers (RPM) is now completed with emphasis on total safety for the operator and increased efficiency of the overall operation. Through the development of telerobotic placement of the RPM, the operator is no longer exposed to vehicle traffic, and the dangerous materials used in RPM placement. The third phase, currently underway, will offer an order of magnitude increase in operational speed and efficiency. Final phases will attempt to eliminate lane closures and/or slow moving trains of protection vehicles currently needed for RPM repair or installation. This will be accomplished as RPM placement nears highway speeds.

Pavement damage caused by cracks are costly, and often cracks tend to reflect through conventional new overlays quite rapidly. Sealing the cracks at an early stage protects not only the pavement surface, but also the road base from water intrusion. UC Davis, Caltrans and Bechtel Corporation have been awarded a contract by the National Strategic Highway Research Program to develop automated equipment to seal pavement cracks, a very labor intensive and slow moving maintenance function. Line scan camera technology will allow the detection of cracks greater than 3 mm while the real-time processor determines a path for the robotic sealant applicator to follow, at vehicle speeds of up to 16 km/hr.

Two projects are currently underway to further increase the efficiency and safety of Caltrans' paint striping operation. The first project, data acquisition and control, attempts to understand the lifecycle of the paint stripes and the optimum application techniques such that control of application rate, paint temperature, pressure, etc. can be automatically adjusted. This automatic, optimal application would help toward achieving longer paint stripe life and depending on conditions, a potential for tremendous savings in paint resources. The second paint stripe project will provide a method for automatic duplication of the original stripe, alleviating the driver of the tedious line tracking currently undertaken, and allowing him to focus on surrounding traffic. Actuators will automatically servo the paint gun outrigger to the correct lateral position and switch the paint gun on and off per the existing stripe at speeds to 65 km/hr. Two different stripe sensing technologies are being investigated including laser and line scan camera technology.

Caltrans picks up and bags approximately 200,000 m^3 of garbage each year from California's highways. The bags are then retrieved by Caltrans personnel within three to five days. This project will see the development of equipment to automatically retrieve the refuse bags through the operation of a single operator driven vehicle. The requirement is that pickup occur at speeds to 16 km/hour, thus reducing exposure of employees while increasing efficiency.

Several automated pavement distress data collection vehicles are now available throughout the world. Komatsu Ltd. of Tokyo, Japan has developed an automated image collection device with on-board processing capability using scanning lasers. This vehicle is used to automatically check a road surface for its condition, seeking out any signs of cracking, wheel rutting and longitudinal unevenness, while traveling at a maximum speed of 60 km/hr. A 2880 hertz laser beam scans the road surface and the reflected signal is processed and stored on videotape. Roads to 4 m wide with ruts and potholes to 0.25 m deep can be evaluated.[Fukuhara]

The French Petroleum Studies Company presented machinery at the Fifth International ATEC Congress in June of 1986. One piece of equipment, the cone dispenser, automatically and quickly sets up and removes one or two rows of warning cones. The cone dispensing unit has 10 vertical cylindrical magazines in which as many as 240 cones can be carried. The cone dispenser can place cones to the right or left of the unit, and collect cones from one side while it is laying them on the other. Thus a lane can be displaced laterally from 0.5 to 3.5 m. The driver can independently program all of these operations. The working speed is 15 km/hr.[Point,88]

The Technique Special de Securite Company (TSS) has developed a new lane separation system. The mobile lane separator, which can place road marker blocks that are similar to the New Jersey type of barrier at either 12 or 30 km/hr. The system includes a series of concrete blocks that are interconnected by an articulated arm around a vertical axis. The machine picks up the blocks on one side and deposits them on the other side, always protected from traffic by its own road marking blocks.[Point,88]

Societe Nicolas of France has developed a multipurpose traveling vehicle (MPV) used for a variety of maintenance tasks. The vehicle was initially intended to mow grass around roadway curbs at speeds between 0 and 20 km/hr, but now has three additional accessories; a coring machine, a brushwood clearer, and a vehicle carrying platform.

For the past several years, researchers at the University of Cincinnati have been developing a mobile robot that can serve as a robot lawn mower. This robot has a novel omnidirectional vision system that can detect between cut and uncut grass and uses ultrasonics for obstacle detection. A teach mode is used initially to lay out the general shape of the area. Automatic operation includes the necessary high-level path planning, path positioning, and collision avoidance capabilities.

3. Summary

As can be seen, many construction and maintenance tasks have been considered reasonable candidates for robotic automation and many more are under consideration. Due to the semi-structured environment of the highways, an autonomous robot working on a specific road tasks may not be seen soon, but the rapid development in robotics assures that we will see such a robot in the near future. In addition, development of robotics applications in space exploration, mining, building construction, and other hazardous environments can be utilized by highway engineers.

References

[Zhou] T. Zhou, "Assessment of the State-of-the-Art of Robotics Applications in Highway Construction and Maintenance", Report to The California Department of Transportation, January 1991.
[Skibiewski] M.Skibiewski and C. Hendrickson, "Automation and Robotics for Road Construction and Maintenance", Journal of Transportation Engineering, Vol.116, No 3, May/June 1990, pp. 261-271.
[Rudall] B.H. Rudall, Reports and Surveys, Robotica, 1989, Vol.7, pp. 183-189.
[Fukuhara] T. Fukuhara, et al., "Automatic Pavement Distress-Survey System", ASCE Journal of Transportation Engineering, Vol.116, No.3, May/June 1990, pp. 280-285.
[Point,89] G. Point, "Some Equipment Innovations to Improve Safety of Highway Users and Maintenance Workers", Transportation Research Record 1183, pp. 78-82.
[Point,88] G. Point, "Two Major Innovations in Current Maintenance: The Multi-Purpose Vehicle and the Integrated Surface Patcher", 67th TRB Annual Meeting,

Applications of Robotics and Automation in Highway Maintenance Operations
B. Ravani[1], member ASME and T. H. West[2]

Abstract

This paper deals with the applications of Advanced Automation and Robotics in highway maintenance operations. The basic components of Robotics and Automation technologies are studied and their appropriate forms for highway maintenance applications are identified. Within each area of highway maintenance operations, maintenance tasks with high potential for the application of the Robotics and Automation technologies are then discussed.

Introduction

One of the major problems in highway maintenance operations is the potential hazard from the on-going traffic to the human workers. In California alone, maintenance personnel suffered over 4,700 work related injuries (45 of which were fatal) between 1974 and 1987 (House Resolution 1989, and Special Hearing 1989). Additional problems in highway maintenance operations include increase in traffic delays and congestions, lack of reliability and efficiency and high operating costs in the use present maintenance procedures and methods. Although Advanced Automation and Robotics (AAR) technologies have become very important in manufacturing, work in space, undersea work, work in nuclear industry and medicine, their applications in automating highway maintenance operations are only recently being exploited.

In many industrial applications, it has been demonstrated that the use of AAR technologies can lead to increased productivity, reduced hazards to human workers, more flexibility, higher reliability and lower operating costs. It is therefore natural to expect similar improvements in highway maintenance operations through the utilization of the AAR technologies. In particular, automating highway maintenance operations has the potential to greatly reduce the number of injuries and deaths to highway maintenance workers.

In response to this, the University of California-Davis and the California Department of Transportation (CALTRANS) have established the Advanced Highway Maintenance Technology (AHMT) program aimed at automating certain highway maintenance operations. This paper first gives a brief description of the basic components of this program. It then provides a functional description and classification of the components of the AAR technologies and develops the functional architechture of robotic systems suitable for highway maintenance applications. Finally a listing of several maintenance tasks, corresponding to each area of highway maintenance operations, with potential for application of AAR technologies is provided.

The AHMT Program

The AHMT program has been established to investigate application of AAR technologies in highway maintenance operation. The program has the following specific goals:
- decreasing the risk of injuries or death of highway workers by reducing exposure time to work area hazards,

---

[1] Department of Mechanical Engineering, University of California-Davis, Davis, CA 95616
[2] Division of New Technology and Research, CalTrans, Sacramento, CA 95819

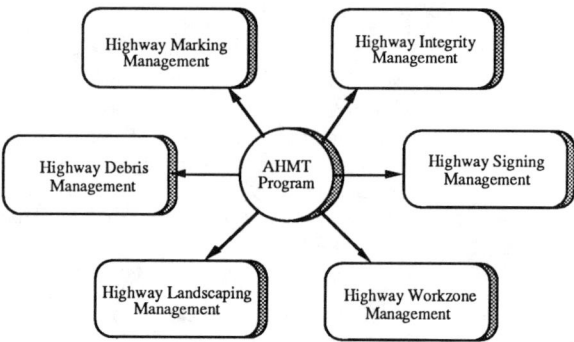

Figure 1: Areas of Highway Maintenance Operations within the AHMT Program

- minimizing hazards to the traveling public due to the slow moving highway maintenance function,

- minimizing maintenance related traffic delays by maximizing the operating speed of various maintenance tasks,

- increasing the reliability, durability and efficiency of highway maintenance functions

The AHMT program has identified several areas of highway maintenance operations for the application of AAR technologies. These areas are shown in Figure 1 and include highway marking management, integrity management, debris management, signing management, landscaping management and highway work zone management. The maintenance tasks for each of these areas and their corresponding automation requirements are described in a later section.

Advanced Automation and Robotic Technologies

The AAR technologies (see, for example, Ferrell and Sheridan 1967, Ravani and Floyd 1988, Webster and Ravani 1988) consist of several component technologies integrated together making up an autonomous, tele-operated or a supervisory controlled system. The component technologies are usually grouped together to provide the basic architecture of a generic sub-system of an automated system. The sub-systems of a present day AAR technology can be functionally characterized to include a manipulation system, a locomotion system, a external sensing system, a control and coordination system, a processing and decision making system, and a communication system including a man-machine interface. Since many highway maintenance tasks require application of other materials such as bitumin, paint, sealant or raised pavement markers to the roadway, a typical robotic system for such applications will also include a material storage and retrieval system. Although the AAR technologies have been applied to applications such as construction (Oppenheim and Skibniewski 1988) or mining (Kassler 1988) which have some features similar to those of highway maintenance tasks, their application in this area is new and only recently evolving. The sub-system architecture of a robotic system suitable for highway maintenance is shown

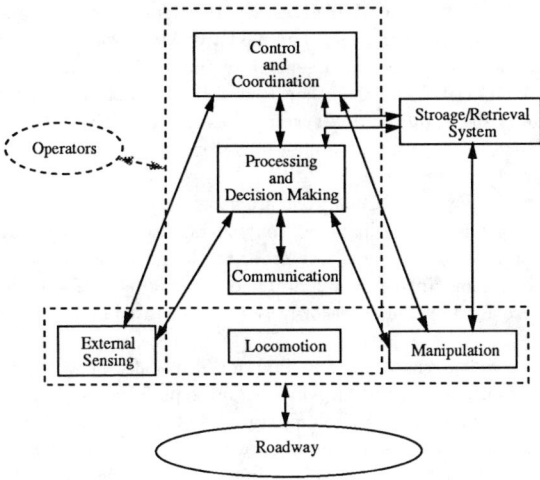

Figure 2: Functional Architecture of a Robotic System for Highway Maintenance Applications

in a functional block diagram form in Figure 2. The first three functions (manipulation, locomotion and external sensing) require interactions with the roadway. The communication function is achieved through the man-machine interface and similar to the material storage/retrieval with the operator. The level of the dependence of the control, processing and decision making functions on the operator depends on the level of the autonomy of the system. These functions become more independent of the operator as the system moves from a manually controlled system to a tele-operated system, next to a supervisory controlled and finally an autonomous system.

In such a robotic system, the manipulation system usually consists of a mechanical positioning device such as a robot arm and an end-effector or tool. The function of the arm is to position the end-effector or tool and the latter is used as an applicator to perform the task. In a robotic system for highway maintenance the manipulator can be, for example, a Cartesian arm (xy table with a servo z axis) and the end-effector may consist of specialized tool used as an applicator depending on the application. In a robotic system the locomotion function is usually achieved by a mobility unit and a navigation system. This system usually consists of sensing and computer equipment and provides mobility guidance for the mobility unit. In highway maintenance, the terrain for mobility is highly structured, a conventional truck or a towed vehicle can therefore be used as the mobility unit. In such a case, the navigation system then consists of the driver and the locomotion system is said to be human guided. The system can also be made more advanced and under supervisory control of the driver by augmenting the driver with advanced displays and sensory control schemes.

The external sensing function is usually intended to identify changes in the work environment and use the information to influence the actions of the manipulation or the navigation system. In highway maintenance, three kinds of external sensors may be needed. The first kind can be used, if necessary, to augment the information available to the driver for navigation purposes. The next two kinds will be used for the maintenance function one providing global sensing of the locations an the roadway where maintenance may be needed and the other used in the form of local sensing to plan the path and actions of the manipulation system. In more primitive forms of an automation system for roadway repair, the human operator may provide all or the navigation and global sensing functions.

The control and coordination functions are usually provided by one or more actuators, the power transmission mechanisms, controllers, feedback sensors and the control software. These functions are usually internal to the manipulation and the material storage/retrieval system. In highway maintenance some level of the control function may have to be provided by the operator in the tele-operated or supervisory controlled mode.

The processing and decision making function is usually provided by a computer system. In a fully automated system, the computer performs all the data processing and decision making functions. In a tele-operated or system under supervisory control only limited or no decision making is performed by the processing system. The human operator is used in the loop for decision making and part of the control function. The processing unit then performs much of the computation and logical operations for some of the control and coordination functions. In highway maintenance tasks, sometime it may be necessary to augment the data processing and decision making function with a data base or a knowledge base providing information on the maintenance process variables.

The communication function for a robotic system is achieved through data communication channels and protocols and man-machine interfaces. In roadway maintenance, at least in many initial implementations, the communication function is only performed through the man-machine interfaces. It will consist of displays, hand controllers, user programming systems and other means of inputting information to the robotic system and observing the actions of the system by the operator in the most transparent manner.

For a robotic and automation system to be effectively used in highway maintenance functions, it should be able to operate for a few hours before it is re-loaded. This means that proper material storage/retrieval is an important requirement for such a system. This would consist of storage tanks, canisters and other equipment integrated with material handling systems such as conveyors, feeders and hoses used to deliver different materials (e.g. adhesives, bitumin, paint, etc.) to the manipulation system.

Highway Maintenance Operations

This section provides a summary of some of the highway maintenance tasks associated with the areas of highway maintenance operations identified earlier. Some of the automation requirements for each task are also identified and provided in the summary.

1. **Highway Marking Management**
    (a) Pavement Marker Replacement
    (b) Paint Re-stripping

2. **Highway Integrity Management**

(a) Crack Sealing

   (b) Pot Hole Filling

3. **Highway Debris Management**

   (a) Litter Bag Pick Up

   (b) On Road Refuse Collection

   (c) Hazardous Spill Clean Up

   (d) Snow Removal

4. **Highway Signing Management**

   (a) Sign and Guide Marker Washing

   (b) Roadway Advisory

5. **Highway Landscaping Management**

   (a) Vegetation Control

   (b) Irrigation Control

6. **Highway Work Zone Management**

   (a) Automatic Warning System

   (b) Lightweight Movable Barriers

   (c) Automatic Cone Placement and Retrieval

Acknowledgement

This work was supported by the California Department of Transportation.

References

[1] House Resolution No. 27, *Assembly Journal*, State Assembly of California, 1989.

[2] *Special Hearing on the Safety and Protection of Caltrans Employees*, California Legislature Assembly on Public Employees, Retirement, and Social Security, Joint Publication Office, Sacramento, CA, 1989.

[3] Ferrell, W. R. and Sheridan, T. B., 1967, *Supervisory Control of Remote Manipulation*, IEEE Spectrum, 4:81-88.

[4] Kassler, M., 1988, *Robot in Mining*, ibid, 2:897-902.

[5] Oppenheim, I. J. and Skibniewski, M. J., 1988, *Robots in Construction*, Int'l Encyclopedia of Robotics, R. C. Dorf ed., John Wiley & Sons, 1:240-249.

[6] Ravani, B. and Floyd, R. E., 1988, *Technological Forecasts for Robots*, Int'l Encyclopedia of Robotics, R. C. Dorf ed., John Wiley & Sons, 3:1702-1709.

[7] Webster, John G. and Ravani, B., 1988, *Teleoperator Control Using Telepresence*, ibid, pp. 1710-1718.

Perception and Control for Automated Pavement Crack Sealing

by Chris Hendrickson[1] M, ASCE, Sue McNeil[2] M, ASCE, Darcy Bullock,[3]

Carl Haas,[4] AM ASCE, Daniel Peters,[5] Darrin Grove,[6]

Kent Kenneally,[7] and Shannon Wichman[8]

Abstract

Pavement maintenance is a major transportation engineering activity that has seen only partial mechanization and virtually no automation. Crack sealing is a typical maintenance task in this area that has not been automated to any degree. In this paper, we report on a program of laboratory experiments to develop perception and control methods to enable effective automation of routed pavement crack sealing. These experiments included:

- Investigation of sensor options and combinations.

- Implementation, calibration and testing of various image processing steps.

- Registration and alignment of video and range sensing data to permit pavement crack modelling.

- Implementation and testing of real time effector controls for tracing pavement cracks.

The series of experimental implementations led to the development of a laboratory prototype system for automated pavement crack sealing. Implications of our experiments for perception of general pavement distress and for control of maintenance equipment will be drawn with respect to automation feasibility, reliability of automated equipment, and the economic efficiency of such systems.

---

[1] Prof., Dept. of Civil Engineering, Carnegie Mellon University, Pittsburgh, PA 15213

[2] Assoc. Prof., Dept. of Civil Engineering, Carnegie Mellon University, Pittsburgh, PA 15213

[3] Res. Asst., Dept. of Civil Engineering, Carnegie Mellon University, Pittsburgh, PA 15213

[4] Asst. Prof., Dept. of Civil Engineering, Univ. of Texas at Austin, Austin, TX 78712

[5] Res. Engr. Undergrad. Fellow, Engineering Design Research Center, Carnegie Mellon University

[6] Res. Engr. Undergrad. Fellow, Engineering Design Research Center, Carnegie Mellon University

[7] Res. Engr. Undergrad. Fellow, Engineering Design Research Center, Carnegie Mellon University

[8] Res. Engr. Undergrad. Fellow, Engineering Design Research Center, Carnegie Mellon University

1. Introduction

Automation of roadway maintenance presents substantial opportunities to reduce labor costs, improve work quality, and decrease worker exposure to roadway hazards. This paper describes a research project investigating an automated method for sealing routed pavement cracks; more information is available in [Hendrickson 91]. A prototype laboratory system was constructed and demonstrated as part of the project. Possible designs for field systems were investigated, and the economic feasibility of automation for this application was analyzed. The major result of the research was demonstrating the feasibility of identifying, mapping and tracking routed pavement cracks in a laboratory setting. Using a combination of video imaging and electronic range scanning, routed cracks 2 cm. (0.9 in.) in width could be located, mapped and traversed with an accuracy of better than 1 cm. (0.5 in.).

In the recommended field system for an automated crack sealing system, a lead vehicle would tow an xy-table assembly or robot arm (Figure 1). The lead vehicle would transport generators, computers, sensor processing hardware, and controllers. Mounted on the rear of the vehicle would be a video camera used for identifying routed cracks. The xy table is similar to a pen plotter with a cart moving within a rectangular frame. Tools mounted on the xy table would be: a depth sensor for verifying cracks, a hot air lance for cleaning cracks, and a sealant wand for sealing cracks. Bulk supplies of sealant material and propane trail behind the xy table.

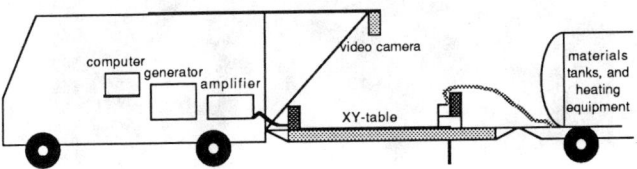

Figure 1: Illustration of a Possible Field System for Sealing Cracks

The laboratory prototype emulated the proposed system in a stationary environment and demonstrated the following steps:
- The vision system identified potential cracks and generated a traversal plan for the depth sensor to collect range data.
- The range data was fused with the video data into a unified map. Potential cracks identified by the vision system, but not corroborated by range data, were dismissed as filled cracks, shadows or oil spots.
- The center lines for the corroborated cracks were identified and a trajectory minimizing the motion of the tools was generated. The cracks were traversed first with a tool cleaning the crack with compressed air. After the cracks were prepared, the laboratory system again traversed the crack and filled the crack volume.

A photograph of the laboratory apparatus appears in Figure 2.

Figure 2: Photograph of the Laboratory Demonstration System

2. Sensing Pavement Cracks

Sensing hardware used in the laboratory prototype system included:

- A commercial VHS camcorder was used to generate the video signal required by the image processing board.

- An image processing board, commonly referred to as a frame grabber, was used to convert an analog video image signal to a digital matrix of numbers representing grey scale. In the laboratory prototype, individual raster cells in the digital image were approximately 4 mm by 4 mm (0.2 in x 0.2 in).

- An infra-red laser range sensor was used to develop three dimensional profiles of the roadway surface. The range sensor had a "footprint" of 15 mm (0.6 in) in the configuration used in the laboratory prototype and range readings were taken every 2 cm. (0.9 in.) across the pavement surface. The range sensor used cost approximately $ 2,000.

Other sensors were considered for the system, particularly sonar as an alternative for range sensing. However, sonar systems were susceptible to environmental disruption from wind and precipitation.

A dual sensor system was used because either video or range sensing alone were not adequate. Surface dark spots or lines could be easily mistaken for cracks in the video image. Range sensing was relatively slow since the sensor had to be moved above each spot being mapped. Range sensing from a single point would be expensive to achieve the required accuracy, and a linear array of multiple sensors would likewise be expensive. Thus, video imaging was used to identify area of potential cracks, and range sensing was used to confirm the location and identity of pavement cracks.

3. Representing Pavement Cracks

A flexible method of representing pavement surfaces was applied in this research based on a two-and-a-half dimensional "quadtree" computer model. The "half" dimension refers to elevation or range data. This representation model accommodated numerous operations for surface perception and modeling such as image filtering, registration of multiple sensor data and generation of crack traversal patterns. The C++ object oriented language was used for software development. The model has been applied to general condition assessment data derived from several commercial sources in addition to the automated maintenance system [Haas 90a, Haas 90b].

4. Sealing Pavement Cracks

Once pavement cracks were identified and mapped, specialized control software was used to traverse cracks on the pavement surface for cleaning and sealing. The motor controls and related software in the laboratory system proved to be quite accurate for single and repeated movements, with an accuracy over a 3m by 3m (10 ft. x 10 ft.) area of 5 mm. (0.2 in.) However, the xy frame was insufficiently rigid to permit rapid movement of the blower and filler. As a result, the sealing operations for cracks on pavement scenes required several minutes. A more rigid frame would alleviate this problem.

5. Alternative System Designs

Several alternative system designs were considered for commercial and field systems. These options included:
- Continuous movement systems with a set of linear nozzles. Unfortunately, control of sealant application would be difficult since instant on-off is required, and the cost of crack perception would be high with such systems.
- Small, mobile, highly maneuverable robots tethered to a field vehicle were considered as an alternative to the xy table. These type of effectors could provide continuous sealing but might increase costs and control problems.
- Alternative sensing systems such as single point laser range sensors were examined. These type of sensors have the potential of dramatically reducing the time required for range scanning, but the existing technology could not provide a fine enough

resolution.

- Alternative effectors such as robot arms could replace the xy-table system. However, these pose difficulty in terms of load bearing capacity, controlling contact forces, spatial constraints and computational intensity.

6. Economic Analysis

A robotic cracksealing system can have significant and substantial economic benefits. These benefits include reduced labor costs, and improved quality and safety. A review of literature and reports related to cracksealing practice and an informal survey of six states and one province provided data on crack sealing practice and expenditures for this project. Although crack sealing practices vary significantly between states, it is a widely used maintenance procedure. It is also relatively labor intensive with labor costs representing over 50% of costs on a per lane mile basis. It is estimated that about $125 million per year is spent by states on crack sealing. This is considered to be conservative as it ignores expenditures by counties and cities as well as airports, turnpikes, other authorities and private organizations. Complete automation of this process would require about 400 crack sealing units nationwide.

Benefits would be realized in the form of reduced labor costs, improved quality and improved worker safety. Assuming a 4 year system life, system acquisition costs of $100,000 and annual operating and maintenance cost of $25,000 per year, a 10% discount rate, productivity rates comparable to existing approaches and the elimination of two laborers, the net savings realized using the automated crack filler may be as high as $60,000 per unit per year or 19% of costs on a national basis. This is equal to $24 million per year in national savings due to labor costs alone.

Acknowledgements

This program was supported by the Strategic Highway Research Program (as part of the New IDEAS program), the National Science Foundation (as part of a Research Experience for Undergraduates program), the National Sciences and Engineering Research Council of Canada, and Carnegie Mellon University. The opinions and views expressed herein reflect those of the authors and not necessarily those of the sponsoring organization.

7. References

[Haas 90a]　　Haas, Carl and Chris Hendrickson, "A Model of Pavement Surfaces," *Transportation Research Record*, Vol. 1260, pp. 91-98, 1990.

[Haas 90b]　　Haas, C. and C. Hendrickson, *A Model of Pavement Services*, Technical Report R90-191, Department of Civil Engineering, Carnegie Mellon University, Pittsburgh, PA, Sept. 1990.

[Hendrickson 91]　Hendrickson, C. and S. McNeil, *Investigation of a Pavement Crack Filling Robot*, Technical Report, Dept. of Civil Engineering, Carnegie Mellon University, Pittsburgh PA 15213, 1991.

Highway Pavement Surfaces Reconstruction by Moire Interferometry

Sidney A. Guralnick[1], Eric S. Suen[2] and Christian Smith[3]

ABSTRACT

The necessary basis for any cost-effective highway and bridge deck pavement management program is reliable, complete and up-to-date information concerning the state of pavement surface. An imaging system using the Shadow Moire method for the quantitative inspection of highway pavement surface in real-time has been developed. The pavement surface reconstruction algorithm presented herein uses two moire interferograms of any point on the surface to obtain overlapping information for correct elevation estimation.

INTRODUCTION

Statistics published by the Federal Highway Administration indicate that maintenance and rehabilitation of highway pavements in the United States requires over $17 billion dollars a year. Despite these vast expenditures the nation's roads and bridges are deteriorating at an alarming rate. Annual maintenance expenditures for roads and bridges will probably not increase during the next few years. Hence, the only option available to reduce or arrest deterioration is to significantly increase the efficiency of maintenance operations through better management practices and the deployment of improved technology.

Conventional road analysis techniques or manual visual inspection techniques are costly and time consuming processes which are inadequate to the task of providing meaningful quantitative information. The need for such information has motivated the development of automated methods for its acquisition. Studies have shown that the shadow moire method is best suited for rather coarse measurements on large surfaces. It is relatively simple to apply and the necessary equipment is relatively inexpensive. The proposed shadow moire method for the real-time inspection of highway pavement surfaces using advanced computer-controlled video and image-processing techniques is capable of providing the needed surface geometry information efficiently.

In this study, two moire interferograms of any point on the surface are captured to obtain overlapping information to provide correct elevation estimation. Hence, almost forty thousand overlapped images would be needed to cover a thirty mile stretch of highway (assuming each image

---

[1] The Perlstein Distinguished Professor of Engineering, Illinois Institute of Technology, Department of Civil engineering, 3200 S. State Street, Chicago, IL 60616.

[2] Assistant Professor, Illinois Institute of Technology, Department of Civil Engineering, 3200 S. State Street, Chicago, IL 60616.

[3] Graduate Research Assistant, Illinois Institute of Technology, Department of Civil Engineering, 3200 S. State Street, Chicago, IL 60616.

captured on the video tape contains information on four feet of highway surface along the direction of travel). The current strategy has been to attack the problem in two phases. The first phase, or screening phase, simply processes the video tape to identify "bad frames" and store their respective frame number pairs in an output file for later analysis. A "bad frame" is a captured moire interferogram that deviates from a specified set of flatness criteria. The frame is also roughly classed for type of failure depending on the severity of its deviation from the flatness criteria. The motivation for screening is based on the knowledge that only those parts of the roadway that are sufficiently deteriorated are of interest to the repair and maintenance manager. The second phase, or analysis phase, reduces the "bad frame" moire interferogram pairs to yield relative surface elevation information which may be used to reconstruct the pavement surface as a three dimensional surface using a commercially available topographical software package. This paper presents an algorithm to achieve the second phase goal.

CONFIGURATION OF THE PROTOTYPE SYSTEM

Moire interferometry, due to its high sensitivity and accuracy, has become one of the most important means of analyzing surface strain and displacement of deformable objects. Because of the tedious and repetitive calculations needed to process moire fringes manually, this technique has not been widely used in the past. Now that relatively inexpensive micro computer based equipment has become available for image processing and image analysis, moire interferometry offers promise as a means to rapidly acquire valuable surface topography information. A block diagram of the automated video image analysis system developed in our laboratory for this investigation is shown in Figure 1. The system consists of a high resolution electronic shutter CCD camera, a high resolution industrial quality 3/4" tape VCR, a video monitor, a real-time image processor subsystem and a host micro computer. This system features a high degree of automation and yields adequate images and accurate results.

DESCRIPTION OF COMPUTER ALGORITHMS

Studies have shown that a single moire image is ordinarily not sufficient to yield complete information about the gradient of a set of contour lines. That is, each and every anomaly on a surface would require a known reference point to establish correct surface elevations. An alternative solution would be to compare two images of the surface taken at different elevations. The surface reconstruction algorithm developed herein consists of a series of programs which perform specific image processing tasks to produce relative elevation values.

Interferogram Acquisition and Digitization

Acquisition and digitization of the interferograms is basically an administrative process because the functions for these tasks are built into the video deck and image processor. The computer directs the initialization of the video deck and the image processor, acquires the bad frame pair from the data file, directs the video deck to the appropriate frame location, and then directs the image processor to store the video image as a digital image in its memory. As the image is digitized it is sent through a Look-Up Table (LUT) which inverts the values of all the pixels in the image. This initial inversion process will reduce subsequent computer processing time.

Binarization of Digital Image

The binarization of the moire interferograms is accomplished by selecting an appropriate threshold for each image. The threshold is determined by the histogram of the digital picture image. Typically, the image histogram for moire interferograms will be a bimodal distribution. This bimodal distribution is fitted with a tenth order polynomial and the valley is selected as the threshold for binarization.

Determination of Moire Fringe Centerlines

The fringe centerlines are, by definition, the contour lines of the given image. These

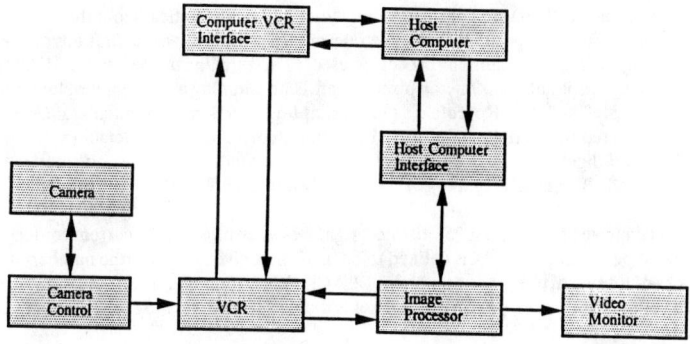

Figure 1. Block Diagram of System Configuration

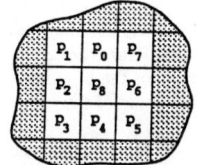

Figure 2. Definition of Near Neighbors of P_8

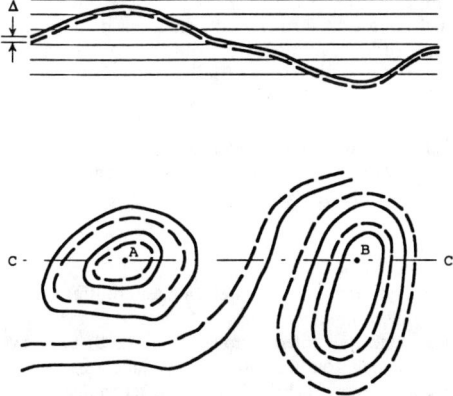

Figure 3. Two Overlapping Thinned Images and the Corresponding Surface Profile

centerlines (or medial lines) are not conveniently described mathematically for arbitrary shapes and curves. Consequently, algorithms have been developed which rely on definitions of the medial lines. For digital images, a thinning process is used. The thinning process simply deletes known boundary pixels until only one line of pixels is left. The thinning algorithm developed herein is adapted from Stefanelli and Rosenfeld's (1971) simplified version of Hilditch's (1969) algorithm (hereafter referred to as Hilditch's modified thinning algorithm). The deletability of any subject pixel p8 in the digital image is dependent on a set of conditions to be satisfied by the near neighbors of p8. The near neighbors of p8 are shown in Figure 2.

The image processor has 256 distinct brightness intensities. In the current context WHITE is defined to be 255 and BLACK is defined to be 0. Hence, given $B(p8)$ as the number of nonzero neighbors of p8 and $A(p8)$ as the number of BLACK WHITE or WHITE BLACK patterns in the ordered set $\{p0, p1, p2, \cdots, p6, p7, p0\}$ a pixel is deleted if the following conditions are met:
1. $510 \leq B(p8) \leq 1530$,
2. $A(p8) = 2$,
3. (a) $p0 \cdot p6 = 0$, or
   (b) $p2 \cdot p4 = 0$, or
   (c) $p4 = 0$, and $p0 =$ previously unmodified p8.

The above thinning algorithm is not able to identify all proper boundary pixels. In some images this leads to a number of spurious branches that are visually known not to be a part of the actual fringe centerlines. These spurious branches are found by an algorithm designed to identify the ends of branches that were a part of the original image boundary and their respective branch node locations. The branches are then deleted from the image. A number of conditions are tested to insure correct identification and deletion of the branches. It is felt that this is a reasonable approach based on the known continuous nature of the moire fringes.

The basis for the algorithm used to detect spurious branch ends is an "inverted" version of Hilditch's modified thinning algorithm. Instead of seeking deletable boundary pixels, this algorithm seeks to delete all the interior pixels of the fringes. Hence, the result will be the outline of the original binarized image. One cycle should be sufficient to identify all interior pixels and delete them. The outlined image is then compared to the thinned image. Pixels which are a part of the outlined image and the thinned image are suspect branch ends. It should be noted that the subject of interest should be sufficiently thick so that the outline is guaranteed not to be a part of the medial line for any fringe in the image. A further set of conditions are imposed on the thinned image to identify branch ends missed by the outlining algorithm:
1. $B(p8) = 255$, and
2. $A(p8) = 2$.

If $B(p8) = 0$ is satisfied for any pixel, it is immediately deleted since it is not attached to anything. Branches that are attached to the fringe medial line form junctions with the medial line. These junctions are identified by the following conditions:
1. $510 < B(p8)$, and
2. $A(p8) > 4$, or $B(p8) > 1020$.

Having identified branch ends and junctions, a cleaning subroutine is used to delete the spurious branches. The cleaning subroutine first labels or "colors" the branch end and junction points with different brightness intensities referred to as END (200) and JUNCTION (80) respectively. It then follows each branch from the end seeking a junction. Branches and medial lines are, by definition, continuous. Hence, a branch following technique has been developed which allows the identification of all pixels belonging to a given branch.

Assignment of Relative Fringe Order
Once the medial lines of the two images to be "overlapped" are obtained, the relative fringe orders can be assigned automatically. The first image will be referred to as the original image. The second image will be referred to as the reference image (both images are images of the same surface at different heights relative to the shadow moire grating). The fringes in the second image are labeled or "colored" GREY (150). Figure 3 shows the two overlapping thinned images and the corresponding surface profile. The original image is analyzed to locate a fringe that spans the entire image. This fringe is labeled as the reference fringe. The rest of the fringes in the original image are then labeled according to their relationship to the reference fringe and the fringes in the reference image.

Calculation of Relative Surface Elevation
After the relative fringe orders are assigned, the relative elevation values are calculated. The image was initially obtained with the shadow moire grating at an angle to the highway surface. Hence, the image of assigned fringe orders is the contour map of a tilted surface with respect to the grating. Relative elevation values are computed for this surface. The relative elevation values of the actual surface are obtained by subtracting the tilt from the tilted surface values. The relative elevation values are computed as

$$\text{Elev} = -j \cdot \tan\theta \cdot x_0 + \Delta h \text{ (shade - lo\_shade / INCREMENT)}$$

where j is the vertical image line position, θ is the angle of the tilted grating to the plane of the pavement, x_0 is the image conversion factor in the vertical direction (in/line), Δh is the contour interval (in), shade is the brightness intensity assigned to the fringe order of the pixel of interest, lo_shade is the brightness intensity of the lowest fringe order in the image, and INCREMENT is the brightness intensity difference between fringe orders. These elevation values may be used to construct a surface profile using currently available topographical software.

CONCLUSIONS

With the system described herein for real-time inspection of highway pavement surfaces, a pavement surface reconstruction algorithm was developed and implemented which uses two moire interferograms of any point on the surface to obtain overlapping information for correct elevation estimation. It is believed that the deployment of the scientific apparatus and the computer software described herein will cost-effectively provide the quantitative information on surface topography that is needed to effect significant improvements in the necessary data collection phases of sophisticated pavement management programs. This improved data will, in turn, permit highway engineers to optimize the expenditure of highway maintenance funds to arrest deterioration without generating significant new maintenance expenditures.

ACKNOWLEDGMENTS

The research presented in this paper was sponsored by the National Science Foundation, Grant No.CES-8814738, and the support provided thereby is gratefully acknowledged.

REFERENCES

1. Buitrago, J. and Durelli, A.J., "On the interpretation of shadow moire fringes", Experimental Mechanics, Vol. 18, No. 6, June 1978, pp. 221-226.
2. Guralnick, S.A. and Suen, E.S., "Real-Time Inspection of Pavement Surface by Moire Patterns", Proceeding of the SPIE's Optical Engineering Midwest Conference, Sep., 1990.
3. Pflug, L. and Oesch, S., "Moire Applied to Real Time Inspection of Highways," Journal of Transportation Engineering, Vol. 112, No. 2, March 1986, pp. 163-171, ASCE.
4. Stefanelli, R. and Rosenfeld, A., "Some parallel thinning algorithms for digital pictures", J. Ass. Comput. Mach., Vol. 18, no. 2, 1971, pp. 225-264.

Design Considerations for Automated Pavement Crack Sealing Machinery

Steven A. Velinsky[1] & Kenneth R. Kirschke[2]

Abstract

The purpose of this paper is to discuss the numerous considerations for designing automated machinery for the sealing of cracks in pavement. Operational requirements are presented and a generic system architecture for such machinery is discussed. Two distinct automated machines are proposed, one to seal only longitudinal cracks, and the other to seal general cracks in pavement. The functions of the component subsystems is also presented.

Introduction

Worldwide, a tremendous amount of resources are expended annually maintaining highway pavement. In California alone, the state Department of Transportation (Caltrans) spends about $100 million per year maintaining approximately 33,000 lane-miles of flexible pavement (Asphalt Concrete - AC) and 13,000 lane-miles of rigid pavement (Portland Cement Concrete - PCC). A portion of these maintenance activities involve the sealing and filling of cracks (approximately $10 million per year). The purpose of crack sealing and filling is to prevent the intrusion of water and incompressibles into the crack, while crack filling is additionally used to hold broken pieces of pavement together. When properly performed, these operations can help retain the structural integrity of the roadway and considerably extend the time between major rehabilitation.

The sealing and filling of cracks are tedious, labor-intensive functions. In California, a typical operation to seal transverse cracks in AC pavement involves a crew of eight individuals which can seal between one and two lane miles per day. The associated costs are approximately $1800 per mile with 66% attributed to labor, 22% to equipment and 12% to materials. Furthermore, the procedure is not standardized and there is a large distribution in the quality of the resultant seal. In addition, while crack sealing/filling, the work team is exposed to a great deal of danger from moving traffic in adjacent lanes.

The crack sealing/filling operation is an ideal candidate for the infusion of advanced technologies in order to automate the process. Automated crack sealing/filling machinery has the potential to:

- Minimize the exposure of workers to the dangers associated with working on a major highway.

---

[1] Associate Professor, Department of Mechanical Engineering, University of California-Davis, Davis, CA 95616.

[2] Civil Engineer, Office of New Technology and Research Management, California Department of Transportation, 5900 Folsom Blvd., Sacramento, CA 95819.

- Considerably increase the speed of the operation.
- Improve the quality and consistency of the resultant seal.

Increasing the speed of the operation will in turn reduce the accompanying traffic congestion since lane closure times will decrease. The combination of the increased speed and the higher quality seal will prove to be extremely cost effective and reduce the frequency of major highway rehabilitations.

In order to have the greatest impact, such machinery should satisfactorily perform the following functions automatically:

- Sense the occurrence and location of cracks in pavement.
- Adequately prepare the pavement surface for sealing/filling with the appropriate methods; for example, any operation that is deemed necessary such as removing entrapped moisture and debris, preheating the road to ensure maximum sealant adhesion, refacing of reservoirs, etc.
- Prepare the sealant/filler for application; i.e., heat and mix the material, etc.
- Dispense the sealant/filler.
- Form the sealer/filler into the desired configuration.
- Finish the sealer/filler.

Additionally, the machinery will have many other more detailed overall functional specifications related to safety, cost, reliability, etc.

Operational Requirements

For purposes of brevity, we will now consider only sealing operations. The first fundamental questions related to sealing of pavement cracks are: which cracks should be sealed and when should they be sealed? These answers to these questions are quite difficult to generalize as they have been the subject of many studies, and thus an automated sealing machine need not be required to make such decisions. However, given information on the general location of cracks and joints to be sealed, such machinery should be able to locate cracks with sufficient accuracy for automated machine operation.

The various types of operations that this machinery should address includes:

- AC/PCC longitudinal crack sealing.
- PCC longitudinal joint sealing.
- PCC transverse joint resealing.
- PCC transverse crack sealing.
- AC transverse crack sealing.

Without getting into too many details of the required operations for each type of seal, one recognizes that the simplest of these operations is longitudinal sealing, and the most complex operation is AC transverse crack sealing. Machinery that is developed to address these two extreme operations (in terms of difficulty) could easily be modified to address any of the others.

Machine System Concepts

Based on the operational requirements for sealing the different types of cracks, we envision two distinct automated machines; one to seal only longitudinal cracks, and the other to seal general cracks in pavement. While these machines would share many of the same component subsystems and components, the overall manner in which these machines will operate are quite different. The development of a machine to automatically seal or fill cracks of sometimes arbitrary geometry (transverse cracks) is quite challenging. The primary challenges arise from:

- The necessity for the sensing system to identify the presence of cracks in addition to locating their position.

- The number of degrees-of-freedom required to follow an arbitrary crack.
- The required unsteady operation of many of the components.

In contrast, a machine to perform the longitudinal operations has several different requirements including:
- Fewer required number of degrees-of-freedom to follow longitudinal joints and cracks.
- Considerably different sensing requirements (i.e., the presence of longitudinal cracks is known and the sensing system is thus required only to locate the position of the crack).
- The steady-state operation of many of the components.

Furthermore, the simpler longitudinal operations are expected to be performed at a much higher rate of speed.

Machine Architecture

The overall system architecture of the two automated sealing machines is identical and is depicted in Figure 1. This system architecture includes four primary systems, and in this figure, each system block includes a partial list of the types of components that may comprise it. The four primary systems are: Crack Sensing System, Applicator Assembly and Peripherals, Positioning System, and Integration and Control System. Of course, all of these systems would in turn be mounted or towed by a support vehicle. The Crack Sensing System will be primarily responsible for locating and describing roadway cracks and joints. The Applicator Assembly and Peripherals includes all the hardware necessary to mix, heat, dispense, shape and finish sealant/filler, and to prepare the pavement including reservoir creation. This system may be comprised of any number of dispensers, valves, cutting tools, heaters, air compressors, etc. The Positioning System will include the hardware necessary to move the applicator assembly end effectors in such a manner that they follow the required path. The Integration and Control System will coordinate the Crack Sensing, Applicator Assembly and Peripherals, and Positioning Systems. It will transform the information from the Sensing System into a desired path for the applicator assembly. The Integration and Control System will then control the motion of the applicator through the Positioning System as well as controlling the individual functions of the Applicator System. Additionally, the Integration and Control System will monitor all of the peripherals to ensure proper sealant/filler supply, sealant/filler temperature, heat supply, etc. More details of each of these systems follows.

The Crack Sensing System

The purpose of crack sensing is to determine the location of cracks and joints on the road in sufficient detail so that crack preparation, sealant/filler application, and shaping can be performed automatically. A wide range of sensing technologies have been investigated in order to select the most appropriate crack sensing system. The Crack Sensing System is comprised of two subsystems; a global machine vision based subsystem with a full lane width field of view, and a local laser range finder subsystem to verify the presence of cracks and joints.

The machine vision based subsystem uses a CCD (Charge Coupled Device) video camera for the sensor as in other recent roadway crack detection studies (Bomar, 1988; Haas & Hendrickson, 1989; Mahler, 1990). We have developed a histogram analysis based image processing algorithm in order to identify cracks in both AC and PCC pavements (Kirschke & Velinsky, 1991). This algorithm has been successful in locating unprepared, and relatively dirty, cracks as small as one-eighth inch in width in both of these types of pavement. However, the global machine vision subsystem cannot distinguish true cracks from apparent cracks which are actually shadows, wet

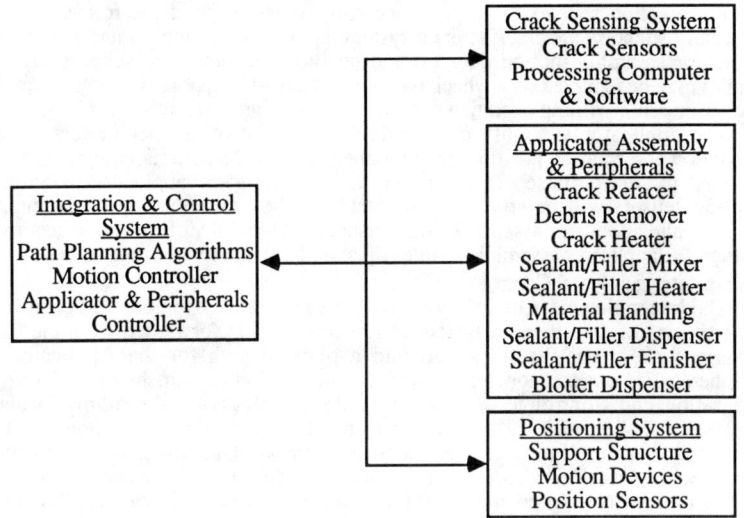

Figure 1. Component Systems of the Automated Crack Sealing Machine

areas, oil spots, or previously sealed cracks. As such, the local laser range subsystem is used to complement the vision system. Preliminary testing and algorithm development has clearly shown the feasibility of this type of sensor for overcoming the vision system's shortcoming. For the longitudinal machine, only the local subsystem will be necessary.

Applicator Assembly and Peripherals System

The key to sealed/filled crack/joint longevity is directly attributable to proper cavity preparation. Additionally, for most hot applied sealants/fillers, a uniform surface heating prior to application is desirable. While there are significant differences between the practices of the various states, for the widest possible applicability, an automated machine must also allow for pavement routing. The Applicator Assembly and Peripherals System prepares the crack/joint and applies and finishes the sealant.

Recent studies (e.g., Rossman, et al., 1988) have shown that the ideal method for crack preparation should include the use of a two phase hot air system. The primary phase of this system should include a source of high temperature and high velocity compressed air to remove entrapped aggregate/vegetation and moisture. The second phase of the heating system should be used to warm, to approximately 280°F (based on recent studies at Caltrans), the surrounding horizontal crack margins to ensure a highly adherent bond between the surface and the sealant material. Once the crack is cleaned, dried and heated, a suitable sealant can be applied. The desired patch configuration requires moderate penetration of sealant material into the vertical crack surface, and sealant penetration can be sharply increased as the temperature differential between the surface and the sealant material is minimized.

Positioning System

The various components of the crack sealing machine are physically connected through the positioning system. The Positioning System will consist of three primary components; the machine support structure, the applicator assembly motion devices, and the position sensors. The machine support structure is the framework that physically supports the crack sensing system, applicator assemblies and peripherals, applicator assembly motion devices, and position sensors. The support structure (frame) may be mounted on a wheel assembly which will support the frame during the crack/joint sealing/filling operation as well as during high speed road travel.

The applicator assembly motion devices consist of stepper motors, drives, translation and rotation positioning components, cables, and other equipment which are used for the purpose of positioning and actuating the applicator assemblies. Position sensors will be employed to determine the exact position of the support structure and applicator assemblies with respect to the road surface and to accurately position the applicator assemblies while adjusting for a variety of disturbances.

System Integration and Control

The Integration and Control System oversees the entire operation and coordinates the activities of the other subsystems. The information forwarded from the Crack Sensing System will be translated into a planned path for the Applicator and Peripherals System components (crack/joint preparation equipment, etc.). Thus, the Integration and Control System will include the necessary algorithms to plan a crack/joint sealing path. This path corresponds to the relative positioning of the Applicator System. If multiple applicators are employed, the Integration and Control System will need to first allocate cracks to the individual applicators and will do so in a manner to maximize speed and avoid interference. This system will keep account of the actual position of the total machine and its components by interacting with sensors on the Positioning System. It will additionally monitor the Applicator Assembly and Peripherals to ensure adequate volume and temperature of sealant/filler, air, etc. Following the planning of the appropriate path(s), the Integration and Control System will control the motion of the applicator(s) and the individual applicator functions.

Conclusions

This paper has discussed numerous considerations for the design of automated machinery for the sealing of cracks in pavement. Such machinery is being developed under the joint UC-Davis/Caltrans project acknowledged below, and prototype machinery will be field tested late in 1992.

Acknowledgement

The authors gratefully acknowledge the Office of New Technology and Research of the California Department of Transportation and the National Research Council, Strategic Highway Research Program who have shared in the support of this work under SHRP H107A.

References

Bomar, L.C.; Horne, W.F.; Brown, D.R.; Smart, J.L. (1988) "A Method to Determine Deteriorated Areas in Portland Cement Concrete Pavements", NCHRP Project 10-28 Report, Gulf Research.

Haas, C.; Hendrickson, C.T. (1990) "A Model of Pavement Surfaces", Dept. of Civil Engineering, Carnegie Mellon University Technical Report #R90-191.

Kirschke, K.R.; Velinsky, S.A. (1991) "A Histogram Based Machine Vision Algorithm for the Automated Sensing and Sealing of Pavement Cracks", submitted for publication.

Mahler, D.L. (1990) "Real Time Image Processing for Pavement Crack Monitoring", presented at 69'th Annual TRB Meeting.

Rossman, R.H.; Tufty, H.G.; Nicholas, L. (1988) "Value Engineer Study-Repair of Transverse Cracking in Asphalt Concrete", FHWA Final report.

COMPARISON OF TRAFFIC SIGNAL SYSTEMS IN AUSTRALIA AND NORTH AMERICA

Kevin J. Fehon[1]

Abstract

Dynamic coordinated traffic signal systems were developed in Australia in the 1970's, and are now installed in many countries. In contrast, strong interest in new generation control philosophy has only recently developed among North American practitioners. The results of applicable comparative studies to are reported.

Introduction

The development of coordinated traffic signal systems in North America has had a rather mixed history. The first electric traffic signals appeared in Cleveland in 1914, and the first electrically connected coordinated signals appeared only three years later in Salt Lake City's "checkerboard" system (ITE, 1971). By 1928 New York City had some 3,000 traffic signals, but very few were under its "block" coordinated control. Denver introduced the computer age in 1952 with an analogue computer using six sampling detectors to select the most appropriate signal timing from a library of programs.

During the 1950's, and 1960's, many cities installed interconnected electromechanical systems, typically offering one to three coordination plans, which define the cycle length, phase splits and offset for each intersection. The first application of a digital computer for traffic signal control was in Toronto in 1963 (Camkin and Sims, 1974), following a pilot study in 1960. However, most developments since then have improved the equipment, more than the control logic.

---

[1] R.J. Nairn and Partners, 214 Northbourne Ave., Braddon, ACT 2601 Australia.

The development path followed in Australia has been rather different. The first major coordinated signal system to be commissioned was in Sydney's CBD in 1963. Programs were selected from a multiple-plan library by time of day, backed up by observer evaluation of system conditions through closed circuit television. Observers could modify the timing of individual signals on line, or manually change to a new plan.

The first Australian application of a digital computer to traffic control was to 17 intersections in Surfer's Paradise in 1970, with a UTCS style system. The commissioning of Digital PDP-11's to control arterial and CBD systems in 1971 signalled the genesis of the Sydney Coordinated Adaptive Traffic System (SCATS).

Signal coordination in Australia is dominated today by SCATS. Initially developed in Sydney, it has been installed in all Australian capital cities except Brisbane, many provincial cities, and in numerous cities in New Zealand, Ireland and Asia. SCATS systems now control almost 5,000 signals in a total of 28 cities (Lowrie, 1990).

Description of SCATS

The control philosophy, development application and hardware of SCATS have been adequately described in the literature. See, for example, Sims and Dobinson (1979), Longfoot (1982), Lowrie (1982) and Sims (1978).

SCATS implements a third generation philosophy. As described in Lowrie (1990), control of traffic is generally effected at two levels ("strategic" and "tactical") which together determine the three principal signal timing parameters of coordination: cycle time, phase split and offset. Strategic control is basically concerned with calculation of suitable timings for the areas and sub-areas, while tactical control operates at the local intersection level, within the constraints imposed by the strategic control.

The grouping of intersections into sub-systems, typically of one to ten signals, is completely free form, constrained only by the boundaries between regions, which typically include up to 120 signals. There is one critical intersection in each sub-system, which can link together to form larger systems. These links may vary according to the prevailing traffic conditions.

The basic traffic measurement used by SCATS for strategic control is a measure analogous to degree of saturation on each approach. This measure is used to determine cycle length and the relative split between phases, and converted to a pcu equivalent to select direction and magnitude of offset.

The degree of saturation is measured by stop line loop detectors typically 4.5 metres long, which allow differentiation between slow moving heavy flows and widely spaced light flows. The associated algorithms are self-calibrating, as are many of the SCATS parameters.

Acceptance of Dynamic Control

For a long time the benefits of dynamic operation were dismissed by adherents to the fixed time philosophy. The Transport and Road Research Laboratory (TRRL) in the U.K. undertook experiments in Glasgow, Scotland, and in Madrid, Spain, during the 1970's. These experiments were similar to those undertaken by FHWA in Washington, D.C., and it was claimed that traffic actuation offered no real advantage over a fully optimized fixed time plan.

This gave rise to a number of scientifically controlled experiments which proved beyond doubt that dynamic operation gave significant benefits compared to systems which use stored timing patterns.

Luk, et al. (1982) reported on a comparison between SCATS dynamic control and fixed time plans derived using TRANSYT, in Parramatta. This showed that for both arterial and CBD type systems, SCATS provided better or similar traffic performance. This and subsequent studies also examined the difference in performance between SCATS and Linked Vehicle Actuated (LVA) signal operation along arterial roads. Again, statistically significant improvements were found to result from SCATS operation.

Nairn (1983) treated the total system performance rigourously, including side street delay and accounting for vehicles travelling along only part of the route. Table 1 illustrates the differences in MOE's measured for an arterial road in Melbourne. The same study estimated the Benefit/Cost ratio of introducing SCATS to 1800

signals in Melbourne at 59. Although the greatest
benefit comes from coordination of isolated signals
(BCR=83), there is still a substantial improvement in
moving from LVA to SCATS control (BCR=21).

Table 1: Percentage improvement of SCATS over other control modes. (Nairn, 1983)

Parameter	AM Peak LVA	AM Peak ISO	AM Bus LAV	AM Bus ISO	PM Bus LVA	PM Bus ISO	PM Peak LVA	PM Peak ISO
Travel time	*	27%	*	11%	*	11%	7%	17%
Fuel	*	15%	*	6%	*	*	*	17%
Stops	34%	123%	*	57%	*	43%	24%	17%
Delay	*	87%	*	59%	*	33%	*	17%

* Difference not statistically significant.

This series of comparisons provided the final impetus for
the development of SCOOT by the TRRL.

Implications for North American Cities

The comparisons of SCATS with LVA operation are
particularly pertinent to many cities today. The many
so-called "closed loop" systems (which often aren't)
operate in a semi vehicle-actuated mode identical to LVA.
For arterial road operation in particular, there would be
substantial vehicle operating cost savings from the use
of SCATS style operation over either time of day or
traffic responsive pattern selection.

A number of the larger cities (e.g., Los Angeles and New
York) are currently installing or expanding 1.5
generation systems such as UTCS Enhanced. Even with such
features as Critical Intersection Control, the overall
performance cannot match dynamic operation. The
shortcomings are recognized by the fact that a number of
sites have been identified to demonstrate small SCOOT
systems. This would be an ideal time to demonstrate the
relative strengths of SCATS, with its longer period of
development and practical approach to coordination.

Conclusions

The increasing emphasis on traffic management in U.S.A.
and Canada will see increasing pressure placed on traffic
signal systems. Traffic engineers are beginning to take
note of the successes of SCATS and SCOOT in improving the

efficiency in traffic operation in Australia and the U.K. The Australian experience is particularly relevant to North America since the style of road network and travel characteristics are so similar.

REFERENCES

Camkin, H.L. and Sims, A.G. (1974), Coordinated Traffic Signals, N.S.W. Department of Motor Transport, Traffic Engineering Branch, Sydney, Australia.

ITE (1971), Traffic Control Devices: Historical Aspects Thereof, Institute of Traffic Engineers, Washington, D.C.

Longfoot, J.E. (1982), SCATS - Development of Management and Operation Systems, in International Conference on Road Signalling, Institution of Electrical Engineers, London.

Lowrie, P.R. (1982), SCATS - Principles, Methodology and Algorithms, in International Conference on Road Signalling, Institution of Electrical Engineers, London.

Lowrie, P.R. (1990), SCATS - Sydney Coordinated Adaptive Traffic System: A Traffic Responsive Method of Controlling Urban Traffic, Roads and Traffic Authority of N.S.W., Sydney, Australia.

Luk, J.Y.K., Sims, A.G., and Lowrie, P.R. (1982), SCATS - Application and Field Comparison with a TRANSYT Optimised Fixed Time System, in International Conference on Road Signalling, Institution of Electrical Engineers, London.

Nairn, R.J. and Partners (1983) Fuel Usage Evaluation of Linked Signal Systems, for Roads and Traffic Authority of Victoria, Melbourne, Australia

Sims, A.G. (1978), The Sydney Coordinated Adaptive Traffic System (SCATS) in Area Traffic Control Workshop, 9th Australian Road Research Board Conference, Brisbane.

Sims, A.G. and Dobinson, K.W. (1979) SCAT The Sydney Coordinated Adaptive Traffic System: Philosophy and Benefits in International Symposium on Traffic Control Systems, Berkeley, CA.

FULLY-DISTRIBUTED CONTROL OF SIGNAL NETWORKS

George List, Associate Member ASCE[1]
Siew Leong[2]
Saaju Paulose[3]

Abstract

A fully-distributed real-time control strategy is described for signalized networks. The signals make compatible, simultaneous decisions about cycle length, offset and splits, without the support of a supervisory master controller. The feasibility of this concept is explored for simple arterials and open grid networks, using NETSIM and a GPSS/H-based simulation model.

Introduction

Fully-distributed control (FDC) is a new concept that has not heretofore been applied to signalized networks. Described initially by List and Pond (1989), it involves the real-time setting of network-wide signal timing parameters by individual signals, in parallel and simultaneously, without the aid of centralized control. It is distinctly different from system-wide real-time control as described by Robertson (1986), Sims (1979), and Luk (1984) in that it lacks centralized control. It is patterned after the type of rule-based decision-making that would be used by an expert system working off the knowledge base of an experienced traffic engineer.

Fully-Distributed Control Concepts and Logic

With FDC, the signals communicate with each other via a databus, sharing information about network volumes
_____
[1]Associate Professor, Department of Civil Engineering, Rensselaer Polytechnic Institute, Troy, NY, 12180.
[2]Graduate Student, Department of Civil Engineering, Rensselaer Polytechnic Institute, Troy, NY, 12180.
[3]Graduate Student, Department of Civil Engineering, Rensselaer Polytechnic Institute, Troy, NY, 12180.

and signal timing. A data cycle is established, with each signal having a block of time for broadcasting data to the rest of the network. Each signal indicates its operating mode - either full-actuated or coordinated - and traffic flows - by movement and/or approach.

List and Pond (1989) describe a version of FDC that involves pattern recognition and timing plan selection, while this paper presents a more robust realization. A complete description of it can be found in List and Leong (1991). The strategy is proactive in that the signals set timing plans in anticipation of major changes in traffic flow patterns and fine-tune those plans, cycle by cycle, to minimize the sum of queue lengths observed at the start of green on all approaches. It is reactive in that, for both operating modes - full-actuated and coordinated, the signals minimize delay and attempt to clear queues on every cycle using normal semi- or full-actuated multi-phase operation.

Case Study Analysis

A GPSS-based model has been developed. GPSS/H is not the authors' first choice for a modelling environment. NETSIM would be preferred, but time and budget precluded such code revisions.

The GPSS/H model has been validated against both NETSIM and TRANSYT-7F (T7F) for situations where comparisons are possible. The GPSS/H model tends to give lower average delay values than NETSIM because it allows the detectors to "move" with the end of queue, meaning that with large flows, there is no detector blockage. Its estimate of queue length is much better than that generated by NETSIM and its ability to distribute green time efficiently is enhanced.

Case Study Findings

Table 1 shows the total delay statistics for four case studies. In Case #1, where the intersection volumes are nearly equal, NETSIM indicates that semi-actuated operation is better than either pre-timed or full-actuated operation, but that the difference is minor. Similarly, T7F indicates a slightly better performance for pre-timed operation (compared to semi-actuated operation). The GPSS model also indicates a better performance by full-actuated operation.

In Case #2, NETSIM indicates better performance for pre-timed than full-actuated operation, and much worse

TABLE 1: Total Network Delay for Case Studies #1 to #3

Case	TRAF-NETSIM			TRANSYT-7F		GPSS/H		
	F-A	P-T	S-A	P-T	S-A	F-A	P-T	F-D
1	108.7	111.0	94.7	97.4	99.0	60.8	110.3	76.4
2	46.0	42.5	98.3	44.4	45.8	32.3	45.8	39.7
3	44.7	19.8	21.0	12.5	18.3	33.8	28.3	35.4

F-A: Full-actuated
P-T: Pre-timed
S-A: Semi-actuated
F-D: Fully-distributed

performance for semi-actuated operation. The GPSS model indicates that full-actuated operation is better than pre-timed. These findings are contrary, but logical. The GPSS/H model does not encounter the situation where its detectors are blocked. Therefore it is more responsive at higher volumes than is NETSIM given the detector positioning assumed. In Case #3, where there are moderately high, directional arterial flows, pre-timed operation seems best.

While Table 1 shows that FDC, in its present realization, does well relative to full-actuated control in Cases #3 and #2, it is not able to achieve the lower level of total system delay afforded by pre-timed operation. Two modeling assumptions seem to be the root cause. One is uniform headways at the external approaches and the other is fixed travel times from one intersection to the next.

Table 2 shows that in Cases #3 and #4, the introduction of random arrivals has a profound effect. It reduces total delay by 26% for Case #3 and 75% for Case #4. The random arrivals allow the FDC to detect the presence of platoons and establish a coordinated timing plan sensitive to the unidirectional flow pattern. It is also important to note that in Case #4, FDC achieves a level of delay much lower than that achieved by full-actuated control, which is logical since the traffic pattern lends itself to one way progression. It is also important to note that the FDC does not achieve a total delay as low as pre-timed control, which is also to be expected since the FDC logic must find the cycle length/offset combination that works best. It is not given a priori.

In Case #2, the change to random arrivals has no impact. This may be due to the fact that the volumes are lower. The headways are greater, and the FDC correctly interprets the situation as calling for fully-distributed operation. Notice that the network delay is nearly identical to that achieved under full-actuated mode. (In fact, the delays are nearly identical for all three control strategies.)

TABLE 2: Total Network Delay for Case Studies #1 to #4 with Various Stochastic Input Conditions

Case	Fully-Actuated			Pre-Timed			Fully-Distributed		
	U-A	P-A	PA-TV	U-A	P-A	PA-TV	U-A	P-A	PA-TV
1	60.8	73.1	74.3	110.3	100.4	104.7	76.4	107.4	112.9
2	32.3	36.2	36.5	45.8	40.6	40.8	39.9	37.9	39.4
3	33.8	27.4	27.5	28.3	22.7	16.8	35.4	26.1	24.8
4	124.9	85.3	85.5	25.0	22.9	23.1	140.6	35.5	39.1

U-A: Uniform arrivals, constant travel times
P-A: Poisson arrivals, constant travel times
PA-TV: Poisson arrivals, travel time variation

In Case #1, the change to Poisson arrivals increases network delay and the change to variable travel times variation increases it further. The randomness in the arrivals seems to partly confuse the present FDC logic. It sometimes sees what it thinks are platoons and transitions to coordinated mode when it may not be required. In fact, it probably is not. Since the same model is used throughout these experiments, leaving the controllers in full-actuated mode for Case #1 might be the best plan since it produces the smallest delays. This would also be logical given the traffic flow patterns. It should be noted, however, that for the PA-TV scenario with FDC, total network delay is no worse than it is for pre-timed operation, and no timing plan had to be prepared.

The significance of these findings is that, even with a very primitive realization of FDC, it is possible to do better than full-actuated control in arterial-like situations and no worse in situations appropriate for full-actuated mode, bearing in mind that the FDC logic should be altered to preclude coordinated operation in situations like Case #1. It is possible to use a very simple set of decision-making rules to get a set of signals to effectively coordinate with one another, making decisions separately and in parallel (without a central controller). That is a major breakthrough. Our task now is two-fold: 1) to double-check these findings and 2) to further refine the decision-making logic so that it functions even better.

REFERENCES

List and Pond (1989). "A Strategy for Real-Time Control of Networks of Demand-Sensitive Lights," in Hendrickson, C. and K. Sinha (eds.), *Proceedings of the First International Conference on the Application of Advanced Technologies in Transportation Engineering*, pp. 367-372.

List, G. and S. Leong (1991). *An Exploration of Fully-Distributed Control for Signalized Networks*, Working Paper, Department of Civil Engineering, Rensselaer Polytechnic Institute, Troy, NY.

Luk, J.Y.K. (1984). "Area Traffic Control Methods: SCAT and SCOOT," *Traffic Engineering and Control*, 14-20.

Robertson, D.I. (1986). "Research on the TRANSYT and SCOOT Methods of Signal Coordination," *ITE Journal* (January), 36-40.

Sims, A.G. (1979). *The Sydney Coordinated Adaptive Traffic System*, in Urban Transport Division of ASCE, Proceedings, Engineering Foundation Conference on 'Research Priorities in Computer Control of Urban Traffic Systems', pp. 12-27.

CARS: A DEMAND-RESPONSIVE TRAFFIC CONTROL SYSTEM

J.Barceló, R.Grau, P.Egea, S.Benedito [1]

Abstract: This paper presents a demand responsive traffic control system for networks based on improvements from a combination of OPAC and ACTS approaches. The system has shown clear advantages over a fixed-time control, tested with the AIMSUN microscopic simulator. A field test is expected in October 1991.

Introduction

Demand-responsive traffic control systems began its development in the early 1960's. The first systems failed mainly due to the use of off-line concepts and inadecuate predictions. SCOOT (Hunt et al. 1982) represents the first clear success, soon followed by SCATS. More recently, GERTRUDE, MOVA, TOL, SAST, OPAC (Gartner et al. 1983) are aimed to isolated intersection control, although the latter has some coordination with the downstream intersections.

Particularly, OPAC is a decentralized control system that acts individually on each intersection, modelling vehicles with constant speeds and vertical queues, and using dynamic programming to determine optimal control policies.

ACTS (Kaltenbach 1983) is a centralised system which uses an underlying model of links divided in blocks with vehicles moving at constant speed and vertical queues. Small variations autoadapt the control policies in each intersection in response to demand changes. Although it hasn't been field tested, it introduces some interesting ideas.

Experience with other systems

We have reproduced the OPAC system and implemented the ACTS approach extending it for complex geometries, give ways and mergings. Our experience with these control systems shows that:

- The modelization with vertical queues is unsuitable to deal with such networks, mainly in conditions close to saturation. It becomes also difficult to find policies of max bandwidth type, for arterials.
- Proper control needs to take into account and measure accurately the free space still available.
- Decentralised control is unsuitable for saturation conditions, where a network-wide knowledge is required.

---

[1] Dept. of Statistics and Operations Research, Universidad Politécnica de Cataluña, Pau Gargallo 5, 08028 Barcelona, Spain. (Tel: 343-4016941, Fax: 343-4017040).

CARS approach

CARS, which stands for "Control Autoadaptativo para Redes Semaforizadas", is a demand responsive control system for networks, arterials and isolated intersections that:

- Shares with OPAC the idea of a rolling horizon to test the policy changes.
- Uses the ACTS (and SCOOT) approach to an adaptive centralised control based on small variations.

It incorporates the following new elements:

- Uses an underlying simulation model that deals with packets of vehicles moving according with an adhoc "packet following" model and horizontal queues. Forbidden zones in the intersecions as well as mergings and give ways are carefully modelled.
- Uses prediction models based on real time measured traffic information.
- Before considering the changes in the control policy, forecasts the network state. The autoadaptive algorithm tests changes in subnetworks, the size of the subnetworks is efficiently reduced by the prediction method.
- Accepts an arbitrary number and positioning of the detectors.

The system comes with a graphical user interface (CARSedi), DecWindows compliant, that allows the user to specify the network characteristics in a friendly and intuitive way. The links are automatically generated, so there is no need for the user to know the actual modelization. Finally, the whole system is written in ANSI C, and so is highly portable.

Control timing

The time is discretized in intervals of delta seconds. The system loops in the system control cycle (Figure 1), of scycle deltas duration, which performs the following tasks:

- send to the regulators the changing phase signals, in each delta of the cycle, when convenient.

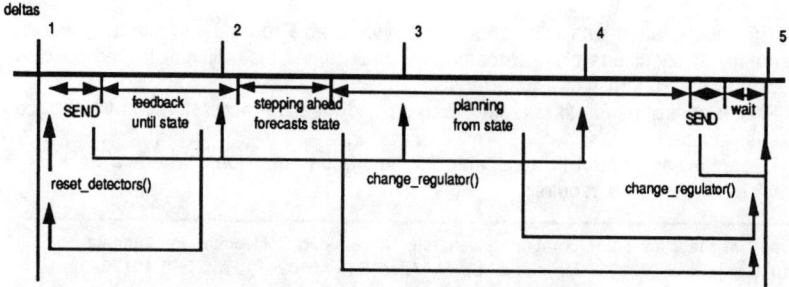

Figure 1. Sample control cycle of 4 deltas

- update the network state with the detector measures in the last control cycle.
- schedules the regulation in each junction from the forecasted network state at the end of the actual control cycle (stepping ahead).

Simulation model

CARS uses the simulator to represent the state of the vehicles in the network and to try control policy changes. The following basic elements are used:

- **lane groups**: same direction and contiguous lanes.
- **links**: a lane group has so much links as possible turns at its end.
- **bridge links**: model the junction's forbidden areas, from a link to the next lane group.
- **junctions**: control the rights of way. Each junction has an arbitrary number of phases and periodes between phases. The control can be inexistent, fixed or demand-responsive.
  The junctions can be characterized as critical and organize in arterials.
- **vehicle packets**: containing a float number of vehicles. They move with a simple packet-following algorithm, experiencing, along the links, grouping and partitioning operations.

Figure 2 shows a junction (A) and how it is modeled in CARS (B).

The updating process acts in two steps, to pass vehicles from one link to the next. The links are ordered to try that, when considering a link, the destination lane group has already been updated and the free space known exactly. Otherwise, a special "between-steps" free space predictor is available.

The objective function employed so far considers the number of vehicles stopped, so the grouping of packets takes special care when the predecessor packet is stopped, performing then a gradual grouping. When a stopped packet is allowed to advance, then it partitions from the front position, creating new smaller packets at a rate specified by the user.

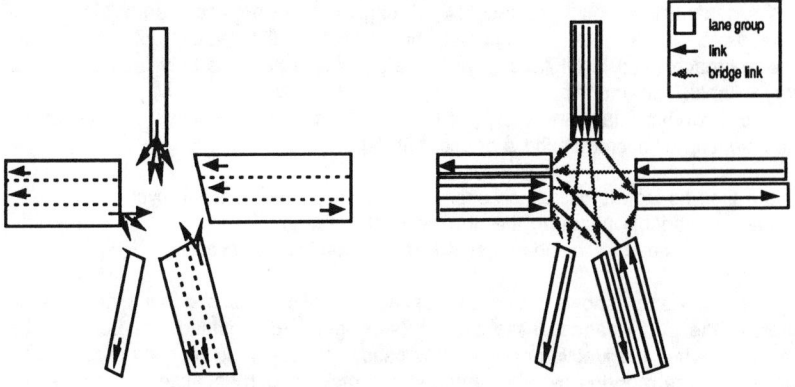

Figure 2. Sample modelization

Feedback

The feedback process takes the state at the beginning of a system control cycle and the measures of detectors in each 1 .. scycle delta, and outputs the state at the end of the cycle. For each delta interval, the model measures are compared to the real ones and the packets corrected. This can produce the generation of new packets, or the partition or deleting of new ones.

Each lane group can have an arbitrary number of detectors, the number and position of them influencing the corrections made in the feedback. If a lane group has no detectors, then its packets are not corrected (al least, until they exit the group).

Stepping ahead

Before considering the changes in the control policy, the system state at the end of the current control cycle is forecasted. This is done by actualizing scycle times, with arrival predictions at the entrances and space free predictions on exits.

The arrivals predictor (Stephanedes, Michalopoulos and Plum 1981) used needs no historical information and adapts to the varying conditions by recalculating the prediction coeficients when the difference between the predicted and the real measure grows.

Planning

The planning task considers changing the control policy in each junction of the network. The changes are tested on a subnetwork around the junction during a time horizon (Figure 3A). The arrival vehicles and the exit free space in the internal links are predicted as a cyclic pattern, made easier because of the knowledge of the control in the origin junctions. This reduces the planning subnetwork to just the entrance lanes to the junction.

Several algorithms for adjusting the control policy have been tested. The binary choice approach, for example, has showed a tendency in producing short periods, and a general inestability. Although there can be some enhancements (Lin 1988), it needs to plan in each time interval, thus restricting the flexibility of the system. The approach used in OPAC of trying every possible policy is only feasible with a very simple(fast) model.

The algorithm finally used is based in ACTS, in that it considers a tendency in incrementing or decrementing a phase, but departs in:

. the time horizon is calculated each time as the current junction cycle.
. the first variable phase in the horizon is adjusted.
. there is no need to consider offsets and common junction cycles.

Our experience shows that incrementing the time horizon doesn't necessarily improve the performance of the system (see figure 3B). Instead, the best results are found when the junction cycle time is used. This seems to be contradictory with the experience reported by other authors, and may happen because of the different model, planification subnetwork and predictors that we use.

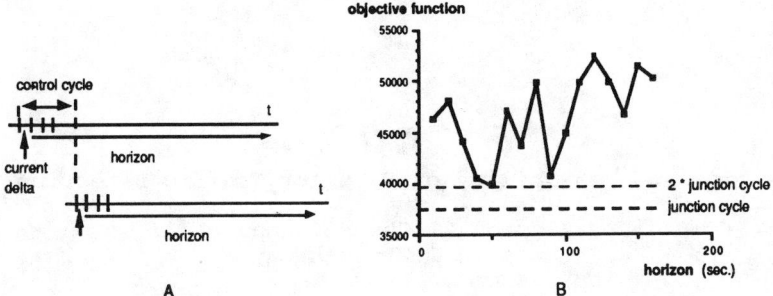

Figure 3. Rolling horizon (A) and sample of horizons (B)

Results

Simulation tests using the AIMSUN (Barceló et al. 1989) microscopic traffic model have been performed. The tests included isolated junctions, arterials and networks with complex geometries. The behaviour of the vehicles being in all cases the same, only one adjustement has been necesary.
The tests show a 10 to 25 % improvement in conditions close to the ones for which a fixed-time control policy was calculated. We have used a delta time interval of 2 seconds and a system control cycle of 5 deltas. With these parameters, a VAXstation 3200 rated at 3 MIPS controls a 10 junctions network using a CPU time equivalent to 20 % of the real time. Thus, with the current availability of more powerful computers, much bigger networks can be considered.

Appendix I. References

Barceló, J., Ferrer, J., Montero, L. (1989). "AIMSUN. Descripción del sistema", Departamento de Estadística y Investigación Operativa, Facultad de Informática, Universidad Politécnica de Cataluña.
Gartner, N.H. (1982). "Demand-responsive decentralized urban traffic control: Part I: Single-intersection policies", Office of University Research, US. Department of Transportation, Rept. DOT-RSPA-DPB-50-81-24.
Gartner, N.H. and Kaltenbach, M.H. et al.(1983). "Demand-responsive decentralized urban traffic control: Part II: Network extensions", Office of University Research, US. Department of Transportation, Rept. DOT-OST-P-34-85-009, 53-97.
Hunt, P.B. et al. (1982). "The Scoot on-line traffic signal optimization technique", *Traffic Eng. Control*, 23(4), 190-192.
Kaltenbach, M. et al. (1986). "Adaptive control of traffic signals", Université de Montréal, Centre de recherche sur les transports, Publication #469.
Lin, F. et al. (1988). "Adaptive signal control at isolated intersections", *ASCE Journal of Transport Engineering*, Vol 114, No 5, 555-573.
Stephanedes, Y.J., Michalopoulos, P.G. and Plum, R.A. (1981). "Improved estimation of traffic flow for real-time control", *Trasportation Research Record*, 795, 28-39.

Testing and Feasibility of VIPS for Traffic Detection
by
Alypios E. Chatziioanou
Stephen L.M. Hockaday
Carl A. McCarley
Edward C. Sullivan

Abstract

The California Department of Transportation (Caltrans) has asked the California Polytechnic State University (Cal Poly) at San Luis Obispo to investigate the effectiveness of existing traffic Video Image Processing Systems (VIPS); as well as the theoretical, hardware and/or financial limitations of VIPS technology. The project is due to be completed in June, 1991. Initial results show that VIPS are both feasible and in some cases well developed as a traffic detection technology.

Three "turnkey" (available for commercial sale) and five "prototype" VIPS detectors from around the world were identified and tested at Cal Poly. The testing procedure was aimed at evaluating the effectiveness of these products under conditions that are likely to be experienced on the congested California highway system. Variables such as camera positioning, traffic volumes, weather, and lighting conditions were examined. Approaches employed by the different systems vary from performing vehicle tracking throughout the video image to analyzing only specific regions of the picture for vehicle presence. Some of the advantages of each approach which became apparent in the tests will give guidance for future VIPS development. The technology as a whole seems to be extremely effective for several categories of detection problems, and the majority of current limitations are likely to be overcome by further development.

Acknowledgement

Success in the research project was made possible by the continuous support of Mr. Randall Ronning, the Caltrans project technical monitor. Staff efforts of the Cal Poly Applied Research and Development Facility (ARDFA) were also fundamental in the success of the project, especially Mr. Dan Need and Mr. Roy Bannon who performed many hours of field data collection and conducted tests in the ARDFA vision laboratory.

Introduction

As part of recent efforts to alleviate the problem of highway congestion through the application of new technologies, Caltrans asked Cal Poly to examine the feasibility of application of Video Image Processing Systems (VIPS) technology to traffic detection through tests of existing commercial and prototype products. The tests were aimed at determining the effectiveness of existing systems in meeting California highway traffic detection needs. The costs associated with possible systems implementation were also examined, together with their potential for use with IVHS/ATMS applications that may become available the near future.

After a thorough worldwide search for systems; a short-list was developed with nine systems that met initial performance capability requirements that might warrant testing on campus. A data-base of related system hardware components (imaging boards, cameras, monitors etc.) was developed. Several vendors pointed out that a specification that met the Caltrans needs was difficult to market at the time of our inquiry, according to investigations that they had conducted. In one case (the only one in the our short-list from Japan), these marketing considerations resulted in the commercializing of a licence plate reader instead of a detector[5] (Takatoo et al, 1989).

Different algorithm formulation approaches[1,2,3,4] (Blosseville et al, 1989 - Cypers et al, 1990 - Inigo, 1989 - Michalopoulos 1990) were identified, and representative examples were examined in a thorough literature search. Formulations of actual product algorithms were not transparent, but major characteristics, like the size of the video picture area of analysis, were clear from the user interface. Correlation of theoretical deficiencies in various approaches to actual system failures have given some indications of the algorithmic formulations.

One additional benefit of the VIPS research project has been the opportunity to become acquainted with some of the extended capabilities of VIPS systems, which are applicable to a variety of other problems beyond the basic highway traffic detection. Such areas already identified in the literature and approached as secondary project targets are: incident detection, license plate reading, and traffic classification. Another area has been identified through the initial investigation on another research project, which is studying safety in connection with the HOV lanes. In addition to conventional statistical analyses of accidents, a real-time analysis of traffic movements leading to conflicts can be performed by a traffic VIPS in a timely fashion with minimal disruption and low costs. Such observations lead to the conclusion that application of VIPS may demonstrate new capabilities to other transportation researchers, and bring to light new and more efficient investigation methods.

The Project Test Plan

The Cal Poly Vision laboratory facility was equipped to perform standardized tests leading to the evaluation of systems. Monochrome MOS/CCD (Metal Oxide Semiconductor and Charge Coupled Device) high resolution video cameras were used to acquire field images of traffic at several locations and over a range of conditions. The Super-VHS video recording format was selected as the standard for all images. All images were acquired and recorded simultaneously in both NTSC (EIA - U.S. standard) as well as PAL (CCIR - European standard) formats. The objective was to collect videotaped traffic images of a sufficient range to test comprehensively the capabilities of each vision system over the range of conditions expected in actual service on California highways, day and night, year round. These test images were edited and reduced to a collection of test video tapes, referred to as the "test suite" for the vision systems. The test suite images were used as video inputs to each candidate image processing system. Table I describes the standard test conditions represented by the test suite.

Table I : VIPS Test Suite

Test #	Parameter Tested
1	Large # of Lanes
2	Small # of Lanes
3	Day to Night Transition
4	Shallow View
5	Steep View, Departing Traffic
6	Shallow View, Departing Traffic
7	Night, Steep View, Approaching Traffic
8	Night, Shallow View, Approaching Traffic
9	Night, Steep View, Departing Traffic
10	Night, Shallow View, Departing Traffic
11-18	Same as 3-10 (above), Offset Camera Mounting
19	Weather - Fog
20	Weather - Rain
21	Weather - Rain at Night
22	Unstable Camera Mount - Shaking
23	Heavy Traffic - Capacity Operation
24	Stop and Go traffic
25	Heavy Shadows from Vehicles
26	Heavy Shadows from the Environment
27-29	Noisy Signal

Most systems tested were designed to and capable of processing monochrome EIA or CCIR video signals to yield estimates of the total number and the average speed of vehicles passing through the field of view in each lane on a cumulative basis. The numbers reported by the systems were compared with known traffic counts and average speeds which were determined by manual analysis of recorded test images. Additional qualitative evaluation was performed concerning system costs, potential problems in field installations, human factors considerations on the system interface, and overall system robustness and reliability.

Initial Results

At the time of preparation of the present paper, only initial results were available. The final report on the project is due at the end of June 1991.

Initial results show that a high percentage (over 95%) accuracy was achieved by most systems on typical low, moderate, and high traffic intensity detection conditions, with minor problems created mainly by occlusion. Some systems had problems with stop and go traffic, particularly if they had to be initialized during the heavy congestion period.

The transition from day to night and vice-versa often reduced the accuracy in the order of 75%. For most systems, this is due to the fact that different algorithms are used under the two lighting conditions. Switching from one algorithm to the other needed refinement and fine-tuning in most systems. This may be singled out as one of the most significant problems, since peak periods (requiring accurate detection), frequently coincide with such transitions for California latitudes.

Some systems were sensitive to camera positioning. Camera optics could correct some of the camera height and angle related problems, but sensitivity to the distance from centerline is more difficult to correct.

Most systems deal effectively with the problem of shadows, particularly those systems that track vehicles. Reflections from a wet roadway have an impact similar to that of a shadow. Count accuracies often range between 80-90%. However, the impact on speed measurement was in some cases severe, stemming from the uncertainty in identified object size that resulted from the inclusion or exclusion of shadows or reflections. The investigation of correlation between shadows and reflections and vehicles is CPU intensive, and it may be one of the reasons tracking systems do not always run at full video speed. Video compression (as a side result of the image processing or CPU limitation) can be helpful, since it can help provide reasonable CCTV picture input for communication over telephone lines. Inaccuracies due to moderate fog are minor since detection is attempted

relatively close to the camera position (normally less than 300ft).
System costs, are expected to be in the range of $10,000 or less per station, for significant quantity orders. Costs are also expected to go down significantly as hardware costs are reduced by specialized hardware manufacturing. Several vendors already use custom-made imaging boards, and we have indications of development of compact weatherproof custom-made integrated processors.

Conclusions

VIPS are ready to enter initial deployment stages. Three systems are marketed as ready to be installed. Accuracies of measurements are over 95% for "ideal" conditions which cover a significant portion of the California traffic detection needs. Lighting transition periods, shadows and reflections may each reduce accuracy by up to 20-30 percent; but system fine-tuning to case-specific conditions is expected to improve this performance.

System costs have decreased in the past two years and are already competitive with other detection systems; and in several cases provide superior features and performance. The ease of, and relatively minor disruption in highway operations for, system installation and maintenance is also a major advantage.

VIPS can be used in a variety of other highway related investigations, such as studies of traffic movement conflicts. One example is the study of HOV lane safety currently underway at Cal Poly.

One of the VIPS project recommendations calls for immediate deployment of a small number of traffic VIPS for field testing and fine-tuning to specific location conditions.

Appendix - References

1. Blosseville J. M., et al, "Titan: A Traffic Measurement System Using Image Processing Techniques," IEE Second International Conference on Road Traffic Monitoring, London, 1989.
2. Cypers L., et al, "CCATS: The Image Processing-Based Traffic Sensor, Traffic Engineering and Control, June 1990.
3. Inigo R.M., "Application of Machine Vision to Traffic Monitoring and Control," IEEE Transactions on Vehicular Technology, August 1989.
4. Michalopoulos P., "Automated Extraction of Traffic Paramenters through Video Image Processing," ITE 1990 Compendium of Technical Papers, Presented August 1990.
5. Takatoo M., et al, " Traffic Flow Measuring System Using Image Processing," SPIE Vol.1197 Automated Inspection and High-Speed Vision Architectures III, 1989.

INCIDENT DETECTION THROUGH VIDEO IMAGE PROCESSING

Panos G. Michalopoulos[1]

INTRODUCTION AND SUMMARY

Extraction of real time traffic data in large-scale street and highway networks has been a major obstacle in implementing practical and reliable Advanced Driver Information Systems (ADIS) as well as Advanced Traffic Management Systems (ATMS). As the instrumentation and data requirements increase, conventional devices become inadequate, expensive and often unreliable for advanced applications such as vehicle guidance and navigation, adaptive control of congested street networks and freeway corridors, real-time forecasting of traffic demand patterns, etc. To address this problem, a machine vision system for vehicle detection and traffic parameter extraction (called AUTOSCOPE) was recently developed at the University of Minnesota. Its advantages lie in its multispot wide area wireless detection capabilities and the ability to extract traffic parameters, such as density and queue length and size, that cannot easily be obtained with conventional devices. The system was extensively tested at the laboratory and later installed at several freeway and intersection locations and compared with loop detectors on a continuous 24-hour basis over a period of eighteen months since the writing of this paper. Following the necessary adjustments, its performance and reliability was demonstrated to be equal to or exceeding that of conventional devices. As a result, the device is currently being implemented at a 3.5 mile freeway section of the I-394 freeway in Minneapolis, Minnesota for automatic incident detection. To further exploit this opportunity a total of 38 cameras were installed in this section for detailed and continuous lane by lane traffic parameter extraction on the freeway and its ramps. In this paper the use of the AUTOSCOPE System for automatic incident detection is briefly presented.

[1]Dept. of Civil and Mineral Eng., Univ. of Minnesota, Mpls, MN 55455.

IMPLEMENTATION OF AUTOSCOPE TO INCIDENT DETECTION

Automated incident detection was selected as the first application of the AUTOSCOPE technology not only for liberating traffic managers from their computer screens and allowing them to concentrate on more critical tasks, but also for reducing the number of TV monitors required for manual monitoring of traffic conditions. At this time a laboratory version of the imaging incident detection system is available while field implementation testing and validation is underway. This breadboard incident detection system is called IDEAS, Incident Detection and Evaluation through the AUTOSCOPE System. The System concept diagram for the deployment of IDEAS is shown in Figure 1. The role of IDEAS in a traffic management system is to provide an alarm and historical log of incidents that occur. IDEAS receives traffic and incident feature detection data from each of the AUTOSCOPE's in the network it is monitoring. The system uses traditional station measurements of volume,

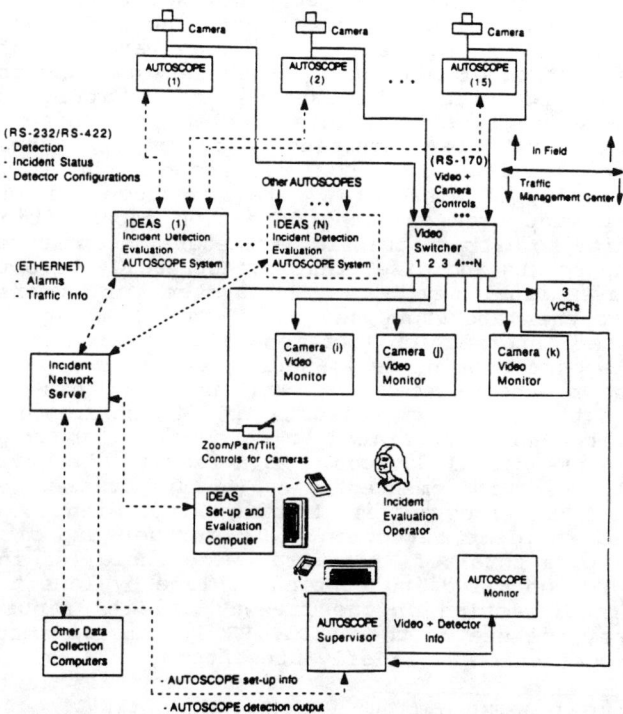

Figure 1   IDEAS System Configuration

occupancy and speed combined with the unique measurements offered by the wide area image processing capabilities of AUTOSCOPE to detect incidents. The incident algorithms are applied to each of the AUTOSCOPE's in the network. When there is sufficient evidence that an incident is occurring, an alarm is set, and follow-up information as to the consequences of the incident (effect on traffic flow, further delay or clearing) is provided to aid the incident management operator. A human operator is required for verification of the incident and determining the appropriate response; the operator can also be assisted by an expert system in this latter task.

IDEAS requires a set-up and evaluation microcomputer to configure, monitor, and control IDEAS. The set-up and evaluation computer loads and saves databases, retains a priority queue of incidents that occur in the network, sets incident report levels and thresholds, displays historical traffic trends to give the operator a measure of how an incident is effecting traffic, displays congestion level metrics of clear, low, medium, heavy, or stopped traffic, provides traffic information of cameras adjacent to the incident site, clears/validates incident reports, and records operator incident results. IDEAS also acts as a communications gateway (feed through) between the AUTOSCOPE Setup/Supervisor computer and the desired AUTOSCOPE in the network.

The set-up and evaluation computer, the AUTOSCOPE supervisor, and any other data collection computers access the IDEAS via an incident network server. The network provides the greatest flexibility to quickly and transparently link additional IDEAS' computers when communications and processing limits of the IDEAS processor(s) are reached. The network also provides clean access to the system for system developers, experimenters, and (outside) data gatherers for an extremely flexible laboratory facility.

There are two algorithmic approaches developed for the detection of incidents. Each approach utilizes the same supporting hardware and software structure (framework), but differ only in the actual sensing mechanism (signals) used to detect incidents, i.e., they use the same AUTOSCOPE's, communication processes, vehicle detection and historical traffic data collection, high level detection logic, and incident reporting mechanisms.

The first incident detection algorithm approach, called the conventional approach, is based on utilizing existing capabilities of the AUTOSCOPE. For each AUTOSCOPE, at least two speed traps are placed in each lane of traffic. The speed traps in each lane are placed as far apart as possible (200 - 300 ft) to indicate speed changes within the field of view. All detection data is retained in a historical database to indicate historically consistent traffic patterns over the last 15 minutes. There are two historical databases maintained for weekday and weekend traffic. Incidents are then detected by comparing the differences in time or volume smoothed speeds against historical data and an incident alarm threshold.

The second incident detection algorithm approach, called the image understanding approach, utilizes computer vision image processing and image understanding techniques that require additional extensions to the "standard" AUTOSCOPE detection capabilities. One such extension involves measurement of a newly defined parameter the traffic pressure that could lead to more reliable incident detection.

As mentioned earlier, currently IDEAS is being deployed on a 3.5 mile section of Interstate 394 in Minneapolis for field testing tuning and verification. While this long term testing is in progress, enhancements to the algorithms are being made to effectively deal with potential artifacts/problems based on measured performance deficiencies. Examples of artifacts to deal with are recurring congestion, compression waves, maintenance lane closures, inclement weather, and special events. The enhancements will include optimization to the existing algorithms to increase processing and communication throughput and implementing promising incident verification or performance improvement techniques that were not as yet incorporated or tested.

CONCLUDING REMARKS

The AUTOSCOPE imaging system for vehicle detection is to the best of the author's knowledge the most mature known device available today.

Unlike other systems, the AUTOSCOPE system provides visual and numerical detection verification. Furthermore, it operates under all weather and lighting conditions (as it has resolved the aforementioned problems) as well as in congested traffic. Most importantly, it does not rely on a particular camera type; this implies that it can be hooked to existing cameras at a minimum cost.

This was manifested in recent live benchmarks in several North American and European cities. Because of this, the system is being implemented in several demonstration projects, two of which began recently. The first of these projects is the largest known machine vision demonstration project for traffic control worldwide. It includes installation of 38 cameras on the I-394 freeway in Minneapolis for automatic incident detection. The second includes AUTOSCOPE installation at an intersection for loop replacement and signal control (also in the Minneapolis-St. Paul Metropolitan Area). Other projects outside of Minnesota which should be underway by June 1991 include large scale real-time network control and incident detection and traffic management at a critical link.

The status of the system today is in the final production line prototype state. This prototype will be available in Fall '91 and it will substantially reduce the size, cost and capabilities of the existing reproduction line prototype which is being produced in limited quantities. The manufacturer of the final product is a firm with extensive experience in the traffic control and is able to supply and support the sale of large quantities of AUTOSCOPE.

In summary, given what we know now it is safe to conclude that AUTOSCOPE is a viable and effective device for multilane-multispot wide area vehicle detection and automatic traffic parameter extraction. Further, it should provide more reliable and effective incident detection. For this reason, large scale production marketing and support was recently undertaken by the industry.

ACKNOWLEDGEMENT

Research for development of the U.S. based system described here was supported by the Minnesota Department of Transportation, the Federal Highway Administration and the Center for Transportation Studies of the University of Minnesota.

REFERENCES

1. Michalopoulos, P.G. and Wolf, B. (1990). "Testing and Field Implementation of the Minnesota Video Detection System," Transpn. Res. Record (in press).
2. Michalopoulos, P.G. (1990. "Automated Extraction of Traffic Parameters through Video Image Processing," Proceedings ITE 60th Annual Meeting pp. 33-38.

TRAFFIC QUEUE DETECTION USING IMAGE-PROCESSING

A Rourke and M G H Bell [1]

1. Introduction

The problem of the automatic detection of congested and queueing traffic has been exercising the minds of traffic engineers for many years. Hoose (1990) provides a concise review of the work in this field. In general, previous attempts to provide an automatic means of queue detection have employed point-based inductive-loop systems, either in isolation or linked into small networks, with statistical analyses of the output data providing the basis for decision-making criteria. The limiting factor on the performance of these systems has been found to be their essentially point-based method of data collection.

The surveillance camera, however, has the potential to provide a much denser level of coverage within its field of view. To exploit this potential, some way of automatically monitoring the video picture is required, since it is impractical to employ a human operator to perform this mundane and boring task continuously. Such a method of picture analysis will be described in the following sections of this paper.

2. Image-Processing for Traffic Applications

A typical digitised image contains 256 kbytes of data, and a video camera will transmit 25 such images per second (30 in USA). For all but the most powerful computers, or highly parallel systems, this amount of data is far too great. Some form of data reduction must be employed. The most commonly used method reduces the number of pixels to be processed in the spatial domain by recognising the fact that vehicles will only occur in well defined areas of the image. The loop-emulator system of Rourke and Bell (1989) for example, reduces the data rate from 6.25 Mbytes/sec to around 2 to 4 kbytes/sec by concentrating processing on a small number of user defined windows. In this way real-time performance can be maintained.

Data reduction in the temporal domain can also be effective, depending upon the application. Queue

[1] TORG, The University, Newcastle Upon Tyne, NE1 7RU, UK.

detection is an example where it is not necessary to process every single frame from the camera. The whole of the image can be processed, exploiting the useful and unique wide area sensing capability of the video camera, at a very much reduced frame rate. Hoose (1990) has described an algorithm for queue/congestion monitoring based on this concept. A number of cells are used to divide up the image providing a high density of 'detectors' in the scene. A cycle time for the algorithm of between 1 and 5 seconds has been found to be useful for most situations, enabling a vast reduction of data in the temporal domain. Even so, the demands of processing the whole image in this time requires the use of some specialised hardware to perform common primitive image-processing functions.

The most demanding traffic application for image-processing is perhaps that of vehicle tracking. Several attempts have been made, one of the most successful being that of Houghton et al (1987). This system uses a number of specialised parallel processing elements to process the video image. For tracking applications, there is very little scope for data reduction in either the temporal or spatial domain, and it is therefore necessary to employ specialised high performance processing elements.

3. Queue Detection Using Low Cost Image-Processing

There is clearly a requirement for an algorithm exploiting the area sensing capability of the video camera whilst also minimising the need for expensive hardware. The method proposed in this paper attempts to achieve these aims by employing a combination of both spatial and temporal data reduction.

Data reduction in the spatial domain is achieved by defining a window, consisting of a single line of pixels, along the traffic lane, parallel with the direction of vehicular motion (figure 1a). Data reduction in the temporal domain is obtained by processing the window only once every few seconds. This coarse sampling rate can be tolerated because of the nature of congestion, which takes seconds to occur.

Three distinct stages of processing are employed to achieve queue detection. Firstly, the pixel data in the window is regarded as a digitised one-dimensional waveform. Each pixel represents a sample of the wave, and its grey value represents the wave's amplitude at that point. This data can, therefore, be processed using the well-known Fast Fourier Transform (FFT) (Brigham, 1974), to reveal its frequency and power spectra.
Figure 1b shows a sample from an empty window. The

Figure 1. Empty window

Figure 2. Queueing in window

frequency spectrum shows a strong dc component, as one would expect from such a sample, along with some low power, higher frequency components, caused by electronic and image noise. Figure 2 shows the effect of introducing vehicles into the window. The frequency spectrum changes significantly, and this provides the basis for detecting the presence of vehicles. Similar changes in spectrum can be caused by the appearance of shadows, which must be differentiated from vehicles. This is achieved by reference to the nature of shadows.

The positions of vehicles are located by approximating the waveform by a series of straight line segments and applying suitable gradient thresholds. This processing results in a measure of the extent of the congestion and the location of the back of the queue (figure 2a). If the scene has been calibrated, these measurements can be related to physical road measurements.

Finally, the amount of movement the vehicles are undergoing is determined by a differencing operation. The absolute difference in the grey levels in the window, on a pixel by pixel basis, at two instances a short time interval apart is found. If there has been no movement this will produce a zero result, neglecting random noise, whilst distinct movement edges will be produced by moving vehicles. These edges can then be matched to the vehicle features and a flow profile over the window obtained. By setting a suitable movement, (or speed) threshold for the window, a queue/congestion alarm can be triggered.

4. Experimental results

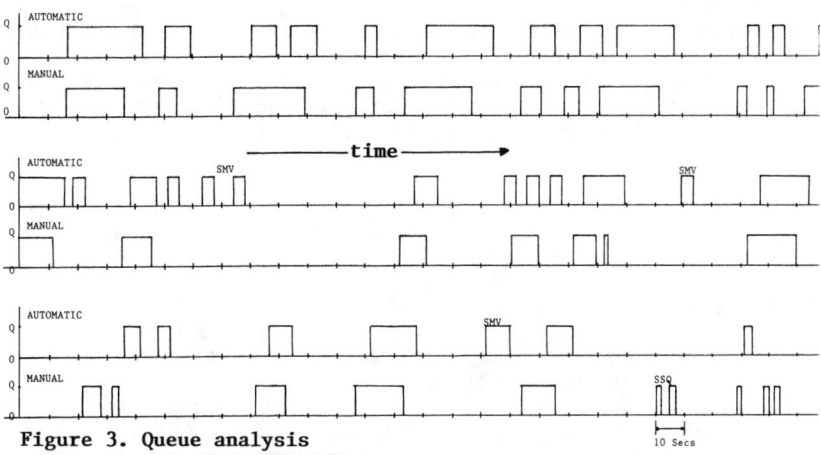

Figure 3. Queue analysis
SMV - Slow Moving Vehicle
SSQ - Stop / Start Queueing

Figure 3 shows a comparison of queue analyses performed both manually and automatically. These data were obtained from the urban traffic light junction shown in figures 1a and 2a. It is evident that good detection is provided by the algorithm. On a small number of occasions the manual and automatic analyses differ. Closer inspection of these periods revealed the automatic analysis to trigger on single slow moving vehicles which were not recorded as congestion by the manual analysis. Similarly, there are periods of stop/start queueing which were not detected by the automatic system. The current implementation of the algorithm requres a cycle time of approximately 3 to 4 seconds. Vehicles stationary for less than this time can be missed.

5. Conclusions

A method of queue/congestion detection has been described using low cost image-processing and computing hardware. It is capable of processing data from a length of roadway within a few seconds, providing distinct advantages over more conventional methods. From figure 3 it can be seen that the current system is capable of providing a response to queueing or congested traffic within 4 to 8 seconds of the queueing occurring. Figure 2b shows that a measure of the extent of the congestion can also be obtained along with the location of the rear of the queue.

Further work is required to improve the speed of execution of the algorithm and to allow processing of calibrated scenes. An increase in execution speed can be achieved by employing a digital signal processing (DSP) co-processor to perform the FFT analysis. This will allow multiple windows to be analysed within the acceptable time-scale. Scene calibration is required to account for minor perspective distortion effects and to enable queue parameters, such as queue length and location of rear end, to become physically meaningful.

6. References

1. Brigham, O E., 1974, "The Fast Fourier Transform". Prentice Hall, Englewood Cliffs, New Jersey.
2. Hoose, N., 1990, "Computer Image-Processing In Traffic Engineering". Research Studies Press, Taunton, Somerset, England.
3. Houghton A D., et al, 1987, "Automatic Monitoring of Vehicles at Road Junctions", Traffic Engineering and Control, October.
4. Rourke, A. and Bell, M G H., 1989, "Applications of Low Cost Image-Processing Technology in Transport". 5th World Conference on Transport Research, Yokohama, Japan.

Report of the DEVLONICS video based traffic detector systems
(DRS-CCATS/DTMSA)
Frans Lemaire[1]
Marc Coussement[1]

Abstract

In traffic monitoring systems many detectors are in widespread use to provide the necessary information. Image and video processing is one of the most defiant newcoming technologies in this domain. Based on the 5 years of field experiences in Belgium, Devlonics Control has commercialised the CCATS (Camera and Computer Aided Traffic sensor) from 1987 on. The traffic information from the sensors can be centralised with the PC compatible DTMSA software package (Devlonics Traffic Monitoring Software Advanced). Recently, Devlonics Control introduced their new system for video image digitalisation, image storage and image transmission, called DRS (Devlonics red light and speed system).

Introduction

This article focus on the results of different evaluation reports of installed CCATS/DTMSA throughout Europe and the first field trials with the DRS for both red light and speed registration.
For the technical background of the video and image processing based traffic sensors, we refer to the publications indicated in the reference list. (1),(2),(3),(4),(6),(10),(11).

The CCATS technology

The CCATS technology is based on video signal and image processing implemented on an enhanced high-speed processing hardware environment.The sensor consists of a camera and a processor box. The sensor hardware digitises the image. The traffic parameters are calculated using video signal processing and image processing techniques within particular programmable windows free definable on the video screen. An additional programmable module enables the definition of typical incident situations, based on a combination of the gathered basic traffic parameters. The processing kernel of the sensor provides individual vehicle information: direction,number,intervehicle gap time, occupancy, speed and length.
----
1 Business Unit Manager traffic department, Devlonics Control NV, Stasegemsesteenweg 110, 8500 Kortrijk, Belgium

These results are collected in minute average packages. The records can be stored in the internal local database, or they can be either transmitted to an existing traffic control centre. On the site, the results can also been collected by connecting a serial link with a portable computer. In addition, a slow scan extention enables to transmit together with the data the digitized image of the camera site.

In the dispatching centre, all the data can be stored for numerical and graphical presentation using an MS DOS driven personal computer and the DTMSA Software package . Both operation data (power breakdown, reboot actions, VMS panel control, alarm detection etc...) as traffic data are collected on the harddisk.

The DRS equipment

In its most simple application, the DRS equipment replaces the fotocamera of a traditional speed or red light offence registration system. Collection of the 'photos' can now be done by collecting the digitized video images on harddisk or by the digital transmission of the images through a telephone line to the central workstation. On this workstation, image processing software tools and dedicated software routines are available for automatic police warrant handling, enhancement of 'bad photos' or pure record keeping.

Actual Application domains of CCATS

CCATS is used for :

- Mobile and fixed traffic detection on motorways, interurban areas and around and inside tunnels.
- Automatic queue detection
- Automatic incident detection
- All purpose traffic detector to control variable message signs and panels
- Analysis of video tapes from different traffic situations

Actual Application domains of DRS

DRS is used for :

- Registration of traffic light offenders
- Registration of speed limit offenders

Summary of the reports of some CCATS installations

Brussels - Schuman tunnel : a CCATS camera unit is installed at the exit of the Schuman tunnel in Brussels to detect queue formation in this area automatically and to automatically switch on a warning panel in the tunnel, to ask the vehicle drivers to reduce their speeds when leaving the tunnel in order to avoid rear-end queue collisions. The system is operational since beginning 1988. Statistics from the police authorities in Brussels confirm that, during an evaluation period of one year, the number of heavy incidents (rear-end collisions) decreased from 7 to 2 (see table I).(7).

Period	Collisions (tail-back)	Accidents (turnings)	injured people
1.87-8.87	7	3	5
9.87-8.88	2	1	0

Ring Brussels - Wemmel : A CCATS prototype is installed in 1987 to evaluate the systems behaviour, especially during queue situations. The table below illustrates the mean accuracies compared with inductive loops during different days and weather conditions:
(volumes : averaged hour values during 3 days)
(speeds : averaged hour values during 3 days)

Traffic situation	mean density rates (veh/h)	CCATS volume loop volume		CCATS speed loop speed	
		min	max	min	max
LOW	0 - 500	0.89	1.01	0.80	1.11
NORMAL	501 - 6500	0.93	1.05	0.91	1.04
QUEUE	2400 - 4300	0.91	1.06	0.74	1.13
ACCIDENT	2400 - 5000	0.81	1.11	0.45	1.29

Entrance to the village of MARTELANGE :Two CCATS cameras were installed during 1988 at the entrance ways of a small village near Luxembourg, called Martelange.(5). The sensors measure volumes, speeds occupancies and classification to detect different threshold values to warn the local police authorities to control a tidal flow system in the small city centre. The installation uses movable cameras to verify the automatic alarms of the fixed CCATS detectioncamerasystems. The system has been successfully used by the police authorities to avoid unnecessary patrol actions and to minimize the total queue time. The official report compared the estimated financial loss due to queue formation and the registrated intervention cost and labour cost of the police patrols. The report showed that the video based system cost was profitable, compared with the benefit in time and intervention cost within less than one year.(8).

City centre of NAMUR : DEVLONICS participates in the DRIVE projects. These projects were organised by the European Commission to stimulate development and study in the traffic domain.(2). In the framework of one of these projects (V1014:IMAURO) , 4 CCATS units were installed in the city centre to evaluate the performance and accuracies in these urban situations. The results will be used to derive all the necessary information to feed into a dynamic traffic model of the city. Official evaluation reports of these tests are available for the EC Commission offices. A comparison was made between loop detection and CCATS Camera counts. The general experience indicates in these urban areas (where the camera installation is beyond its normal specs) the relation between accuracies and the distance between the camera and the surveilled surface (lane). Maximum error rates of 5% were achieved on the lane close to the camera. Error rates up to 25% occured on the thirth lane. A mathematical module to integrate in CCATS to improve these effects is in development.

Ring Antwerp : A set of 4 CCATS camera's have been installed on the Kennedy tunnel entrance motorway. The objective was to detect automatically queues, to locate them and to detect the tale and head of the queue. The system operates during day and night with the same accuracies. The obtained accuracies tested and approved by the Belgian Government were sufficient to succeed in the project objectives. The table below compares the results from loops and CCATS during several 24 hour periods.(9).

Day	Volume		% error
	CCATS	LOOPS	
1/7/90	47564	50005	4,88
2/7/90	60608	64787	6,45
3/7/90	58761	62721	6,31
4/7/90	60318	64409	6,35
5/7/90	61550	64667	4,82
8/7/90	49475	52036	4,92

Evaluation conclusions

Some expected product features are achieved and integrated after the engineering experience. Some other product features and user needs were achieved by additional engineering. Some product features will not be implemented in a short term and they must be considered as features of a second generation video based traffic sensor.(e.g. vehicle tracking).

In any way, the next paragraphs summarize the actual sensor aspects :

Installation camera : - height motorway and urban sites 9...16 m
- height tunnel sites : min 4.6 m
- camera angle : 35..70 °
- Site length : max 200 m
- illumination for simular day/night performance : min 30 Lux

Shadows and different weather conditions

- Elimination of horizontal moving schadows : 60 - 100%
- Elimination of fixed shadows : 100 %
- Slow light variations : filtered out
- Sun and cloud game : filtered out
- Rain,fog,dawn,wind,snow : worst case errors less than 20%
- Day to night changes : no influence
- Environmental illumination : min 30 LUX
- Night without illumination : no possible length measurements
- Heavy snow : maximum error rate up to +10%
- Normal circomstances : error rates -4% to max -10%

References

(1) Blosseville,Krafft,Lenoir,Motyka,Beucher. TITAN : new traffic measurement by image processing, IFAC CCCT'89 Conference, Paris, 1989
(2) Clauwaert,Vervenne,Lemaire. Control and data acquisition of traffic based on image processing technologies. Route et Informatique, Conference Paris, 1990
(3) Cypers,Kolacny,Poncelet,Vervenne,Lemaire,De Jaegere. CCATS: the image processing based traffic sensor. Traffic Engineering + Control, June 1990
(4) Cypers, Goossens, Van Campenhout, Lemaire, Vanderschaeve, De Jaegere. VERKEERSTECHNISCHE STUDIE ANTWERPEN. Ministry of Public Works Belgium-Antwerp. Evaluation Report 1990.
(5) Fabrimetal. CASE STUDY MEMORANDUM ELECTRONISCHE SYSTEMEN in de weginfrastructuur - Case Martelange - gebruik van CCATS camera's. Symposium Fabrimetal, 1990.
(6) Hoose. IMPACT. IFAC CCCT'89 Conference Paris, 1989
(7) Maes, Lemaire. Incident reduction at the exit of the Schuman tunnel at Brussels. Ministry of Public works Belgium. Evaluation report 1988.
(8) Poncelet,Bellens,Bigoni,Lemaire. Traitement d'images video pour la detection automatique des incidents. Raport Interne Ministère des travaux publics Belge, 1989.
(9) Van Campenhout, Verbeek. Vergelijkende studie manuele tellingen: video, lussen en CCATS. Vlaams Ministerie van Leefmilieu en infrastructuur, 18 september 1990.
(10) Versavel,Lemaire,Van der Stede,Maes,Mortelmans. CCATS - Camera and Computer Aided Traffic Sensor. CCCT'89 IFAC, Conference Paris, 1989
(11) Wan,Dickinson. Road traffic monitoring using image processing - a survey of systems,techniques and applications. IFAC CCCT '89 Conference, Paris, 1989

An Application of Advanced Technologies for Freeway Traffic Management - An Indiana Case Study

Michael J. Cassidy[1]
Kumares C. Sinha[2]

ABSTRACT

The Indiana Department of Transportation is in the process of implementing an electronic surveillance and control system on the Borman Expressway (I80/94). This paper presents a summary of technologies being proposed for the system. The system would consist of three elements: 1) Traffic Surveillance, 2) Motorist Information and 3) Congestion Management.

INTRODUCTION

Research described in this paper was conducted in an effort to identify appropriate technologies for a proposed electronic surveillance and control system to be implemented on the Borman Expressway in Northwest Indiana [Cassidy and Sinha 1990]. The study itself was a synthesis of existing information on appropriate technologies. Information was collected from 1) literature detailing relevant research findings, 2) discussions with professionals in the area of freeway traffic management and 3) visits to several freeway surveillance and control centers [Harmelink et al. 1984; McDermott 1980; MnDOT 1979]. To the extent possible, technologies selected for implementation represent state-of-the-art electronic systems. It is hoped that information in this paper can assist highway agencies who are presently considering the implementation of advanced freeway traffic management systems.

BACKGROUND

The Borman Expressway is a 12-mile freeway linking Gary, Indiana with Chicago, Illinois. The expressway, which serves an average of over 140,000 vehicles each day, represents the highest volume facility in the state of Indiana. These heavy traffic flows, combined with a high

---

[1] Assistant Professor, Purdue University, School of Civil Engineering, West Lafayette, IN

[2] Professor & Head of Transportation Engineering, Purdue University, West Lafayette, IN

proportion of trucks in the traffic stream (approximately 30%), contribute to frequent roadway incidents. Thus, efforts have been undertaken to design a surveillance and control system capable of mitigating operational problems through the timely detection and removal of expressway incidents.

In identifying the general elements to be incorporated into the proposed system, an important consideration was recognized. The proposed system will be implemented in the near future and must function in a "real world" environment. All elements in the surveillance and control system must be implementable and should not rely upon technologies which require additional development. The study has therefore sought to identify technology advancements which have proven to be effective and appropriate.

SYSTEM COMPONENTS
The proposed surveillance and control system is to consist of three components. 1) Traffic Surveillance, 2) Motorist Information and 3) Traffic Management.

1) Traffic Surveillance
The function of the surveillance system is to detect and verify congestion problems caused by incidents or other capacity constraints. This information is used for incident response purposes and to supply input data to motorist information and traffic management systems. Although a number of technologies are available for performing freeway surveillance, this study has sought to incorporate system elements which have already proven to be effective and appropriate.

Inductive loops have traditionally been the technology used for traffic surveillance. Loop detection systems provide a reasonably cost-efficient approach to monitoring roadway operation. Rather than single loop detection points, paired loop stations (i.e. two loops located 15 to 20 feet apart in the same lane) will be used to facilitate surveillance. With single loop detection, vehicle speed can only be derived by using an assumed (i.e. average) vehicle length. Where loop detectors are located in pairs, vehicle length is no longer an input variable. Moreover, vehicle classification, or more specifically, vehicle length can be determined once speed is reliably measured. Specific locations for all loop stations are to be determined as part of the system's final design process. On average, stations will likely be located at distance intervals of one-half mile (in all travel lanes). The installation of loops has also been recommended at expressway on-ramps to facilitate ramp metering [Hibbard et al. 1990].

To minimize data transmission requirements, detector data will be processed on-site using Type 170 microprocessors (or the equivalent). Thus, processed data will be sent to the control center and to other elements of the freeway traffic management system.

A **Closed Circuit Television System (CCTV)** provides improved incident detection, identification, verification and management. Where an incident is automatically detected by loops, CCTV cameras will be used by personnel in the traffic operations center to verify and identify the nature of the incident. Moreover, CCTV will provide incident detection capabilities. Periodic scanning of the cameras can be used to locate operational problems. Under certain flow conditions (e.g. low flows), CCTV may prove to be the primary detection system.

A nominal camera spacing of one-half to three-quarter mile should be appropriate. Cameras will be located so as to provide full view of the expressway, as well as a view of all ramps and critical changeable messages signs. Each camera will have full pan, tilt and zoom capabilities, and will be remotely controlled by personnel in the Traffic Operations Center (TOC).

The most desirable camera technology appears to be a Silicon Intensive Tube camera. Although the camera is black and white, this technology produces the best image during darkness and/or inclement weather conditions. Cameras will be housed in environmentally protected casings and mounted on self-standing poles at heights of 40 to 50 feet above the surface of the expressway.

2) Motorist Information

The objective behind motorist information systems is to provide drivers with real-time information concerning prevailing traffic conditions. Alerting motorists to downstream operational problems essentially serves three functions:
1) **Improved Safety:** By informing motorists of downstream roadway problems, drivers are likely to approach problem locations with greater caution.
2) **Improved Operation:** A motorist information system can provide drivers with alternate routes during congested operation.
3) **Improved Public Image:** A motorist information system is visible evidence of a state highway agency's commitment to improving highway safety and performance.

The **Changeable Message Sign** (CMS) can be an effective means of providing information to motorists. A number of CMS technology options exist. The most reliable existing technology appears to be the "rotating drum" CMS. This type of CMS typically consists of three rows of six-sided rotating drums. Thus a number of possible preset messages are possible. Moreover, text lines can be replaced or updated when appropriate.

Light Emitting Diode (LED) technology is another CMS option. This technology provides superior flexibility and legibility. Moreover, LED signs have graphics capabilities, can display messages in different colors and automatically adjust LED intensity for sunlight and nightfall conditions. The effectiveness of such technology, however, is not without costs. Beyond the rather high capital costs, LED signs require continuous power to operate.

A final selection of CMS technology to be used on the Borman system will be made as part of the final design process.

Other technologies can be employed to supplement motorist information provided by the CMS system. One such information source is **Highway Advisory Radio** (HAR). Like the CMS system, HAR broadcasts can alert motorists to operating problems and provide detour information. A number of strategies exist for implementing and operating an HAR system. A so-called "leaky" (coaxial) cable can be installed in the facility's median barrier. It is also possible to operate an HAR system using an exclusive commercial AM radio channel. The Borman's HAR system will likely operate via a public radio channel controlled by Indiana's Lake County Tourism Bureau.

3) Traffic Management

Beyond detecting and informing motorists of roadway problems, technologies can be employed to manage corridor operation during incident conditions. By minimizing congestion and delays caused by incidents, these management techniques improve freeway safety and performance.

In the past, **ramp metering** has been a commonly used technique for managing congested freeway operations. Although issues such as equity and surface street queueing have historically been a concern, operating conditions within the Borman Corridor appear to be particularly conducive to ramp metering. On-ramp motorists wishing to avoid delays caused by metering, do generally have alternate routes available to them. Thus, metering may encourage motorists, particularly short-trip motorists, to use alternate surface streets and thereby reduce individual and system delays.

To better facilitate trip diversion, roadway improvement measures, such as arterial signal optimization, will be implemented. Ideally, signal optimization will be self-adaptive [JHK 1984]. Future research efforts related to this project will investigate techniques for distributing diverted vehicles uniformly across the roadway network.

Traffic Operations Center

Operation within the Borman corridor will be supervised from a single remote Traffic Operations Center (TOC). The TOC will serve to 1) monitor traffic operations, 2) supervise traffic control and management strategies and 3) issue information on operating conditions to neighboring transportation agencies, emergency response agencies and the media.

Communication Linkages

Leasing phone company utility lines appears to be the most cost-effective manner of transmitting information for the proposed system. Leasing lines provides the flexibility to change communication routing strategies in the event that the TOC is relocated. Transmission service can be canceled at any time should alternative communication strategies become attractive in the future.

Linkages from inductive loops and changeable message signs to the on-site microprocessors can be accomplished using traditional coaxial cable. Fiber optic cables, which can better accommodate information sent by the CCTV system, will be used for all trunk lines. Transmitting information via satellite will be explored as part of a pilot study related to this project.

SUMMARY & CONCLUSIONS

The objective of this paper has been to outline technologies to be implemented as part of a freeway traffic management system in Indiana. The proposed system will also include management strategies such as mobile service patrols and commercial radio traffic broadcasts. The scope of this paper, however, is confined to the discussion of technologies. Thus, the details of the proposed traffic management strategies are excluded.

REFERENCES

1. Cassidy, M.J. and Sinha, K.C. [1990], <u>An Electronic Surveillance and Control System for Traffic Management on the Borman Expressway</u>, Final Report, Purdue University, JHRP-90/14.

2. Harmelink, M. et al. [1984], "Freeway Traffic Management on Highway 401, Toronto, Canada", ITE Journal (March).

3. McDermott, J. [1980], "Freeway Surveillance and Control in the Chicago Area", ASCE Journal of Transportation Engineering (May).

4. Minnesota DOT [1979], "I-35W Traffic Management System".

5. Hibbard, J. et al. [1990], "An Overview of the Ramp Metering Subsystem for the Phoenix Freeway Management System", Compendium of Technical Papers, ITE 60th Annual Meeting.

6. JHK & Associates [1989], <u>Smart Corridor for the City of Los Angeles</u>, Demonstration Project Conceptual Design Study.

Evaluation of Reliability of Road Network
for Better Performance, Advanced Management
and Future Network Design

Hiroshi Wakabayashi[1] and Yasunori Iida[2]

Abstract

This paper proposes reliability as a new indicator for advanced management and future construction of road networks; and also presents a new reliability analysis method. This article contains three sections. First, the significance of road network reliability and the benefits of increasing reliability are discussed. Second, a method for increasing reliability through traffic control is demonstrated. Third, an efficient and practical method for calculating an approximation of terminal reliability is developed. Numerical operations will be provided verifying that the method provides stable and satisfactory approximations. The CPU-TIME is 1/200 to 1/20,000 times more efficient than both the precise method in calculating the exact value, and the Monte Carlo method in obtaining the approximate value.

Introduction

In traffic management and new road construction, road planning requires an indicator of present road network performance levels in order to explain potential benefits derived from the construction and management of new and existing roads. Conventional indicators such as the total length of road per unit area or average travel time have been quantitative or static. Now a new indicator reflecting the quality of a road network and the constancy of its smooth performance is required.

This paper proposes terminal reliability as a new indicator for road network management and construction, discusses the significance of increasing reliability, and also proposes a new reliability analysis method.

The terminal reliability of a road network is defined in this study as the probability that specific two nodes

---

[1] Associate Professor, Dr. Eng., Dept. of Civil Eng.,
   Osaka Prefectural College of Technology,
   26-12, Saiwai-cho, Neyagawa 572 JAPAN.
[2] Professor, Dr. Eng., Dept. of Transportation Eng.,
   School of Civil Eng., Kyoto University,
   Yoshida-Hommachi, Sakyo-ku, Kyoto 606 JAPAN.

are connected for certain traffic service levels for specific time periods.

The reliability indicator illustrates the stability of the road network performance where service level is defined as daily smooth traffic without a heavy delay. A high reliability road network provides sure and unfluctuating traffic service by offering drivers alternative routes even when traffic accidents occur or road maintenance is taking place. Thus, not only general drivers, but ambulances and fire engines can arrive at their destinations smoothly. Therefore, in today's cities of heavy traffic congestion, it is important to increase the reliability of road networks in order to promote quick medical care and the prevention of disasters, as well as fluid daily transit.

Increasing Reliability through Traffic Control

This section demonstrates that when there are two parallel roads between two given nodes and heavy traffic is producing congestion on both roads, traffic control can increase terminal reliability. It will be shown that systems perform more efficiently with traffic control than without, from the point of view of reliability.

The assumptions in this section are as follows:

(1) Link reliability is the result of variations in link flow. Link reliability is defined as the probability that the demand flow on the link does not exceed the capacity of the link for a given time period.

(2) Demand flow range obeys a normal distribution. The mean of the demand flow is $\overline{v}_a$.

(3) Variances in the normal distribution σ_a^2 are functions of the demand flow and link capacity(C_a) on link a. Coefficients of variation (COV) $\sigma_a/\overline{v}_a$ are given using the following function. (α, β, γ and δ are parameters).

$$COV = \sigma_a/\overline{v}_a = \alpha \cdot \exp\{-(\overline{v}_a/C_a+\delta)\cdot \beta\}+\gamma \quad , \quad (1)$$

With these assumptions, link reliability can be determined from demand flow and link capacity. We can, therefore, make a model that a change in link flow affects link reliability.

A model network is shown in Fig.1. One road is a highway and the other is an access-controlled road(e.g., an express-way). For simplification, road specifications and "zero-flow" travel time are the same for both roads. Traffic capacity is 2,000(cars) per hour and "zero-flow" travel time is 10 minutes for each link. Travel time is attained using the modified B.P.R. function.

The Origin-Destination demand flow rate consists of four cases; that is 3,500(cars), 4,000, 4,500, and 5,000. Terminal reliability R between node 1 and node 2 is given using

$$R = 1.0 - (1.0 - r_1)(1.0 - r_2). \quad (2)$$

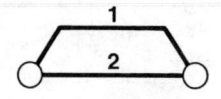

FIG.1. Model Network

When no traffic control is carried out, traffic is assigned in an equilibrium manner to result in the same flow on both links. We investigate the change in reliability when inflow control is carried out on the access-controlled road. We calculate the terminal reliability by (2) shifting the path flow from link 1 to link 2 through traffic control. The results are illustrated in Fig.2. The horizontal axis depicts the difference of flow between link 1 and link 2 (= $\bar{v}_2 - \bar{v}_1$) and the vertical axis indicates the terminal reliability.

By shifting the path flow from link 1 to link 2, the reliability of link 1 increases and the reliability of link 2 decreases. Consequently, the terminal reliability increases. Thus, for increasing terminal reliability, traffic control is valid. However, congestion on link 2 increases with traffic control. Therefore, we should consider this trade-off relationship. A simple index for explaining the benefits of traffic control on both routes is introduced using the following integrated index;

$$\text{Integrated Index} = \sum_a r_a \times \bar{v}_a . \qquad (3)$$

This index resembles the expected traffic flow value that does not encounter congested traffic. Figure 2 illustrates this index for each demand flow with a dotted line. The vertical axis is on the right-hand side of the figure. When the total demand flow does not exceed the total capacity of both links(i.e., demand rate is 3,500 or 4,000), traffic control is not effective since the index does not increase when the path flow is shifted flow from link 1 to link 2. When the total flow exceeds the total capacity(i.e., demand of total flow rate is 4,500 or 5,000), both the terminal reliability and index increase through shifting the path flow. Therefore,

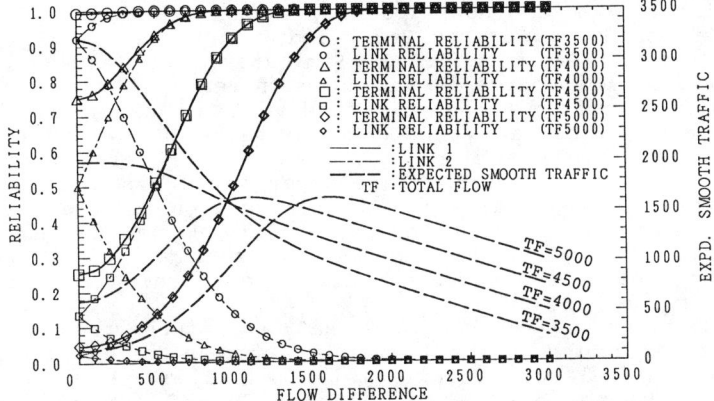

FIG.2. Change in Reliability and Integrated Index through Traffic Control

traffic control is worth introducing. When the path flow is shifted more heavily from link 1 to link 2, however, the index decreases. Of course, when no traffic control is carried out, both routes collapse with congestion. Thus, there is a desired level of traffic control for operating the system most efficiently as well as for avoiding a complete system failure.

We can conclude from this analysis that routine traffic control is necessary for providing a system that is reliable. In future studies, we need to develop a more appropriate index that can be applied to actual traffic control.

The Intersection Method

In general, network reliability analysis requires a great amount of computational work since computational work increases exponentially with network growth. This section presents an efficient and practical new method based on Reliability Graph Analysis for estimating an approximate value of reliability between two given nodes in a road network (Iida and Wakabayashi 1989). We introduce the two reliability functions as

$$R_p = 1 - \prod_{s=1}^{p'}(1 - \prod_{a \in P_s} r_a) ,\qquad(4)$$

and

$$R_k = \prod_{s=1}^{k'}\{1 - \prod_{a \in K_s}(1 - r_a)\} ,\qquad(5)$$

where P_s and K_s represent the s-th minimal path set and the s-th minimal cut set, and p' and k' are the number of partial minimal path sets and cut sets respectively.

These two functions are characterized as follows: R_p is the increasing function of the number of path sets, and the value R_p increases monotonically with an increase in the number of path sets from the lower bound value of reliability to the upper bound value of reliability. This upper bound is known as Esary and Proschan's upper bound (Barlow and Proschan 1965). Similarly, the value of R_k decreases monotonically with the increase in the number of cut sets from the upper bound value of reliability to Esary and Proschan's lower bound value of reliability. Thus, the two functions will cross at a certain point between Esary and Proschan's upper and lower bounds. We propose that the value at the intersection is a satisfactory approximation for the reliability value. The use of partial minimal path sets and cut sets allows for extremely efficient calculations. We call this the intersection method.

Next, we will demonstrate that the intersection method provides sound approximations and the necessity for computational work is extremely limited. Three different grid-type networks which consist 24, 40 and 85 links respectively are investigated. The node pair investigated is diagonal. The reference standard for the comparison is

TABLE 1. Comparison of Reliability Values and Cpu-times

The number of links contained in network		Intersection method		Exact method		Monte Carlo method	
		Reliability (Error from Reference Standard)	CPU-TIME (Sec.)	Reliability (Error from Reference Standard)	CPU-TIME* (Sec.)	Reliability (Error from Reference Standard)	CPU-TIME (Sec.)
24	Case 1	0.97717 (+0.00212)	0.033	0.97505 (Ref.Std.)	738.092	0.97542 (+0.00037)	6.698
	Case 2	0.19732 (-0.00112)	0.032	0.19844 (Ref.Std.)	737.886	0.19818 (-0.00026)	5.545
40	Case 1	0.97612 (+0.00095)	0.031	Uncalculatable		0.97517 (Ref.Std.)	16.194
	Case 2	0.11447 (-0.03709)	0.045	Uncalculatable		0.15156 (Ref.Std.)	12.425
85		0.87038 (+0.04342)	0.049	Uncalculatable		0.82696 (Ref.Std.)	61.175

*This cpu-time is from a vector processor(FACOM VP-400E) and other cpu-times are from a general purpose processor(FACOM M-780).

given using the exact method. When the system expands, however, it is impossible to calculate the exact value since the computational work increases exponentially with network expansion. Thus, approximations attained using Monte Carlo method in conjunction with the restricted sampling technique (Kumamoto et al. 1977) is used as the reference standard instead.

Table 1 shows the results. Approximation values attained using the intersection method are accurate in comparison with the exact values. The intersection method cpu-times are extremely small and the computational efficiency is 200 to more than 20,000 times greater than that of the exact method and Monte Carlo method. The increase in the Monte Carlo method's computational time is greater than the increase in the network scale in spite of its high accuracy. When reliability is used as indicator for traffic management and new road construction, this accuracy is suitable for practical use. In addition, the partial minimal path sets used by the intersection method correspond to actual traffic routes, and cut sets reflect screen lines. Thus, we can execute reliability analyses of actual traffic conditions using the intersection method. We can conclude that the intersection method is an appropriate method for practical use in road network reliability analysis.

References

Barlow,R.E. and Proschan,F.(1965). "Mathematical theory of reliability." John Wiley & Sons,Inc., New York.
Iida,Y. and Wakabayashi,H.(1989). "An approximation method of terminal reliability of road network using partial minimal path and cut sets." Transport Policy, Management & Technology towards 2001, WCTRS,4,367-380.
Kumamoto,H.,Tanaka,K. and Inoue,K.(1977). "Efficient evaluation of system reliability by Monte Carlo method." IEEE Trans.on Reliability, R-26(5),311-315.

Public Agency Issues in the Implementation of IVHS

S. Edwin Rowe[1]

Abstract

Public agencies will be extensively involved in the deployment, maintenance, and operation of IVHS infrastructure. The existing fragmentation of transportation responsibilities will require development of new approaches to multi-jurisdictional coordination. Certain IVHS measures will be perceived as contrary to local interests and will arouse political opposition. Many public agencies do not currently have the technical expertise to maintain and operate IVHS equipment. New funding mechanisms need to be devised to assure an adequate level of public agency support of IVHS.

Introduction

The transportation community has initially focused its attention on the technological requirements of Intelligent Vehicle Highway Systems. This is appropriate, since advanced technology is unquestionably at the heart of IVHS. It is important to keep in mind, however, that our ability to effectively deploy and use emerging vehicular and highway technologies will depend on the active participation of public agencies. These governmental entities will have responsibility for the design, construction, operation, and maintenance of a significant portion of the infrastructure necessary to support IVHS.

A role as important as this deserves some very careful consideration. Many questions need to be

---

[1]General Manager, Los Angeles City Department of Transportation, 200 North Spring Street, Room 1200, Los Angeles, CA 90012

addressed: How do we go about organizing a large number of public agencies into a coordinated nationwide IVHS network? How do we gain political acceptance of certain IVHS strategies? How do we ensure that public agencies develop the necessary technical skills? How do we make available to public agencies the funding required to deploy, operate, and maintain IVHS?

Answers to these questions should be developed in parallel with the technology if we are to be ready for large-scale IVHS deployment. We should also expect that many of the IVHS technology trade-offs will be influenced by such considerations.

Multi-Jurisdictional Coordination

A national IVHS Program will require expanded cooperative relationships between state and local governments. The states operate the Interstate Highway System and other state highways linking metropolitan areas. Jurisdiction over the highway network within metropolitan areas is usually fragmented between a state DOT and a multitude of local governmental agencies. In Los Angeles County, for example, there are 86 cities which control operations on highways within their boundaries except designated state highways.

The practical consequences of this jurisdictional fragmentation are a lack of uniformity in traffic control practices, equipment, and capabilities throughout the metropolitan area. The problem is particularly acute at the boundaries between cities, since usually there are no longer any large open spaces separating them. The metropolitan area is one seamless web of highways Balkanized by artificial jurisdictional boundaries. At the extreme, some jurisdictions adopt parochial traffic management strategies designed to discourage regional traffic from using their arterial streets. Obviously, such cities will be cool to IVHS strategies that might divert freeway traffic to their highways. They would also generally view regional traffic management goals as threatening to their "home rule" prerogatives.

Embryonic efforts are underway in several states to deal with the problem of multi-jurisdictional coordination of regional traffic management programs. State and local agencies have joined together on the Los Angeles Smart Corridor Demonstration Project and are overseeing all phases of the project from conceptual design to the development of traffic management plans. The Texas Department of Highways and Public Transportation has organized multi-jurisdictional

Traffic Management Teams to coordinate improvements in traffic operations within nine metropolitan areas and freeway corridors.

Future federal and state highway funding programs should contain provisions that will encourage cooperative approaches to traffic management and the establishment of organizational structures for metropolitan-wide coordination. The public agency organizational infrastructure is as critical to the success of IVHS as is the equipment infrastructure and should be established early in the system development cycle.

Political Acceptance

Most of the IVHS goals and strategies should not arouse opposition of local governments, since they are consistent with local concerns over traffic congestion, safety, and air quality. There are some IVHS strategies, however, such as ramp metering and route diversion, which can be expected to generate strong local opposition unless introduced with a great deal of sensitivity. These measures can be perceived to improve freeway travel at the expense of added congestion on local surface streets. With regard to ramp metering, there is a considerable amount of actual implementation experience that can be drawn upon to allay local concerns. Route diversion strategies that would advise drivers to leave the freeways and use alternative routes on local surface streets are another matter. These are sure to be viewed as a degradation of local street traffic conditions unless local agencies are granted a degree of control over the timing and magnitude of diverted traffic.

Technical Capability

Many public transportation agencies do not currently have the technical capability to manage the implementation, operation, and maintenance of IVHS. Engineering staffs have their major experience in highway and traffic engineering and are unfamiliar with the complexities of command/control systems that involve the integration of computers and communication networks. Technicians are commonly selected from the ranks of journeyman electricians capable of installing wiring, but lacking in experience related to digital electronic systems.

Such a state of affairs has impeded the adoption of currently available computer and electronic technology for traffic management systems and would severely inhibit a national IVHS Program utilizing even more advanced technology. Deployment of IVHS nationally will place severe technical demands on public agencies, which must be dealt with before large-scale commitments are made.

First, it will be necessary to keep the technical managers of public agencies informed of IVHS developments so that they can begin planning for their future involvement. This can be done in a variety of ways including workshops, courses, and professional societies such as ASCE.

Once an IVHS project begins, special training programs for engineering and technical staff should be provided at an early stage so that they can become progressively familiar with the new technology during design and installation. Crash training courses before system delivery should be avoided.

Organizational changes will be required in most agencies to provide adequately for the wide range of new activities required for an IVHS program. It also will be necessary to review position descriptions and classifications to insure that appropriate skilled technical personnel are available.

Serious consideration should be given to "turn-key" projects that are designed, implemented, operated, and maintained under contract. Private systems contractors have much more flexibility in acquiring and maintaining the necessary technical skills required for a successful IVHS program than most public agencies.

To insure against gradual degradation in system performance, federal and state funding agencies should establish standards for both operation and maintenance of new IVHS projects. Enforcement of these standards could be tied to IVHS and other transportation funding programs.

Cost Barriers

Initial costs to deploy IVHS infrastructure are beyond the capability of most city governments. For example, the Advanced Traffic Management System (ATMS) element of IHVS can cost as much as $80,000 per signalized intersection to implement. A city with 1,000 signalized intersections would thus be faced with an $80 million capital project. New funding sources will be necessary for projects of this magnitude at the city level.

Funding of operations and maintenance costs of IVHS projects also will be a major issue, especially with cities. Many cities that implemented computer traffic signal control systems over the last 30 years with federal and/or state gas tax funds found that they did not have the necessary resources from local taxes to fund adequate operations and maintenance of these high tech systems. Consequently, the systems often did not perform up to expectations.

Given the intense competition for general fund tax revenue at the local level, it is not likely that adequate funding will be available to operate and maintain IVHS infrastructure. This is likely to be true even though the increase in local expenses over what is currently allocated for maintenance of existing traffic signal systems may not be large. The adverse results for IVHS would be even greater than was the case with computer traffic signal systems. System reliability and accuracy of information on traffic conditions are absolutely essential to a successfully functioning IVHS Program.

Future consideration should be given to a more flexible approach to the allocation of gas tax funds that would allow for use in operations and maintenance activities related to IVHS. This would provide some assurance that the investment in equipment and highway infrastructure would return the expected benefits. More importantly, the funding mechanisms also could be applied in a way that provided a greater uniformity of operations and maintenance levels over the entire system network. Funding incentives also might be devised that would encourage participation of local jurisdictions in joint IVHS operations and maintenance programs.

Conclusions

Clearly, public agencies will be major players in any future national IVHS program. It is not too early to begin to address the special problems that they face in participating in such a program. We must take the time to learn from past public agency experiences implementing new highway transportation technologies and create the organizational, educational, and funding mechanisms necessary for future success.

STEREO VISION ONBOARD A VEHICLE FOR OBSTACLE DETECTION

J.-L. BRUYELLE and J.-G. POSTAIRE[1]

Abstract: We show how two linear cameras, mounted behind the windscreen of a car, can be used to detect the obstacles on the road in real time, even with small computer architectures. Results of simulations and field tests are presented, that demonstrate the efficiency of the described system.

I. INTRODUCTION

In recent years, the increase of road traffic has made necessary the improvement of safety. The aim of the research described in this paper is automatic visual obstacle detection. Much work in this field has been concerned with image sequence analysis and with stereo vision. However, these techniques, when used separately, lead to unacceptable constraints (only one moving object, fixed camera, rigid motion...) or are much too slow for our purpose.

Dynamic stereo vision is a relatively new technique for addressing depth and motion estimation problems in complex, unstructured environments [7]. Performing stereopsis simultaneously with temporal matching is a new way, in which motion information is used to limit possible matches between the image pairs while temporal tracking can be improved by 3D knowledge of the scene. Hence, disparity analysis between the two images of each pair and tracking of prominent features provides many ways of determining motion [1,3,6,11] as well as depth maps [4,5,8,9].

All the approaches outlined above have been developed to work under as various environments as possible. They result in slow, complex algorithms, and are not suitable for real time implementation on small computers, as required for road vehicles. Fortunately, it can be assumed that the motion occurs on a planar road [2,10]. This is the basic assumption of our approach. Linear cameras can thus be used instead of video cameras, so as to reduce strongly the data flow.

---

[1] Centre d'Automatique de Lille, Bât. P2, USTL-FA, 59655 Villeneuve d'Ascq Cedex, France. This research is supported by the EEC's *Eurêka-Prometheus* project, and by a joint *INRETS-regional council Nord-Pas de Calais* grant.

II. STEREO VISION WITH LINEAR CAMERAS

II.1. Cameras geometry and evaluation of disparities

Suppose the two cameras are mounted behind the windscreen of a car so that their optical axes are parallel and separated by a distance E. These two axes define the "optical plane" which intersects the planar surface of the road about 50 metres in front of the car (fig. 1). Let us choose the base line connecting the lens centres as the x-axis and let the z-axis be in the optical plane, parallel to the optical axis of the cameras, so that the origin of the $\{x,z\}$ coordinates stands midway between the lens centres. The intersection of the optical plane with the scene is focused on the linear CCD sensors, which are parallel to the x-axis, in the optical plane. Let the image coordinates of a point in the left and right linear images be x'_l and x'_r, respectively (fig. 2).

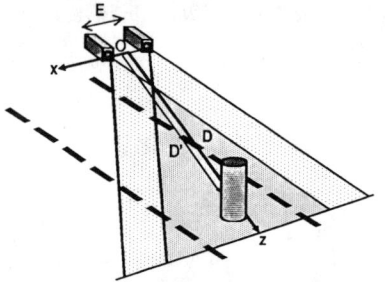

Figure 1 : Linear stereo vision set-up   Figure 2: Formation of the stereo image

Using the pinhole camera model, the disparity $d = |x'_l - x'_r|$ between the left and right images of a point $P(x, z=D)$, of coordinates x , z=D in its environment, can be easily computed as: $d = E f / D$, $\forall$ x, where f is the focal length of the lenses, E the base line length and D the distance between the point P and the base line. This result indicates that the disparity is the same for the images of all the points lying at the intersection of the planar surface of the road and the optical plane. When the road is clear, its two linear images can then be registered by means of a simple shift of amplitude d . When a vertically extended obstacle is seen by the cameras, its distance D' from the base line is smaller than the distance D of the background. After shifting one of the images by d , the remaining disparity of the object images is: d-d' = Ef(D-D')/(D.D')

II.2. Obstacle detection scheme

When the way is clear, the remaining disparities between the two images after global matching of the background are negligible. Now, if any vertically extended obstacle stands in front of the car and intersects the optical plane, it causes a noticeable remaining disparity (fig. 3). The analysis of disparities after global registration of the background is thus an appealing solution for deciding whether an obstacle is present in front of a vehicle.

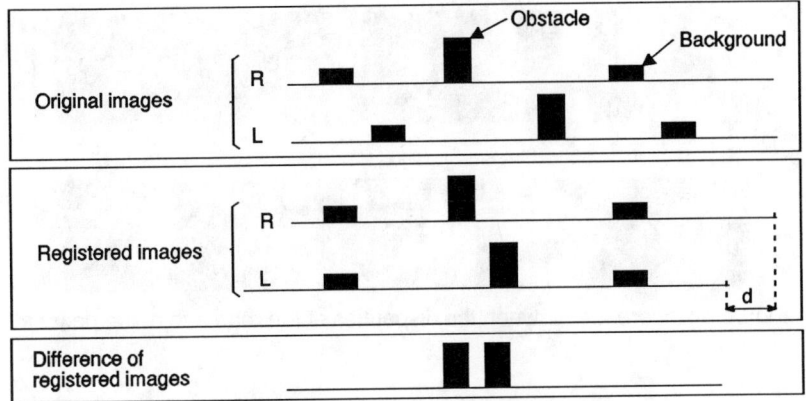

Figure 3: Obstacle detection from registered images

The above described set-up has been simulated with computer images, so as to determine the performances that can be expected. Figure 4 (a) shows a synthetic pedestrian crossing a road, as seen by one one the cameras, as well as a user-selected window. The same window, enlarged in figure 4 (b), shows the effect of the sampling by the sensor, and provides a qualitative idea of the possibility of recovering the pedestrian's image from its sampled representation.

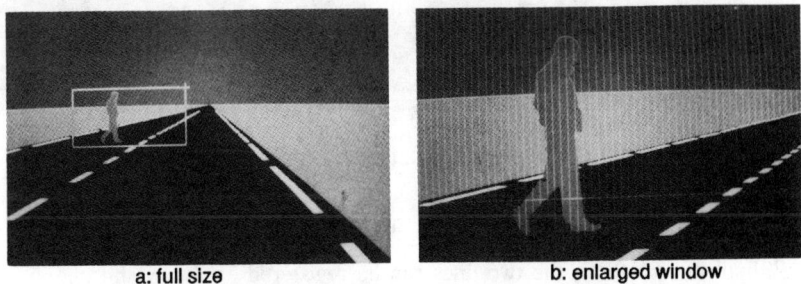

a: full size   b: enlarged window

Figure 4: Computer simulation of the linear stereo set-up

The central horizontal lines of the two 2D images correspond to the linear images. Figure 5 shows these lines, vertically widened for a better legibility. The disparity between the left and right images, with respect to the background, is perfectly visible and quantizable.

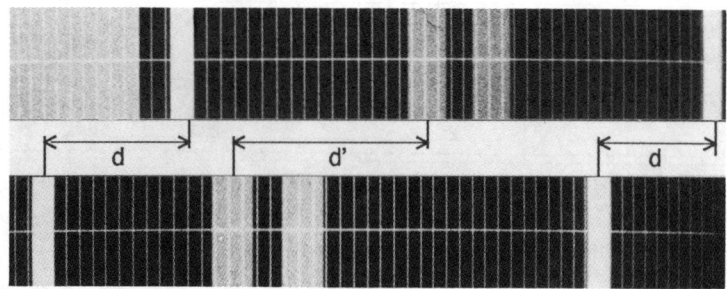

Figure 5 : difference between the disparities of the road and of the pedestrian

IV. EXPERIMENTAL RESULTS

The set-up resulting from the above described simulations has been mounted on a car. It consists of two linear cameras and a PC-AT 386 computer. A sequence, in which a pedestrian walks towards the car, is shown in figure 6.

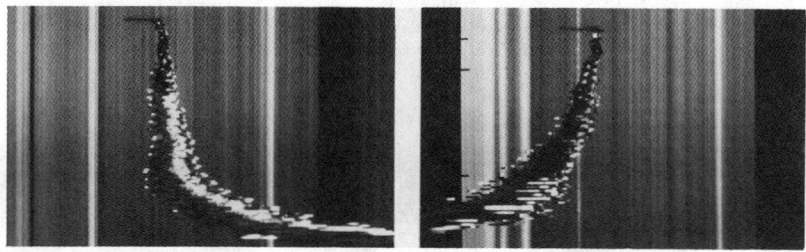

a: Left sequence   b: Right sequence
Figure 6: Pedestrian shot by the linear stereo set-up

Linear images shot simultaneously are displayed, from top to bottom, as corresponding lines on the left and right images of figure 6.
Without any obstacle, the two lines can be registered easily, by shifting one of them along the other until the peaks in grey level corresponding to the white markings are superimposed. The quality of the match is measured by the sum of the absolute values of the pixel-to-pixel grey level differences between the two registered images. Let $g_l(i)$ and $g_r(i)$ denote the grey levels at the ith pixels of the left and right line images, respectively. If the right image is shifted by d pixels with respect to the left image, the quality of the match is measured by:

$$Q(d) = \sum_{d=1}^{R-L+1} \sum_{i=L}^{R} | g_l(i) - g_r(i+d) |$$

where L and R are the indexes of the leftmost and rightmost pixels delineating the road surface on the left image. With a pedestrian, whatever is the shift d, the quantity Q(d) remains at a high level, thus indicating the presence of an obstacle in front of the car.

V. CONCLUSION

The encouraging results so far obtained show how an obstacle can be detected by analyzing linear stereoscopic image pairs. Further developments are presently in progress in order to extract some information about the direction of displacement and the velocity of the obstacles, with respect to the vehicle. By combining detections from several consecutive stereo pairs, the performances of the procedure should be increased while more information about the detected objects should be extracted. Some efforts are also devoted to the identification of the markings on the road, in order to improve the robustness of the global registration process.

REFERENCES

[1] AYACHE N. and FAUGERAS O.D.: *Building, registering, and fusing noisy visual maps*, Int. Journal of Robot. Res., Vol. 7, n°6, 1988

[2] CARLSSON S. and EKLUNDH J.O.: *Object detection using model based prediction and motion parallax*, Proc. IEEE Int. Workshop on Intell. Robots and Syst.,pp 51-56, 1990

[3] DICKMANNS E.D.: *An integrated approach to feature based dynamic vision*, Proc. IEEE Compt. Vision and Pattern Recog., pp 820-825, 1988

[4] FORSTNER W.: *Precision of geometric features derived from image sequences in High Precision Navigation: Integration of Navigational and Geodetic Methods*, Linkwintz and Hanleiter, ed. Springer-Verlag, pp 313-329, 1989

[5] GENERY D.B.: *Modelling the environment of an exploring vehicle by means of stereo vision*, Stanford Univ., PhD thesis, 1980

[6] GENERY D.B.: *Stereo vision for the acquisition and tracking of moving three-dimensional objects, in* Tech. for 3D Machg. Perception, Rosenfeld, ed., Elsevier, pp 53-74, 1986

[7] JENKIN M.: *Tracking three dimensional moving light displays*, Proc. Workshop Motion Representation Constr., Toronto, Canada, pp 66-70, 1983

[8] MAROQUIN J.L.: *Probabilistic solution of inverse problems*, MIT, PhD thesis, 1985

[9] MATTHIES L.: *Dynamic stereo vision*, Carnegie Mellon Univ. Tech. Rep. CMU-CS-89-195, 161pp, 1989

[10] SCHAASER L.T. and THOMAS B.T.: *Finding road lane boundaries for guided vehicle navigation*, Proc. IEEE Int. Workshop on Intell. Robots and Syst.,pp 101-109, 1990

[11] YOUNG G. and CHELLAPA R.: *3-D motion estimation using a sequence of noisy stereo images*, Proc. IEEE Compt. Vision and Pattern Recog., pp 710-716, 1988

AUTOMATIC OBJECT IDENTIFICATION IN ROAD

Martin Lipičnik [1]
Danijel Rebolj [2]
Tomaž Tollazzi [3]

ABSTRACT

The methodology for automatic object recognition in real road traffic anvironment is presented. The proposed data acquisition methods have been tested on site. In addition, the fundamental solution for the recognition of object involved in road traffic is presented.

1. INTRODUCTION

Traffic participants involved in road traffic have their own needs and requirements. As soon as their number or the number of their requirements exceeds a certain limit conflicts start occurring. In well organized road traffic the possible conflicts are more or less known, and so is their location - the conflict point.

The main task of traffic experts is to prevent the occurrence of conflict situations - conflicts, and if they nevertheless occur, to moderate their consequences.

As our paper is limited to road traffic, we shall discuss the road as a traffic surface on which we have direct traffic participants (pedestrians) and indirect traffic participants (drivers of motor cars and bicyclists). A road traffic expert considers pedestrians and vehicles as objects involved in road traffic.

We can consider man as a top-level intelligent technical system consisting of sensors and a processor. The sensors are the eye (it measures visual impressions), the inner ear (it measures acceleration), the skin - the sense of touch (the dynamometer), and the ear (it measures auditory signals). All the mentioned so-to-say sensors operate in real time: they send the data directly to the brain - the processor. There the data are processed into an information which tells how to respond to the actual situation.

Speaking in terms of technical systems man can be considered a top-level achievement whose capabilities have not yet been exceeded by any product of most advanced technologies. However, despite his »technical superiority«, man remains imperfect and unpredactable. Wrong decisions often result in a catastrophy (the last mistake in one's life).

This is the reason why we are working intensively on the development of such technical systems which will assist the driver in making the correct decision in different traffic

---
[1] Professor, Head of the Road and Traffic Centre, Faculty of Technical Sciences
Smetanova 17, YU 62000 Maribor, E-MAIL: LIPIČNIK@UNI-MB.AC.MAIL.YU
[2] M. Sc., Research Engineer, Road and Traffic Centre, Faculty of Technical Sciences
Smetanova 17, YU 62000 Maribor, E-MAIL: REBOLJ@UNI-MB.AC.MAIL.YU
[3] Assistant, Research Engineer, Road and Traffic Centre, Faculty of Technical Sciences
Smetanova 17, YU 62000 Maribor, E-MAIL: RDRTC@UNI-MB.AC.MAIL.YU

situations. Technical systems that help people involved in road traffic are of two kinds: direct and indirect systems.

Direct systems are, for instance, systems that would be capable of recognizing the sharpness of the curve and would thus prevent the driver from running into the curve at a too high speed. Another example is a system that would guide the driver directly to a free parking place.

Indirect systems enable a rational and safe general organization of the traffic.

In our paper we shall present one of our systems that offers indirect assistance to traffic participants, which is done by traffic course monitoring in real time and place by means of video-systems.

2. VISUAL DATA ACQUISITION SYSTEM IN ROAD TRAFFIC

Video systems enable to organize the control of traffic in large road networks. They also help to acquire numerous data which are indispensable to road traffic experts and necessary for a safe and rational traffic organization aiming at the reduction of possible conflicts. As there are numerous factors that affect traffic events, just as many data that describe these factors are necessary. Visual recording of traffic events is one of the possible ways of acquiring this quantity of data. In our opinion visual recording is most efficient and best adjusted to man because it imitates the human eye, i.e. the most developed human sensor.

The monitoring system can be based on two principles. Recording stations are connected to the central station whose role in the system is either more or less important. In this respect, two kinds of communication between the recording station and the central station are foreseen depending on the kind of information that is being transmitted.

In the first case, the information of the type »picture of the object« is transmitted from the recording station to the central station where it is processed and identified. This approach requires good communication means and a powerful central station capable of rapid receiving, processing and storage of data. Intelligence of recording stations is not necessary.

In the second case, the communication is based on intelligent recording stations, which comprehends data acquisition as well as the ability to process the »picture of the object« and to send the attributes of the identified object to the central station. In this case the central station has a minor role. Its major function is statistical processing of results and their presentation. The intelligence of the recording station is of primary importance. The factors that decide which data transmission approcah to use are the quality of communication means and the time necessary to identify the recorded object.

Both approaches require special software (for the central station in the first case and for all recording stations in the second case). We have developed three methods for data acquisition, data storage and data processing. These methods result from the development of an idea about possible ways of acquiring data from the real environment.

The first method is the simplest because the main part of the procedure is done manually, the second method is more up-to-date, and the third method makes excellent use of the latest achievements in the development of electronic instruments.

3. LAYOUT OF THE RECORDING NETWORK

A network of recording stations is placed into the road network. The density of recording stations is a function of the required accuracy of recording results. Recording stations are placed at at the beginnings and ends of all important road sections. Recording stations are connected to the central station which is responsable for the processing and presentation od results. Data acquisition at recording stations is continuous throughout the year. The central station conveys the data in cycles or constantly, depending on the needs of the user. The central station also stores the data and performs updating. The connection between the central station and individual recording stations is constant.

The described basic recording system is fixed. Each recording station has a subsystem that works constantly, in cycles or according to the needs. This subsystem is meant for detailed data acquistion. The location of the subsytem is also changed according to the needs. Each subsystem conveys the data to the central station through its recording station. Each road section on which the real conditions are to be recorded should be equipped with at least two recording stations. Recording stations are located either at the roadside or above the road, depending on local circumstances.

4. DATA ACQUISITION AND DATA PROCESSING METHODS

4.1. DATA ACQUISITION BY A VIDEO CAMERA AND MANUAL DATA PROCESSING

Recording stations are placed at the beginning and at the end of a road section. Recording stations are equipped with the recording apparatus and belonging activation mechanisms (if the activation of the recording apparatus is mechanical). The recording station is equipped with a video camera, the activation mechanism and the cable.

The activation mechanism enables to activate the camera if a vehicle is approaching the recording station. The camera is activated for a certain period of time, which means that the recording is cyclic. In this way the tape of a four hour video cassette is sufficient for one day of recording provided that the camera is activated for one second. The cameras at the beginning and at the end of the road section should be well timed (absolute time).

This method is not expensive and requires no sophisticated technology because all subsequent phases are performed manually.

The data acquisiton system consists of two video cameras and the display of absolute time. The data are stored on the video tape. The processing system includes the video player with which the video cassette is reproduced.

The data obtained by this method are the following: the time when the vehicle passes the recording stations at the beginning and at the end of the road section; the license plate number (used for vehicle identification and the calculation of the average vehicle speed); type of vehicle, vehicle make, country of origin, and the number of passengers in the vehicle The data are entered manually, further processing is automatic.

4.2 DATA ACQUISITION BY VIDEO-STILL AND COMPUTER PROCESSING

The location of recording stations and their density are the same as with the first method. The difference is that a video- still is used instead of the video camera. This method is more complex because a digital record is obtained which can letter be inserted into the computer by means of a graphic card of great resolution. The data acquisition, storage and processing are entirely computer aided.

When a traffic object passes the recording station the video-still is activated , and the the information or the picture of the object is recorded into the memoy unit. A special software enables to transmit the picture into the computer. The stored record is subsequently processed by special graphic identification programs (Picture Recognition System). All further processing is no longer manual but computer aided. The results obtained by the vehicle identification program serve as input data for further statistical processing.

4.3 VIDEO CAMERA DATA ACQUISITION, COMPUTER PROCESSING

The location of recording stations and their density are the same as with the first two methods, but there are some differences as regards data acquisition, storage and subsequent processing. It should be noted that a hardware sensor for activating the camera is not necessary because the software simulates the sensor. The recording station is equipped with a video camera, but we also have a PC and a graphical card of great resolution in the configuration. The computer has programs for vehicle identification (PRS) which digitally record the data into the computeržs memory.

The recording with the camera is continuous, and records are made only when objects pass (Figure 1). This method is the most complex one. It requires the sophisticated software and hardware available on the market.

The computer gets the recorded data by means of a picture transmission adapter and processes them by vehicle identification programs. The pictures are on the video tape and the records are stored in the computer. They can also be stored on diskettes or diskette tapes. As in the previous method, the results obtained by the vehicle identification programs serve for further statistical processing.

Figure 1: The result of recognition

5. RECOGNITION OF OBJECTS INVOLVED IN ROAD TRAFFIC BASED ON GEOMETRICAL CHARACTERISTICS

The software system is of essential importance for the recognition of objects from digitized videopictures of the traffic situation on a certain road section. As the surrounding is very specific, our solution of the problem is very unusual. The methods known in computer vision are used in a simple way and only at the basic level of picture recognition (for instance edge detection by Sobelov's operator 1) 2).The matching of some key words occurring in the paper with the terms used in object oriented programming is not a coincidence as the method has been efficiently used. However, it is not the programming technique that is so important, but the object in road traffic.

5.1 OBJECTS

The presentation of objects in a system is of primary importance. The surrounding has first been divided into static (traffic environment) and dynamic objects (traffic participants). Static objects have only geometrical parameters, whereas dynamic objects (simply called objects in this paper) are characterized also by dynamic time dependent parameters. Static parameters of objects (or attributes) are comparable with attributes of objects known in object oriented programming, while dynamic parameters can be understood as the results of methods of these objects.

The inverse principle of object oriented programming has been used for the recognition of objects. Known are the objectźs attributes and the results of the methods the object enables (i.e. its behaviour), but the object itself is not known. At the beginning of our research we worked just on simple functions. Nevertheless, this idea proved to be very useful for the drafting of the system. Its full applicability is expected in the recognition of objects which do not belong to the same class and therefore have different attributes and methods that define them.

At present, all dynamic objects are defined by the following attributes: Static - the basic shape (visible surfaces, the body) and size. Dynamic - function of velocity function of acceleration, and the trajectory.

5.2 OBJECT RECOGNITION

Objects are recognized in several phases in real time. The scanning subsytem (SSub) first perceives the presence of the object and determines its static attributes on the basis of simple graphic algorithms. (Static attribues also vary in real time, but the SSub registers these variations only to improve their accuracy). SSub adds the name to each perceived object and sends the data (the name, visible surfaces, the contour in real lineal measure) in equal time intervals to the Dynamic Object Reciognition System - DORS. This system can receive blocks of data from several subsytems at the time. In case we have several SSubs the naming of objects is synchronized, so that the object that is observed by several SSubs at the time is correctly identified. In addition, the data about one object are synthesized in case the object is observed from several recording stations. This approach makes all the data and subsequent processing much more accurate.

DORS monitors each object in space and establishes its behaviour. Gradually it determines its dynamic parameters and simultaneously tries to rank it into the closest possible subclass to which it belongs, thus recognizing it in detail as far as possible. Due to the incorporated rules, DORS can form new subclasses, which contributes to its knowledge base. In this phase it acts as an expert system.

The next step done by DORS is the recognition of the situation. Basing on the the attributes of objects, the system can predict events in the observed area and send messages with corresponding probability factors and priority degrees of predicted or past events.

5.3 LAYOUT OF THE DISTRIBUTED SYSTEM

Observing subsystems are the basic elements installed at observation stations together with the unit for picture receiving and digitizing. In this way the quantity of data sent through the network is substantially reduced. If required, the local subsystem can store the entire visual information into its own memory, just the number of frames is limited.

If we have just one observation station, DORS can be installed on the same location, but it uses its own processor. It can also be installed at a special place or in the control station if it is not too far away.

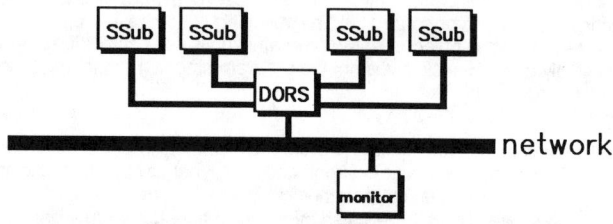

Figure 2: Schematic presentation of the system for object recognition system in road traffic

A monitor can be connected to every DORS (physically and logically if the system is independent, and logically from any location in the network to which DORS is connected). It enables communication with other DORSes. At present, synchronous operation of several DORSes, which would convey information about objects and monitor them through the entire network, is not planned.

5.4 SYSTEMS OPERATION

When the installation is over (installation of hardware, connection into the network, loading of software), the teaching of the system begins 4) 6). The teaching refers predominantly to local characteristics because the basic knowledge about static and dynamic objects as well as the knowledge base and the rules for recognition are already installed. We describe the static objects for each SSub separately and enter the dimensions of characteristic objects (the width of the pavement in a road intersection for instance). Global objects are marked separately.

The teaching of DORS is more demanding. DORS takes the basic knowledge about the environement from local SSubs (global static objects). However, we must indicate which static objects overlap or mark the fields which overlap (minimal shifts of cameras after installation are permissible because the SSub is not bound to the sight field itself but to the objects inside it). Greater shifts are crucial for the system. They mean new teaching of the affected subsystem and DORS.

Careful and precise installation is of highest importance for correct recognition and for correct determination of dynamic object parameters. In the opposite case, DORS will not find the right subclass and will consequently start creating its own.

After the installation, the system operates independently. DORS completes its knowledge base and tries to narrow subclasses and adjust the limits between them to the accuracy of data acquisition 3) 5). The operation of DORS can be monitored and we can intervene into the knowledge base if necessary. DORS also elaborates reports about its activities and about events in the observed area and stores them. If connected into the network, it sends them to desired addresses.

6 LOCATING THE RECORDING LOCATION

When locating the recording station the laws of optics should be strictly observed, as well as all other obstructive circumstances Obstructive conditions are due to:
- the maximal sizes of the maximal vehicles driving along the road section
- free road profile requirements
- the winter weather conditions and snowploughing
- the protection of the recording apparatus against precipitations
- the protection of the recording apparatus against stealing
- optical performances of the recording apparatus
- the need for high quality resolution of graphical outputa

CONCLUSION

Our paper presents the research work performed at the Road and Road Traffic Centre at the Faculty of Technical Sciences, University of Maribor. In our opinion our research efforts should continue as the acquired information is of high quality and most applicable, and the hardware and software are developed to a promising degree.

LITERATURE:

1) D. H. Ballard and C. M. Brown, Computer Vision. Prentice Hall, New Jersey 1982
2) H. Wang, Edge Detection in a New Dimension, University of Leeds, Department of Computer Studies, Technical Report, University of Leeds, Leeds, UK 1987
3) P.H. Winston, Learning Structural Description From Examples, in The Psychology of Computer Vision. Ed.P.H. Winston, McGraw Hill, New York 1975
4) Y. Ohta, Knowledge Based Interpretation of Outdoor Natural Color Scenes. Pitman, London 1985
5) H. J. Levesque, Knowledge Representation and Reasoning in Annual Review of Computer Science. Ed. J. F. Traub et al., 1986
6) I. Bratko, I. Mozetič, N. Lavrac: KARDIO. A study in deep and qualitative knowledge for expert systems. The MIT Press, Cambridge, Mass. 1989.

STEREO VISION WITH SCAN-LINE CAMERAS FOR INTRUSION DETECTION ON L.R.T. TRACKS

L. DUVIEUBOURG[*], J.-P. DEPARIS[**] and J.-G. POSTAIRE[*]

Abstract

In this paper we present a new method for the detection of intrusions by means of stereo vision in order to infer 3D information from two linear sensors. This method can be used for security enhancement of underground transport systems. This new concept which consists in combining two parallel surveillance planes has been validated in an experimental station of the R.A.T.P. tube in Paris.

Key Words

computer vision - linear cameras - stereo vision - intrusion detection - space time image - transport system security.

1- Introduction

Guided transport systems with fully automated driving have been under development for about fifteen years. Generally, those systems are built on dedicated guideways so that interactions with the users are limited to station areas. The safest solution, used in Lille, France, for the V.A.L. stations is to prevent the users to fall onto the tracks by means of landing doors. This results in high building costs (Boutry, 1987). Research for the automation of transport systems without landing doors is therefore very important.

Others solutions have been proposed using detecting devices to stop the trains and switch off electric power on the rails, in case of a fall on the tracks. Many technologies were considered, namely: pressure carpets, hyperfrequency barriers, video-sensors, passive infrared detectors and active infrared beams (Brachet, 1989).

Among these technologies, the concept of video stereo vision is appealing for intrusion detection because computer vision systems are continuously vigilant and may react within a very short time lag. The use of image processing to detect objects moving in three dimensional space generally calls for analysis of two images obtained by means of two matrix cameras aiming at the scene from different point of view. The problem is then to establish correspondence of points in multiple views while taking into account size changes and displacements (Barnard and Thompson

[*] Centre d'Automatique
Université des Sciences et Techniques de Lille Flandres Artois
59655 Villeneuve d'Ascq Cedex FRANCE

[**] CRESTA-INRETS
20 rue Elisée Reclus
59650 Villeneuve d'Ascq Cedex FRANCE

1980 ; Ohta and Kanade, 1985). The analysis of such stereo images requires from several seconds to hours of computer time.

In this context, we propose a new detecting device using linear video-sensors. Considering that the problem is to detect a person falling from the platform onto the tracks, we have designed a device with a linear camera in which the image of a line of fluorescent tubes is focussed on the linear sensor of the camera, so as to define a surveillance plane. This set up is very effective to detect the presence of any object between the camera and the line of fluorescent tubes (section 2).

In order to discriminate small objects crossing the surveillance plane (birds, insects, dust, snow, etc ...) from humans entering the protected zone, we use two twin parallel surveillance planes. In this configuration, the two cameras, as well as the two lighting tubes, are facing each other so that stereovision triangulation techniques yield the position and the size of any object crossing the double surveillance plane (section 3).

The corresponding software has been implemented on a specific hardware architecture in order to derive spatio-temporal information about the objects crossing this double surveillance plane. The system has been tested in an experimental station of the R.A.T.P. underground in Paris . Experiments in field conditions have demonstrated that the experimental set up is able to distinguish in real time between humans and nuisance alarms produced by the intrusions of small objects (section 4).

2 - Definition of a surveillance plane

In our system, each linear camera shoots a straight line of fluorescent tubes which defines, with the optical centre of the camera, a surveillance plane (cf. Fig. 1).

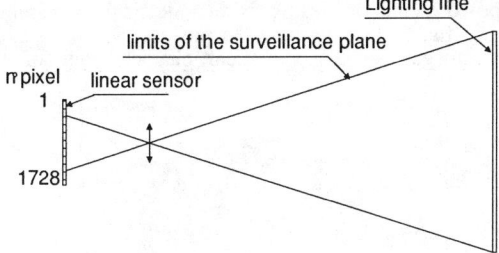

Fig. 1 : Definition of the surveillance plane

The camera produces a line image which is quite different from standard T.V. frames. When nothing crosses the surveillance plane, the lighting line is entirely seen and no significant information can be extracted from the image line (cf. Fig. 2). On the other hand when an object crosses through the surveillance plane, it leaves a shadow on the recorded line image and can be easily detected (cf. Fig. 3).

However, it is important to note that objects of different sizes may give rise to identical shadows while objects of same sizes may produce different shadows in the line image (cf. Fig. 4). Although such a simple device does not give information about the position and the size of the objects crossing the surveillance plane, it is

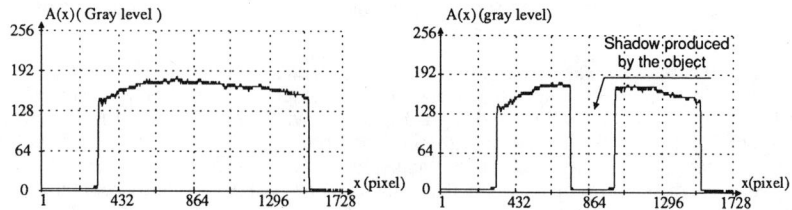

Fig. 2 : Line image of the ligthing tube   Fig. 3 : Line image with an object in the surveillance plane

used as the basic element of the stereo detection system presented in the next section.

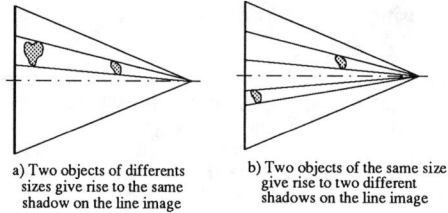

a) Two objects of differents sizes give rise to the same shadow on the line image

b) Two objects of the same size give rise to two different shadows on the line image

Fig. 4 : Effects of the projection of different object

3 - Recovering object position and size

The proposed method makes use of two identical and parallel surveillance planes, as shown in fig. 5, to determine the position and the size of any object crossing this optical gate. The geometrical characteristics of this particular

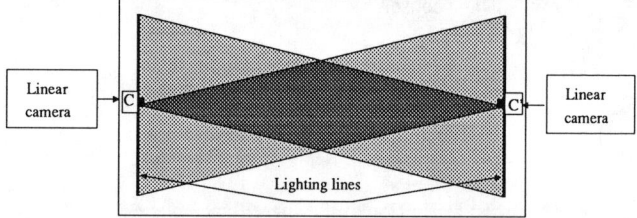

Fig. 5 : Relative position of the two identical planes

stereovision system yield to substantial simplifications with respect to the basic triangulation principles of stereovision.

In this configuration, the distance between the two elementary planes is made as small as possible, so that they can be considered as merged in a single plane. Figure 6 shows the stereo geometry in which cameras are separated by a distance 2D. We assume that the two optical axis of the cameras Oz and Oz', are colinear and as for matrix stereovision, we define a reference plane for each camera.

Any object crossing the optical gate is bounded by the quadrilateral defined by the shadows left on each line image. Let us consider the lower point A of this

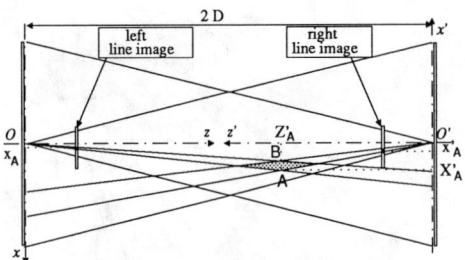

Fig. 6 : Stereo camera geometry

quadrilateral. Its position is defined in the reference plane (O'z',O'x') associated with the right camera C' by its two coordinates X'$_A$ and Z'$_A$.

These scene coordinates can be computed from the abscissas x_A and x'_A of the images of the point A on the two line images, by the relation :

$$X'_A = \frac{(x_A . x_A' . 2.D)}{(x_A' - x_A)} \quad \text{and} \quad Z'_A = \frac{(x_A . 2.D)}{(x_A' - x_A)},$$

provided the point A does not lie on the optical axis of the cameras.

We can obtain in the same way the scene coordinates of the upper point B. Let M be the middle of the segment AB and T its size, respectively. The position and the size of the outline of the object in the surveillance plane at the time of the analysis of the two lines are defined by means of the coordinates (X'$_M$, Z'$_M$) of M and by T.

4 - Performance evaluation

To demonstrate the effectiveness and the performance of the proposed intrusion detection scheme, a prototype based on a Digital Design Visionix linear image processing system equipped with two 1728 pixels Micronix linear cameras has been developed. This 68000 multiprocessor based system was programmed in Motorola machine language. The two line images are shoot simultaneously thanks to specifically designed camera interface and the time lag between two consecutive shots was set to 5 ms. This system has been mounted on a laboratory experimental platform. This test facility was used to evaluate the detection reliability over long periods of time and for a variety of targets. Several kinds of objects, such as sheets of paper, matchboxes, tennis balls, handbags, etc ... have been thrown at different speeds through the experimental double surveillance plane. Such objects were always detected, provided their speed did not exceed 36 Km/h. Humans crossing the twin plane at different speeds have also always been identified. A simple criterion on the size T of the reconstructed outlines has proved to be very effective to distinguish between humans and small objects. To demonstrate the usefulness of this technique, the prototype has been installed in an experimental station of the R.A.T.P. underground in Paris. In this station, the distance D between each camera and the corresponding lighting line was 70 m (cf. Fig. 7). 200 mm telephoto lenses were used to aim at the lighting lines which were 2.4 m high. During a two days long test period, all intentional human intrusions have been detected, which no false alarm have been recorded.

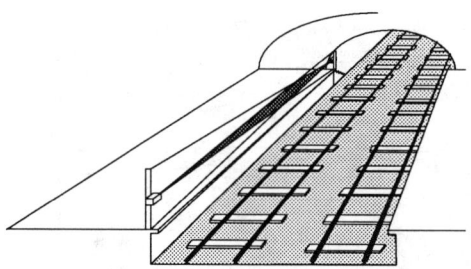

Fig. 7 : Configuration in station

5 - Conclusion

In this paper we have proposed a simple solution for the detection of moving objects in three dimensional scenes. The use of linear cameras instead of matrix sensors for stereovision leads to real time determination of the size and the position of any object crossing a surveillance plane. This new scheme was found to be applicable to the detection of intrusions on the tracks of an L.R.T. system. We are presently working on the problem of recovering motion information about the objects crossing the surveillance plane and on the enlargement of the active area of the system.

Acknowledgements

This work has been supported by a grant from the Institut National de la Recherche sur les Transports et leur Sécurité (INRETS) and from the Centre Régionnal d'Etude Sur les Transports Automatisés (CRESTA). The authors want to thank Y. David for his continual support.

References

Aoki M (1984) : " Detection of moving objects using line image sequence ". Proc. 7the Int. Conf. Pattern Recognition, pp 784-786.

Barnard S. and Thomson W. (1980) : " Disparity analysis of images ". I.E.E.E. Trans. Pattern Analysis and Machine Intelligence, Vol. PAMI-2, n° 4, pp 333-340.

Boutry F , Khun F , Deparis J-P and Postaire J-G (1987) : " New tendencies in L.R.T. systems : Automatic driving by means of computer vision ". Int. Conf. on Local Public Transport, Liverpool , G.-B.

Brachet J-C : "Fall detection system in a metro line " Control, Computers, Communications in transportations, Sept 1989, Paris, pp 143-147.

Ohta Y. and Kanade T. (1985) : " Stereo by and inter scanline search using dynamic programming " I.E.E.E. Trans. Pattern Analysis and Machine Intelligence, Vol. PAMI-7, n° 2, pp 139-154.

The Heavy Vehicle License Plate Program
and Crescent Demonstration Project

C. Michael Walton [1]
Member ASCE

Abstract

The HELP Program and Crescent Demonstration Project is a bi-national multi-state cooperative study and demonstration project involving government and industry participating to investigate new technologies that have the potential to provide an integrated heavy vehicle management system with applications to both highway and vehicle systems. This initiative has been described as a leading example of the Intelligent Vehicle Highway System (IVHS) initiative focusing on commercial vehicle operations (CVO).

The study elements of this project focus on selected technologies that can be integrated into a heavy vehicle management system: (1) automatic vehicle identification (AVI), (2) weigh in motion (WIM), (3) automatic vehicle classification (AVC), (4) data communication networks including systems integration. The program, initiated approximately seven years ago, consists of three phases which include assessing the feasibility of the concept, technical studies involving laboratory and field tests, and, lastly, the demonstration phase. Perhaps the most significant of the activities to be derived from this project to date are those focusing on institutional arrangements afforded by the emerging technology in this experiment.

The demonstration element of the program, referred to as the Crescent Demonstration Project, began in 1991 involving six states within the United States and in one province of Canada and will be phased into a full scale operation over the next three years. It is estimated that over the life of the program and the project that more than $20 million will be expended or contributed by both industry and government in testing this landmark program. This paper provides an overview of the program and the components of the implementation plan.

Introduction

With the advent of programs aimed at advancing the efficiency and productivity of the highway transportation system, the major "umbrella" initiative in the USA is referred to as intelligent vehicle highway systems (IVHS) has emerged. Among the major IVHS program elements recently defined by IVHS America is one focused on commercial vehicle operations (CVO).

In defining the significant components of a national CVO program in IVHS, one program surfaces as the leading activity - the Heavy Vehicle Electronic License Plate (HELP) program. The HELP program is comprised of a number of research, technical, administrative, and demonstration projects with the Crescent Demonstration project being the "flagship" activity.

----
[1] Bess Harris Jones Centennial Professor and Chairman, Department of Civil Engineering, University of Texas at Austin, Cockrell Hall, Suite 4.2, Austin, Texas 78712.

Background

The HELP/Crescent initiative grew from a mutual interest for fostering CVO efficiency from representatives of the states of Arizona and Oregon. Their motivation was to explore the emerging electronic technologies for potential application in a highway environment. The intent was to focus on one class of users in which various government agencies had operational and administrative responsibilities and in which intuitively, there seemed to be an opportunity of mutual interest with the user - commercial vehicle fleet owners and operators.

From this initial interest, circa 1983, the program has been designed to bring to bear emerging technologies in an integrated system to facilitate heavy vehicle management (HVM). This research and demonstration project is intended to design and test the integrated technologies including automatic vehicle identification (AVI), automatic vehicle classification (AVC), and weigh-in-motion (WIM) technology using low-cost technology, systems integration via a private operator, communication linkage networks, and institutional arrangements within and between state agencies.

The HELP initiative has involved fifteen US states, one Canadian province, the US Department of Transportation Federal Highway Administration, Transport Canada, the Port Authority of New York and New Jersey, and respective representatives of the trucking industry from each governmental entity. A forum has been created to examine what lies ahead for improvements in heavy vehicle management technology and to what extent a cooperative effort between public and private sectors can be effective in fostering this initiative.

A variety of integrated system scenarios have been developed including some which might be able to facilitate trucking systems using simple coded messages sent by two way communications with enhanced AVI, on-board computers, or, ultimately, satellite systems. Using roadside AVI/WIM facilities, messages could be sent or received by trucks when they pass an AVI/WIM site. Users would need to install special equipment on their trucks. The system might provide the motor carrier industry with information that they may desire and government agencies may be provided information for planning and management in addition to monitoring routes for specific movements of special cargos such as hazardous materials.

Program Goals

The goals and objectives of the program are to define the feasibility of the HELP program and the Crescent Demonstration as determined by progress in achieving the following four goals:

1. Assess the Viability of the Technology in the Highway Environment (eg. reliability, accuracy, etc.)

2. Improve Institutional Arrangements (eg. one stop shopping, pre-clear at weigh stations and ports of entry, border transparency)

3. Measure Efficiency and Productivity (eg. improve safety, enforcement, reduced administration, data collection, etc.)

4. Identify Additional Applications of the Technology (eg. public/private sector applications)

The Crescent Demonstration Project element of the program began in mid-1991. The demonstration project should last for a minimum of one year once fully displayed and will be followed by an evaluation. The demonstration is intended to test as many applications of the HELP technologies as possible within the budget constraints and should provide useful information and data on the utility of the technologies involved as well as opportunities afforded by systems integration. The challenges of working within state institutions as well as multi-state arrangements are built upon other initiatives underway within the National Governors Association, the IRP, IFTA, and others in an effort to provide regulatory uniformity thereby enhancing productivity and efficiency.

The HELP Program

As indicated, the program, initialed in 1983 by representatives of Arizona and Oregon Departments of Transportation, began with the development of concept papers. The FHWA provided grants to Arizona DOT to undertake a conceptual feasibility study and to Oregon DOT to perform a proof of conceptual demonstration. These efforts considered merging of new technologies into an automated vehicle monitoring system. The results were encouraging regarding the technical feasibility and there appeared to be potential benefits to states and truckers alike worthy of further pursuits.

In early 1985, a core group of western states joined with a number of other interested states to form a joint-funded, multi-state development and testing program. The project was originally call the Crescent Study because of the shape of the proposed demonstration route along the west coast and east to Texas. With the inclusion of many participating states outside of the Crescent, the name of the overall study was changed to HELP (Heavy Vehicle Electronic License Plate), although the major demonstration component maintains the Crescent name.

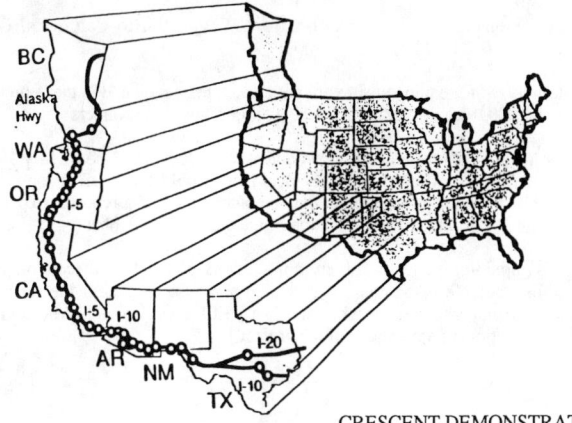

CRESCENT DEMONSTRATION

Program Description

As previously stated, the HELP project is a North American, bi-national, multi-governmental and industry, cooperative study and demonstration project whose objective is to investigate current technologies that have the potential to provide an integrated heavy vehicle management system with applications to both highway and vehicle management.

The study elements of the HELP program focus on four technologies that can be integrated into a management system:

1. Automatic-Vehicle Identification (AVI)
2. Weigh-In-Motion (WIM), using low-cost technology
3. Automatic-Vehicle-Classification(AVC)
4. Date Communications Networks and systems integration

The entire HELP program can be divided into three phases:

<u>Phase I</u> involved the early work by Oregon and Arizona to assess the feasibility of the concept. This effort was performed during 1983 to 1985.

<u>Phase II</u> has consisted of a number of technical studies involving each of the individual

technologies. These studies have investigated each technology through research, field and laboratory testing and development of equipment performance specifications, where appropriate. This phase began in 1985 and was essentially completed in 1989 with continuing studies regarding on-board computers, refinement of some of the other technologies, and further efforts into a universal reader for various AVI systems.

Phase III is the Crescent Demonstration, which will test the integration of the technologies and several institutional aspects in an operating environment.

The Crescent Demonstration

The demonstration project will provide approximately forty (40) vehicle monitoring locations and 5000 participating heavy commercial vehicles to test an integrated system concept in an actual highway environment. Detailed planning for the demonstration is being handled by an internal group referred to as the Crescent Implementation Group which is comprised of the affected state government agencies and participating trucking interests. The following is a brief overview and status:

- The planned demonstration corridor is I-5 from Washington to California, I-10 from California to Texas, and I-20 in Texas.

- Approximately forty (40) equipped sites are anticipated to be included in the demonstration. A minimum of 5,000 trucks are expected to be equipped with transponders.

- The current schedule calls for all equipment and systems to be in place by July 1991. The actual Crescent Demonstration, once fully deployed, would then run for one year. An evaluation will be undertaken throughout the demonstration, and a report is expected to be completed within a year following completion of the demonstration phase.

- The project is intending to test as many applications of the HELP technologies as is possible within available resources. The final report will document the results of the demonstration, the extent to which each of the four HELP goals was achieved, and possible national and international implications of the project.

Applications

Equally important as the functioning of the technology in an operating environment is the testing of institutional arrangements within and between states which will allow trucks operational and administrative opportunities beyond what is currently possible. Indeed many participants in the HELP program view the institutional issues to be far more challenging and, potentially, more rewarding than advances in the use of these technologies. However, it is obvious that the operational efficiency and reliability of the hardware is a complement to the institutional issues. The outcome of any one aspect of the program could have a significant effect on the entire concept. Therefore, a series of applications has been identified as illustrative of the institutional issues to be confronted. These applications are summarized as data collection, weight enforcement, and automatic clearance.

Evaluation

The goals and objective were carefully defined so as to facilitate the evaluation of the HELP program as reflected in the Crescent Demonstration Project. Internal committees were formed and comprised of state agency and industry representatives. A committee was assigned to each program goals and corresponding objectives and charged with devising a methodological framework for evaluating the degree to which the demonstration project satisfied their assigned goal and objectives. In some cases this may involve a "before and after" approach, others a detailed performance critique based on measurements of accuracy and reliability, for example, while in other cases it may be an assessment of how well various agencies worked together in accepting an alternative means of achieving their particular mission (i.e. transparent borders).

An overall evaluation framework will be created which will encompass the results of each committee assignment and provide for integration into a final determination as to the viability of individual technologies, institutional issues and the like.

Conclusion

The HELP Program and Crescent Demonstration represents a regional application of a CVO activity in IVHS. This public and private initiative is unique in many respects. One such spinoff has been the appreciation and mutual understanding each participant has gained of the needs and concerns of industry groups and government agencies represented. These attributes are only a beginning of what may, in the long run, be more significant in overall productive gains than envisioned in the formative days of the program.

Using Real-Time Location Information
for Hazardous Materials Shipments

Mark A. Turnquist[1]

Abstract

Real-time location information on shipments of hazardous materials can be used to update routing decisions while the shipments are in transit. The effects of unforeseen delays, accidents on the network, and time-varying network characteristics can be taken into account, in an attempt to reduce risks, especially for shipments of "ultrahazardous" materials.

Introduction

In the U.S. there are nearly a million shipments of hazardous materials daily, and most of these occur with little direct consideration of risks in routing and scheduling. However, for some "ultrahazardous" materials, like radioactive wastes, risk considerations are paramount. The growing national concern with risks due to hazardous materials movements, and the recognition that risks may be time-of-day related, imply that routing and scheduling decisions are closely interrelated.

Global positioning satellites make it possible to monitor the position of shipments continuously, and this real-time information opens the possibility of creating adaptive routing strategies which could change the directions given to a driver while the shipment is en route, if such changes could reduce risks to population along the route, or result in avoidance of an accident scene along the originally planned route.

Adaptive routing strategies can be characterized by "windows" within which the routing decision remains unchanged. As long as the vehicle remains "within the

---
[1] Professor, School of Civil & Environmental Engineering, Cornell University, Ithaca, New York 14853-3501.

window" as it moves along the chosen route, no update is necessary. However, if the vehicle is delayed excessively, or an event occurs which changes the window, and the vehicle's position is outside the window's limits, a revised routing should be considered. Options for rerouting depend on the vehicle's current location and the current time. The methods used in this paper are based on extensions of earlier work on multiobjective, dynamic routing/scheduling for hazardous materials shipments (Turnquist, 1987).

This paper illustrates dynamic rerouting on a hypothetical network, focusing on two example analyses. In the first, the primary concern is with a single routing objective -- to minimize total population exposure along the route -- but it is recognized that the population along a link is time-of-day dependent. Thus, the choice of route may be affected by time-of-day. For a given destination in the network, a dynamic shortest-path procedure can determine, for each node in the network, a time-dependent "next link" en route to the destination. Thus, as the shipment approaches each node along its route, it can be checked to determine whether a change in its currently projected path is desirable, in order to minimize population exposure. As long as the shipment "stays on schedule," its originally planned route will remain optimal, but if the shipment is delayed (or gets ahead of schedule), rerouting can occur at the next node in its route.

The second example focuses on multiple objectives and rerouting as a result of an incident somewhere along the planned route ahead of the shipment. In this case, the route guidance system creates a set of alternative routes from the current location of the shipment to its destination, reflecting both the current time (and hence any time-of-day variations in population, for example), and the temporary obstruction of one or more links in the network. The set of routes is generated to include all "non-dominated" alternatives for a set of criteria, which may reflect multiple objectives. This set would be presented to a dispatcher, to allow him/her to select an alternative and transmit instructions to the driver.

The Example Network

Figure 1 shows the network used in these two example applications. It includes 41 nodes and 146 directional links. The area around node M in the network represents a major urban area, with a "beltway" around it (E-F-G-N-Q-P-O-K-E), radial freeways (F-M-P and K-L-M-N), and an "inner belt" (F-I-L-O). Node E represents the destination of the shipments in both examples.

On the links in the urban area, population exposure is represented as a function of time-of-day. This reflects a combination of residential population in the areas through which the links pass, employment in those areas, and traffic volume on the links themselves. Each of these three constituents has a different temporal pattern, and the resulting combination for a typical link is shown in Figure 2. This temporal pattern is expressed as a percentage of the mean population exposure for the link, by hour of the day.

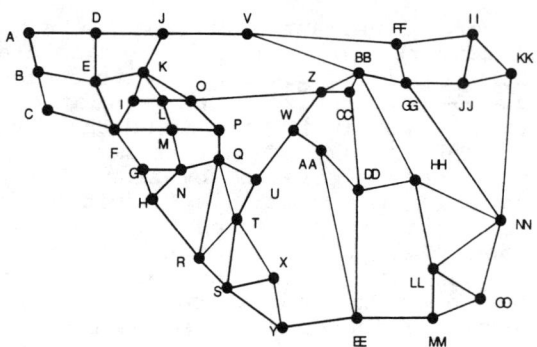

Figure 1. Hypothetical network for examples.

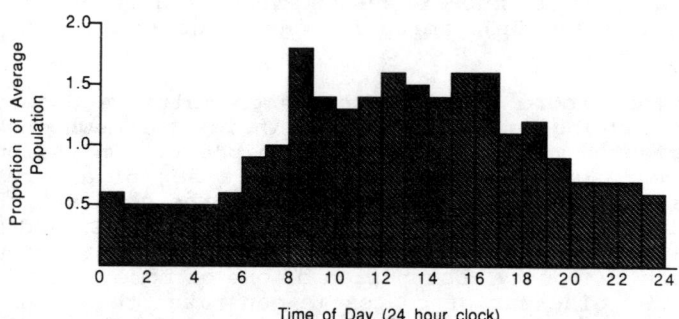

Figure 2. Illustrative temporal variation in population exposure along a link.

An Example of Schedule "Windows"

Figure 3 shows schedule "windows" for nodes OO and AA, for shipments en route to node E, to illustrate the concept. For example, a shipment arriving at node OO between 1500 and 2000 (using 24-hour clock notation) should proceed to node MM (with an anticipated route OO-MM-EE-AA-W-Z-O-K-E) to minimize population exposure. However, if that shipment arrives at node OO before 1500 or after 2000, it should proceed to node NN (with an anticipated route OO-NN-KK-II-FF-V-J-D-E). These "windows" are being determined by the time-varying population exposure on the "beltway" links O-K and K-E. If the shipment can cross those links in the early morning hours (before 0600), the route through the "middle" of the network is preferrable. However, at other times of the day, the population exposure along those links is higher, and the result is that the minimum-population route is the longer route around the perimeter of the network.

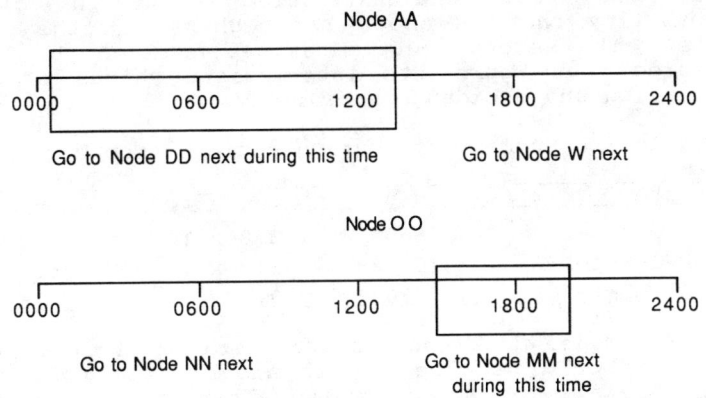

Figure 3. Schedule "windows" for two nodes.

The schedule "windows" at node AA are determined by the same links as for node OO. Thus, if a shipment leaves node OO at 1800, for example, and goes to node MM, as long as it stays on schedule, it will arrive at node AA at 2218, and proceed to node W as planned. If, however, the shipment is delayed before reaching node AA, and arrives after 0030 the next morning, it would be rerouted to DD, with an anticipated route to the destination AA-DD-HH-BB-V-J-D-E. This route skirts around the smaller urban area and makes its way to the "perimeter" route into node E. The reason for this is that, as a result of the delay, the shipment would likely arrive on the "beltway" links O-K and K-E during the morning rush hour if it continued on its present route, with attendant large population exposure.

An Example Involving Incident Avoidance

As a second example, suppose that the shipment passed node AA at 2218, on schedule, but while it is traveling along link AA-W, a major traffic accident occurs on link W-Z, along the intended route for the shipment. To avoid the accident scene, it is clear that when the shipment reaches node W, it must be rerouted toward node U, but what are the alternative routes to the destination? Supposing we are interested in minimizing: 1) population exposure, 2) probability of an accident and 3) travel time, we can construct the set of non-dominated routes from node U, based on the shipment's current location and the current time.

Table 1 lists the routes identified, with their characteristics. The first route listed minimizes both population exposure and travel time, but has an accident probability that is about 12% higher than the third route. The second route minimizes none of the three measures, but represents intermediate values of both population and accident probability.

Table 1. Summary of routing options from U to E.

Route	Population	Acc. Prob.	Time (hrs.)
U-Q-P-O-K-E	25650	3.8×10^{-4}	4.0
U-Q-P-O-K-I-F-E	45365	3.7×10^{-4}	5.0
U-Q-P-O-L-I-F-E	56365	3.4×10^{-4}	4.6

The model does not dictate which of these routing options should be chosen. It simply determines that these are all the non-dominated alternatives available at the time and place in the network where the shipment is (or will be when a choice is required), and presents them to the dispatcher (or system manager) for a choice based on relative evaluation of the three criteria.

Conclusions

Availability of real-time location information for shipments of ultrahazardous materials, through satellite tracking or other means, offers the opportunity for dynamic rerouting of such shipments in the event of unforeseen delays or incidents on the network. Multi-objective methods have been developed which can take advantage of such information, and create new routing alternatives in real-time, based on time-varying link characteristics and current data on link availability. The example in this paper demonstrates proof-of-concept, and the methods described here can provide the basis for further development of an implementable system.

Reference

Turnquist, M.A. (1987). "Routes, Schedules and Risks in Transporting Hazardous Materials," in *Strategic Planning in Energy and Natural Resources*, B. Lev, *et al* (eds.), North-Holland, Amsterdam, 289-302.

An Analytical Framework For Minimizing Freeway Incident Response Time

Kostas G. Zografos[1], A.M. ASCE and Teti Nathanail[2]

Abstract

The reduction in incident delay is an important aspect of any freeway management program as it results in substantial savings to roadway users. This paper presents an integrated decision making framework for the reduction of the incident delay through the optimum deployment of Traffic Restoration Units (TRU).

Introduction

Freeway traffic congestion is a problem of growing public concern affecting the socio-economic activities of the residents of large metropolitan areas. Traffic congestion is divided into two components: 1) recurring, and 2) non-recurring congestion. Recurring congestion is caused by the combined effect of heavy traffic volume and inadequate capacity. Non-recurring congestion is caused by incidents such as vehicle breakdowns, accidents, roadway maintenance/construction activities etc., which reduce the capacity of a given freeway section. While recurring congestion is predictable and follows a well defined temporal and spatial pattern, the incident congestion is random in terms of occurrence and duration.

Incident delay represents substantial percentage of urban travel delay (Lindley, 1986). An FHWA study (Lindley, 1986), suggests that 60% of the overall delay is due to the occurrence of incidents; the same study projects that by the year 2005 incident delay will account for 70% of the total delay.

The reduction in incident delay is an important aspect of traffic management programs (TRB, 1989) to the extent that it results in substantial savings for the roadway users.

---
[1] Assistant Professor, Department of Civil and Architectural Engineering, University of Miami, Coral Gables, FL. 33124

[2] Graduate Student, Department of Civil and Architectural Engineering, University of Miami, Coral Gables, FL. 33124.

The objective of this paper is to develop an integrated decision making framework for reducing incident delay, through the optimum deployment of traffic restoration units.

Definition of the Problem

The demand for freeway emergency response services is created by incidents causing traffic flow interruptions. These incidents occur randomly in time and space. When an incident is detected a Traffic Restoration Unit (TRU), is dispatched to restore the traffic flow.

For a given freeway, the delay caused by incidents depends on the severity of the incident (i.e. how many lanes are blocked by the incident), the traffic flow rate, and the traffic flow restoration time (i.e. the time elapsed between the incident and the restoration of the traffic flow).

The traffic flow restoration time consists of the following four components: 1) incident detection and identification time, 2) dispatch time, 3) travel time, and 4) incident clearance time. Figure 1 shows schematically the traffic flow restoration time and its components.

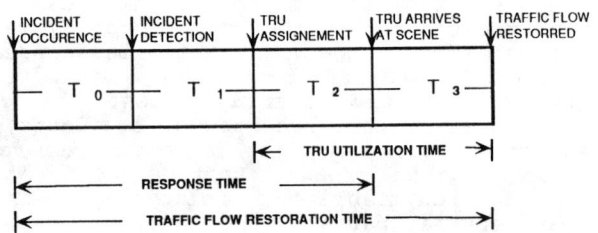

Figure 1: Sequence of Events in the Traffic Flow Restoration Process

The detection time is the time interval between the occurrence of the incident and the identification of the location and nature of the incident causing capacity reduction. The dispatch time is the time elapsed between the identification of the incident and the dispatch of the first available TRU. The travel time is the en-route time between the origin of the response unit and the scene of the traffic incident. Finally, the incident clearance time is the time required to clear the roadway from the incident.

The traffic flow restoration time is equal to the sum of the incident detection, dispatch, travel, and incident clearance time. Therefore, reduction in the duration of any of the four components may result in a reduction of the overall incident duration time and consequently to a reduction of the incident delay.

Although substantial attention has been directed to the reduction of the incident detection time (Payne, 1976) there has been little emphasis on the next two components that define the response time to an incident. This paper introduces a new approach for the reduction of incident delay through the minimization of dispatch and travel time.

In its general form the traffic flow restoration problem can be stated as follows:
Given the expected number, type, frequency, spatial, and temporal variation of incidents in a transportation corridor; define the number, the location, and the dispatch strategy of emergency response units in such a way as to achieve a threshold value of traffic flow restoration time.

Methodology

The traffic flow restoration problem is a complex resource allocation problem which involves the following decisions: 1) Determination of the required number of TRU's, 2) designation of emergency response areas for the TRU's, and 3) establishment of the dispatch policy.

An iterative procedure combining computer simulation and optimization is suggested for the solution of the traffic restoration problem. The proposed approach consists of four major steps.

The first step deals with the optimum partition of the corridor into service districts for the TRU's. The objective of the districting problem is to create districts that are contiguous, have similar workload, and minimize the total weighted travel time of the TRU's.

The mathematical expression of the districting problem is an extension of the p-median formulation suggested by Toregas (Toregas et. al., 1971), and can be written as follows:

$$\text{Minimize } F = \sum_{i \in N} \sum_{j \in M} W_i t_{ij} X_{ij} \qquad (1)$$

S.T.

$$\sum_{j \in M} X_{ij} = 1 \qquad \forall \; i \in I \qquad (2)$$

$$X_{jj} \geq X_{ij} \qquad \forall \; i,j \qquad (3)$$

$$\sum_{j \in M} X_{jj} = N \qquad (4)$$

$$X_{jj}(1-\varepsilon)\overline{W} \le \sum_{i \in N} W_i X_{ij} \le (1+\varepsilon)\overline{W} X_{jj} \quad \forall\ j \in J \qquad (5)$$

$$X_{ij} \ge 0 \quad \forall\ i,j \qquad (6)$$

where:

I = the set of demand nodes i = 1,2,...N

J = the set of potential TRU locations j = 1,2,...M

W_i = workload of node i

t_{ij} = shortest travel time from node i to node j

X_{ij} = $\begin{cases} 1 \text{ if workload of node (i) is assigned to TRU} \\ 0 \text{ otherwise} \end{cases}$

$\overline{W}$ = average workload

ε = allowable deviation from the average workload

The second step of the proposed procedure uses a simulation model to estimate the traffic flow restoration time for the incidents generated within each service area. The estimation of the traffic flow restoration time requires: 1) the generation of freeway incidents according to their temporal, spatial, and severity characteristics, and 2) the simulation of alternative dispatch policies for the TRU's.

The third step of the proposed approach uses a dynamic macroscopic simulation model, KRONOS, (Lee et.al., 1990) for the estimation of the freeway incident delay. The traffic flow restoration time for each incident, the location and time of occurrence of each incident, and the V/c ratio prevailing on the freeway are the inputs to the traffic flow simulation module.

The fourth step of the algorithm compares the freeway incident delay obtained by KRONOS with a threshold value of acceptable delay (D*). If the estimated delay exceeds the threshold value, the number of the required emergency response units is increased by one and the procedure is repeated until the delay constraint is satisfied. The basic steps of the proposed method are shown in Figure 2.

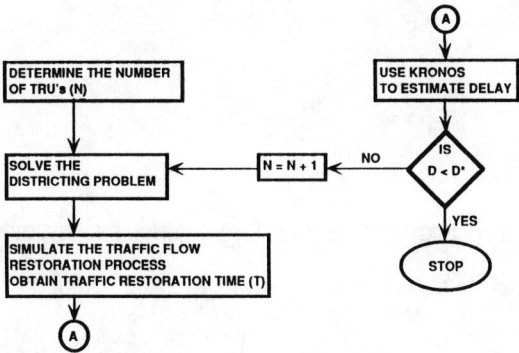

Figure 2: The Proposed Analytical Framework

Concluding Remarks

An analytical framework for managing freeway emergency response services was presented. The proposed approach can be used to determine the number of TRU's, their service territories, and the dispatching policy, required to achieve a threshold value of freeway incident delay. The trade-off between the number of TRU's and freeway incident delay also can be examined.

References

Lee, C.F., R. Plum and P. Michalopoulos (1990) <u>KRONOS-6: An Interactive Freeway Simulation Program for Personal Computers</u>, Center for Transportation Studies, Department of Civil and Mineral Engineering, University of Minnesota.

Lindley, J.A. (1986). <u>Quantification of Urban Freeway Congestion and Analysis of Remedial Measures</u>, U.S. Department of Transportation, FHWA/RD-87/052, Washington D.C.

Payne H.J. (1976). <u>Development and Testing of Incident-Detection Algorithms: Vol.1 Summary of Results</u>, FHWA, Washington D.C.

Toregas, C., R. Swain, C. ReVelle and L. Berguen (1971). The Location of Emergency Service Facilities, <u>Operations Research</u>, Vol. 19, pp. 1366-1373.

Transportation Research Board. (1989). <u>Freeway Incident Management</u>, National Cooperative Highway Research Program Synthesis of Highway Practice 156, TRB, Washington D.C.

COST SAVINGS FROM AVMC SYSTEMS IN TRUCKING

Susan F. Hallowell[1], Edward K. Morlok[2], Member, ASCE
and Lazar N. Spasovic[3]

Abstract

The paper presents a brief overview of a methodology by which the cost savings of advanced vehicle monitoring and communication (AVMC) systems in long distance or intercity trucking can be quantified. The model can be used for two purposes: (1) for initial feasibility studies of AVMC systems, and (2) for evaluation of systems that are in place. The model is formulated as an engineering or unit cost model, in which various capital and operating costs are determined on both a before-tax and after-tax basis, before and after introduction of the AVMC system. The model has been applied to various situations, and results indicate that the AVMC systems are beneficial over a wide range of circumstances.

Introduction

The purpose the methodology was to quantify the cost savings or benefits of advanced vehicle monitoring and communication (AVMC) systems in the long distance or intercity trucking industry. These AVMC systems may be based on satellites, meteor-burst reflection, or mobile radio. The ways in which these systems could be used in various types of intercity trucking systems and the benefits to be derived therefrom, for carriers, shippers,

---

[1]Research Fellow, Department of Systems, University of Pennsylvania, Philadelphia, PA 19104-6315.

[2]UPS Foundation Professor of Transportation and Professor of Systems Engineering, Department of Systems, University of Pennsylvania, Philadelphia, PA 19104-6315.

[3]Assistant Professor, School of Industrial Management and Transportation Center, New Jersey Institute of Technology, Newark, N.J. 07102.

and more generally, the public at large is discussed in detail in (Morlok et al., 1989) and (Morlok and Hallowell, 1989). Our focus here is on the benefits to be realized in the form of cost savings to trucking companies, for these are most readily translated into increases in profit.

Truck Cost Model

The model calculates the cost impacts of introducing an AVMC system in irregular route trucking in two ways. First, it computes the total cost of providing an entire system. Then, these results can be modified to yield the average cost per tractor, per load, or other average cost. Categories of cost, such as operating costs, can also be calculated.

The model is an engineering unit cost model, in which we first estimate the resources required to produce a given transportation product, in this case a given amount of cargo movement, and then price these resources to yield the cost (or expenditures) required. The overall structure of the model is shown in Figure 1. Details of the model are presented in (Hallowell and Morlok, 1991). The estimate of resources required is determined by the underlying technology of transportation, and it is of course this technology which is being changed by the introduction of an AVMC system. Thus, the inclusion of technology as an explicit variable is significant. The transportation product input describes the cargo movement that is to occur.

The output of the production process model consists of both physical measures of resources used, or surrogates for resources (such as vehicle-miles). The resource variable estimates and performance measures are derived using an extensive set of equations developed to accurately model trucking operations and the changes in driver and fleet utilization resulting from introduction of an AVMC system.

The cost outputs consist of annual operating expenses and capital costs, either before taxes or after taxes. The various cost outputs are listed in Table 1.

Example Application

Numerous example applications based on the typical unit cost and performance values for intercity truck load (TL) motor carriers, have been performed. The input data were obtained from reports filed with the Interstate Commerce Commission, American Trucking Association data, trucking journals, and interviews with motor carriers. In computing the costs, 1989 prices were used. Satellite-based communication and positioning equipment's initial cost included hardware, installation, training and a one-year warranty. It was assumed that a compatible computer

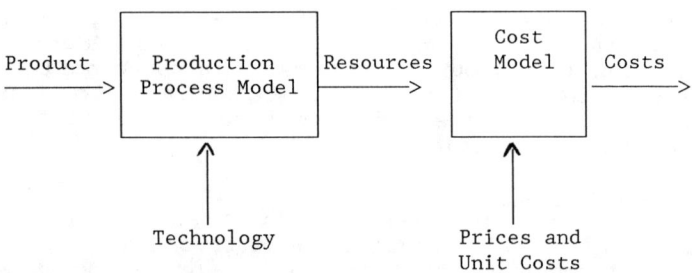

Figure 1. Model for estimating cost savings of AVMC systems in irregular intercity trucking

Table 1. Cost outputs of the model.

Operating expenses
    Driver wages and benefits
    Tractor operating supplies and
        maintenance (may include trailer, also)
    Trailer operating supplies and maintenance
    User fees and taxes
    Telephone communication
    Hiring and training new drivers
    Insurance
    Dispatcher wages and benefits
    AVMC operation
    Overhead
Capital ownership
    Tractors
    Trailers
    AVMC equipment
Ordinary income-tax impacts
Capital ownership-tax impacts
    Tractors
    Trailers
    AVMC equipment
Annual total cost after taxes
Annual total cost before taxes

with required software existed. The minimum attractive annual rate of return was assumed to be 15%.

To illustrate an application, for one example typical of our analyses, the total costs were about $35.0 million per year (before taxes) for a system that consisted of 275 tractors, 550 trailers, and operated 30.9 million tractor miles, with no AVMC system. After installation of the AVMC system, to accommodate the same cargo only 245 tractors and 490 trailers would be needed, and tractor mileage would drop to 29.3 million for a cost of $32.2 million per year before taxes. This is a reduction in the total cost of about 7.97%. Based on the annual average of 112,233 miles per tractor before the AVMC is installed, the average cost per tractor mile before taxes would be $1.14 After the AVMC equipment is installed the average cost per tractor-mile before taxes would be $1.04 per "before" AVMC tractor-mile. The total after tax costs decreased by 8.05% as a result of the introduction of the AVMC system. It should be noted that the reduction of average cost is due to an increase in equipment utilization, a decrease in communication time and a reduction in driver turnover. This is consistent with the benefits reported by TL carriers, as discussed earlier.

A variety of cases were analyzed, and in all cases there was at least some reduction of cost resulting from introducing the AVMC system. These analyses included, changes in the product (cargo and traffic) characteristics, fleet utilization, load factor, and driver communication activity.

Summary and Conclusions

The general model of trucking costs of irregular route intercity trucking can be used in: (1) feasibility studies and planning of an AVMC system, and (2) in post-implementation evaluation of an installation, to assess the actual economic effect and help to pinpoint areas of success as well as problem areas. The model is formulated to be sensitive to the types of changes that accompany installation of an AVMC system and traces through their impacts on operating and capital costs, both before and after taxes. While our results showed that cost savings are more than sufficient to justify these systems, clearly productivity changes will naturally vary from one carrier to another, due to different operating conditions, markets, personnel and organizations. Thus each case must be analyzed carefully, before investment in an AVMC system, to ensure that it is justified and to plan how the improvements will be achieved. Furthermore, a company may take advantage of the AVMC system to change its price-service position in the freight market.

Although this model has been developed for general

irregular route intercity trucking operations, it can also be applied to specialized services such as the movement of hazardous materials cargo and movements in support of just-in-time manufacturing and distribution.

Acknowledgement

This research was supported by the Advanced Technology Center of Southeastern Pennsylvania under grant no. 07.402NU with additional support from the Mobile Satellite Corp., the American Mobile Satellite Consortium, and the UPS Foundation Professorship in Transportation at the University of Pennsylvania. Dr. Russell B. Capelle, Jr. of the Regular Common Carrier Confrence also provided much useful data and contacts, too numerous to mention. This support is gratefully acknowledged but implies no endorsement of the findings.

Appendix: References

1. Hallowell S.F. and Morlok E.K. (1991). Estimating cost savings from advanced vehicle monitoring and communication systems in irregular route trucking, Working Paper, Department of Systems, School of Engineering and Applied Sciences, University of Pennsylvania, 19104-6315.
2. Morlok E.K., Bedrosian S.D., El Zarki M. and Hallowell S.F. (1989). Vehicle monitoring and telecommunication systems for enhancements of trucking operations. Proceedings of the IEEE Conference on Vehicle Navigation and Information Systems, Toronto, Sept. 11-13.
3. Morlok E. K. and Hallowell S. F. (1989). Reported benefits of advanced vehicle tracking and communications systems, Working Paper 89-8-1, Department of Systems, School of Engineering and Applied Sciences, University of Pennsylvania, Philadelphia, PA 19104-6315.

A Fleet Management System for Road Transportation

P.R. Schrijver and H.G. Sol[1]

Abstract

In this paper, an architecture for a fleet management system for road transportation companies is being proposed. This fleet management system integrates planning support with mobile communications to solve the problems found within trip planning and trip execution.

Introduction

Road transportation companies nowadays have to deal with a diversity of problems during vehicle scheduling and trip execution. The increased amount of congestion of the road network, the shorter intervals between acceptance and execution of the orders, the necessity for a higher level of customer service, and the need for more accurate real-time information about trip execution, both for the transportation company and for the customers, make it more and more difficult to handle vehicle scheduling and trip execution in an effective and efficient manner. In the European situation, these problems are even more obvious because trucks carry their load over shorter distances and yet have to go through customs more frequently.

According to De Jong and Sol (1991), throughout this paper, we will use the term *trip planning* to address the issue of vehicle routing and scheduling. By *trip execution* we mean the process of delivering goods from origin to destination.

When we take a closer look at trip planning and trip execution, several problems can be identified (Schrijver and Sol 1991). Most of these problems result from a lack of direct insight into the trip execution process and the inability to communicate directly with the driver. Besides, the considerable amount of time-consuming admin-

---

[1] Department of Information Systems, Delft University of Technology, P.O. Box 356, 2600 AJ Delft, The Netherlands.

istrative activities play an important role in causing problems at the planning department.

Mobile Communications

Mobile communications is a mean for solving them mentioned problems in trip planning and trip execution. Recent developments in mobile communications offer a variety of new possibilities for road transportation companies. One of these developments has been the introduction of mobile satellite data communications. The main advantage over the conventional mobile communications technologies, such as the mobilophone, is that the area in which a mobile unit can be reached is much larger. Another advantage of satellite communications is that automatic vehicle positioning is possible.

Several authors already have stressed the potential importance of utilizing mobile communications to support trip planning and trip execution (Assad 1988; Morlok et al. 1989; Powell 1990).

Architecture for a Fleet Management System

For solving the above mentioned problems, we should regard the trip planning process as a decision process (Sol 1991). Supporting this decision process can be established by developing a *planning support environment* (De Jong and Sol 1991; Schrijver and Sol 1991). In our case, such a planning support environment is based on the presence of mobile communications and will be called a *fleet management system*.

The mobile communications system we will use in our fleet management system should meet some requirements: message exchange between driver and dispatcher is realized by data communication, it is possible to use "preprogrammed" messages (macro's), and automatic location reporting is available within the mobile communications system.

In figure 1, the proposed architecture for the fleet management system is presented. Support components are depicted by rectangles and arrows represent the data flow between two support components.

A more complete overview of the architecture is given in Schrijver and Sol (1991). The proposed architecture for the fleet management system consists of the following support components:

1. ORDER ACQUISITION AND REGISTRATION. The registration of order and customer data takes place in this component. Beside this function, the order acquisition and registration component supports the dispatcher by making the decision whether or not to accept an order.

2. *EQUIPMENT MANAGEMENT*. The equipment management keeps track of the movements of the transportation units and coupled equipment. It also takes care of the registration of locations and status of uncoupled equipment and stationary transportation units. The registration of the assignment of drivers to the equipment also is performed within this component.

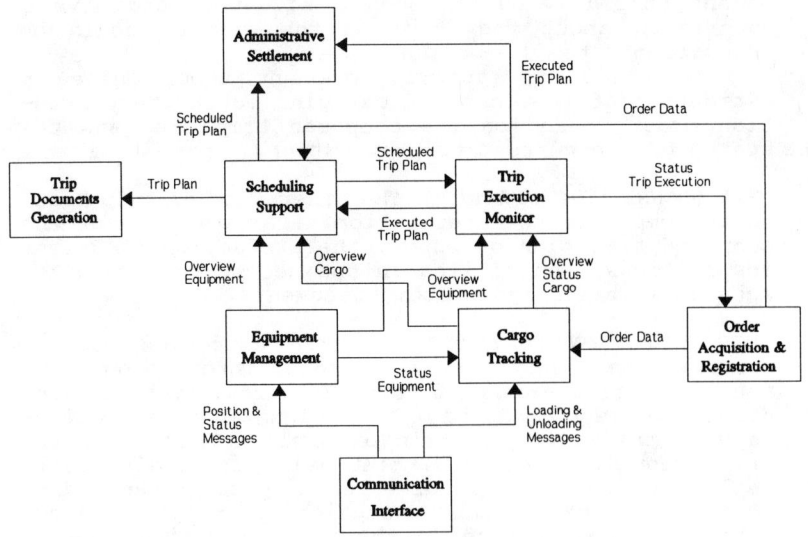

Figure 1. Architecture for the Fleet Management System.

3. *CARGO TRACKING*. The cargo tracking component keeps track of cargo by determining locations and status of cargo, and performs the registration of cargo related data, such as waiting times during loading and unloading.

4. *SCHEDULING SUPPORT*. Scheduling support deals with supporting the trip planning problem itself. This component should support the dispatcher in composing the trip plan. Powell (1990) distinguishes four subproblems within the trip planning problem:
- Driver work assignment.
- Empty repositioning of drivers.
- Load acceptance/rejection.
- On-line pricing.

The assignment of work to drivers can be supported by presenting an overview of the unassigned loads and the positions and points of time at which drivers will become available. The support component should be able to present suggestions on the assignments of loads to drivers, based on certain optimization criteria. When a driver has finished his current job so far, he should receive a new order. Sometimes, when there no loads to assign him to, it is necessary to reposition the driver empty to another location. The support component should recognize such a situation and give a suggestion about the location the driver should be repositioned to.

The scheduling support component should give an estimation of the costs of carrying out a transportation order, based on the composed trip plan and the status of the current trip execution.

5. *TRIP DOCUMENTS GENERATION*. The generation of trip documents comprises the preparation of an overview of the composed trip plan on paper, the production of driver instructions, and filling in the necessary documents, such as waybills and customs documents.

6. *TRIP EXECUTION MONITOR*. The trip execution monitor supports the dispatcher by providing information about the status of the current trip execution. Normally, the dispatcher is interested only in significant deviations between trip plan and trip execution. Therefore, the trip execution monitor continuously performs a "soll-ist"-comparison, and informs the dispatcher, according to the "management-by-exception"-principle, about noticed irregularities in the trip execution and, on the other hand, about deviations that probably will appear.

7. *ADMINISTRATIVE SETTLEMENT*. The administrative settlement component prepares information, derived from trip planning and trip execution information, for administrative purposes. This information will be made available for other departments within the transportation organization.

8. *MANAGEMENT INFORMATION*. This support component is meant for providing trip execution information for the management of the planning department. Although this information is not used for direct feedback to trip execution, it is included in the architecture.

9. *COMMUNICATION INTERFACE*. The communication interface establishes a link between the overall communication system and the other components of the fleet management system and takes care of the message traffic and message management.

Final Remarks

In this paper, we presented an architecture for a fleet management system. Our research will continue with the development of a prototype for the proposed fleet management system, after the architecture has been refined. With this prototype, experimentations will be carried out within transportation companies.

References

ASSAD, A.A., "Modeling and Implementation Issues in Vehicle Routing", in B.L. Golden and A.A. Assad (Eds.), *Vehicle Routing: Methods and Studies*, Elsevier Science Publishers (North-Holland), Amsterdam, The Netherlands, 1988, pp. 7-45.

DE JONG, R. AND H.G. SOL, "Vehicle Routing and Scheduling Support: Coping with Strict Time Windows", in H.G. Sol and J. Vecsenyi (Eds.), *Environments for Supporting Decision Processes*, Elsevier Science Publishers (North-Holland), Amsterdam, The Netherlands, 1991, pp. 277-293.

MORLOK, E.K., S.D. BEDROSIAN, M. EL ZARKI AND S.D. HALLOWELL, "Vehicle Monitoring and Telecommunication Systems for Enhancement of Trucking Operations", in D.H.M. Reekie, E.R. Case and J. Tsai (Eds.), *Conference Record of Papers Presented at the IEEE Vehicle Navigation and Information Systems Conference (VNIS '89)*, IEEE Vehicular Technology Society, New York, NY, 1989, pp. 356-360.

POWELL, W.B., "Real-Time Optimization for Truckload Motor Carriers", *OR/MS Today*, Vol. 17, No. 2 (April 1990), pp. 28-33.

SCHRIJVER, P.R. AND H.G. SOL, "Integrating Mobile Communications and Planning Support: Design of a Fleet Management System for Road Transportation Companies", Presented at the *IFORS-SPC 1 Conference on Decision Support Systems*, Bruges, Belgium, March 26-29, 1991.

SOL, H.G., "Information Systems to Support Decision Processes", in H.G. Sol and J. Vecsenyi (Eds.), *Environments for Supporting Decision Processes*, Elsevier Science Publishers (North-Holland), Amsterdam, The Netherlands, 1991, pp. 115-127.

OPTIMAL RAMP-METERING CONTROL FOR FREEWAY CORRIDORS

Yorgos J. Stephanedes, A.M. ASCE[1]
and
Kai-Kuo Chang[2]

1. INTRODUCTION

Ramp-metering control is one of the most effective Advanced Traffic Management System (ATMS) methods for relieving congestion in freeway corridors. However, current control strategies are not truly dynamic, are not optimal or consider only parts of the freeway corridor. The focus of this work is on alleviating congestion in all interacting corridor components, including freeway, parallel arterials and connecting streets, through dynamic, optimal ramp-metering in real time.

For determining optimal ramp-metering rates on a freeway, an input-output ramp flow model was first solved by linear programming [1] and improved [2,3] to treat congested situations. However, since frequent state transitions and rapid flow variations are observed in most real traffic systems, the steady-state strategies determined by the model are of very limited use even if updated periodically.

Based on the vehicle conservation principle, a nonlinear flow model for optimal ramp control was linearized to a linear programming problem [4]. The conservation equation was simplified and a quadratic objective minimized by Frank and Wolfe's method [5]. Since the flow model is linearized, the control strategies are hard to adjust locally, and since adjacent streets are not considered, the control shifts congestion to these streets when ramp demand is high. Further work with nonlinear modeling is promising but remains mostly theoretical (see summary in [6].)

---

[1] Professor; [2] Graduate Student, Dept. of Civil & Mineral Engrg, University of Minnesota, Minneapolis, MN 55455

2. OPTIMAL CONTROL FORMULATION

In our optimal ramp-metering control model, a freeway corridor is considered an open loop system. The corridor is discretized into small road sections, with density as the state variable and ramp metering rates, the control variables. In each section the density dynamics are estimated by a continuum model developed with the conservation equation, and formulated as a state equation. The optimization process adjusts the metering rate $R_i(n)$ of each ramp i according to corridor traffic conditions such that total corridor travel time T_T during a certain period is minimum:

Find control trajectories $R_i(n)$ which minimize the objective function $T_T = T(K_j(n))$ subject to the state constraints $K_j(n+1) = f_j(K_j(n), R_i(n))$ and $K_{j,min}(n) \leq K(n) \leq K_{j,max}(n)$, and the control constraints $R_{i,min}(n) \leq R(n) \leq R_{i,max}(n)$ for all ramp meters i and road sections j in the optimization period, where $K_j(n)$ is the density of section j at time $t_n = t_0 + n*\Delta t$, and $R_i(n)$ is the metering rate of ramp i for time period t_n to t_{n+1}.

The traffic continuum model is discretized [7] as a function of Q, flow rate; G, generation flow; x, length of section j; and t, time increment:

$$K_j(n+1) = \frac{1}{2}\left\{K_{j-1}(n) + K_{j+1}(n)\right\} + \frac{\Delta t}{2\Delta x_j}\left\{Q_{j-1}(n) - Q_{j+1}(n)\right\} + a*G_j(n)$$

In order to consider the control variable in the state equations, we separate each ramp area into two sub-systems defined by the meter stop line. The actual flow rate crossing the line links the two sub-systems, and is a generation term, G(R), in the state equations of the upstream and downstream sections. We similarly link sub-systems for constructing intersection state equations satisfying the vehicle conservation principle and the intersection constraints.

We convert the optimal control problem into an unconstrained optimization, and solve it by the Conjugate Gradient method. Substituting the state equations in the objective function, the objective becomes a function of the initial conditions and the control variables, and the domain constraints are satisfied. A necessary condition is that the gradient of the objective function with respect to the control variables is zero.

3. APPLICATION

We first tested our traffic flow model against the TRAF program in the I-94 freeway corridor in St. Paul, Minnesota with encouraging results [8]. In this paper we present a test of the optimal metering for a simple one-entrance-ramp segment of the freeway corridor, illustrated in Figure 1. The small network consists of a 4700-foot freeway and a 3400-foot section of local streets. Fixed signal timing, upstream demands, diversion rates, turning percentages, and initial densities are known and the optimization period is three minutes.

In order to execute the optimization process, the initial control strategy for this example is set at 400 veh/hour. After fifteen iterations, the optimal ramp metering rates for the first, second, and third minutes are 805, 1151, and 980 veh/hour, averaging 979 veh/hour. We may evaluate these results by considering the simulated state surface generated by a set of average controls, which remain constant for the simulation period. Such a surface indicates a desired average control of approximately 1000±50 veh/h, close to the real-time optimal findings, as illustrated in Figure 2. Table 1 summarizes the control strategies and corresponding traffic performance at each iteration, with convergence to the optimal control values. The control method has been extended to longer corridor sections within a hierarchical control structure [8] with encouraging results.

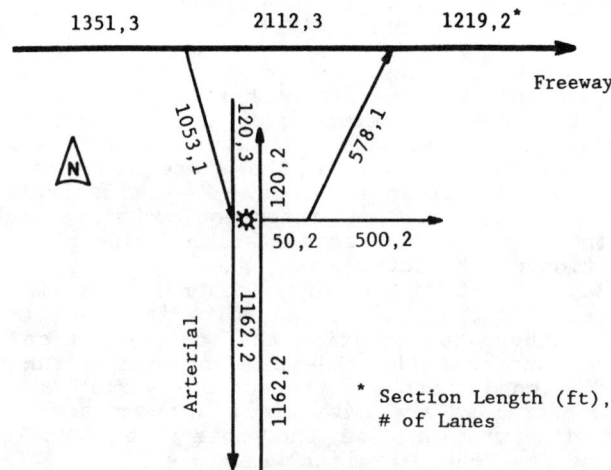

Figure 1. Simple One-Entrance-Ramp Corridor Segment

Acknowledgements - This study was supported by National Science Foundation project NSF/CES-8713277. The Minnesota Supercomputer Institute provided partial support through projects mg26701 and mg29901. The Center for Transportation Studies, Department of Civil and Mineral Engineering, University of Minnesota is acknowledged for its support.

Table 1: Results from Each Iteration in Optimization Process

Iteration Number	Ramp-Metering Rate (veh/hr)			Total Travel Time (Hours)
	1st Min.	2nd Min.	3rd Min.	
Initial	400.00	400.00	400.00	24.479397114638
1	536.70	855.46	400.00	24.345324162568
2	544.53	863.05	400.29	24.336675756149
3	545.19	866.10	399.15	24.336644794118
4	868.85	863.98	561.78	24.235307704717
5	933.40	886.40	610.89	24.214272160862
6	796.72	918.21	637.58	24.193151558044
7	796.72	918.22	637.59	24.193145819074
8	915.27	1156.74	949.91	24.074648379917
9	914.29	1161.54	957.88	24.074606590889
10	908.42	1160.07	958.40	24.073449981991
11	896.50	1157.97	960.09	24.071441196931
12	805.06*	1151.03*	980.38*	24.055511845143
13	805.06*	1151.03*	980.38*	24.055511810840
14	805.06*	1151.03*	980.38*	24.055511810839
15	805.06*	1151.03*	980.38*	24.055511810839

* Only two-digit precision shown.

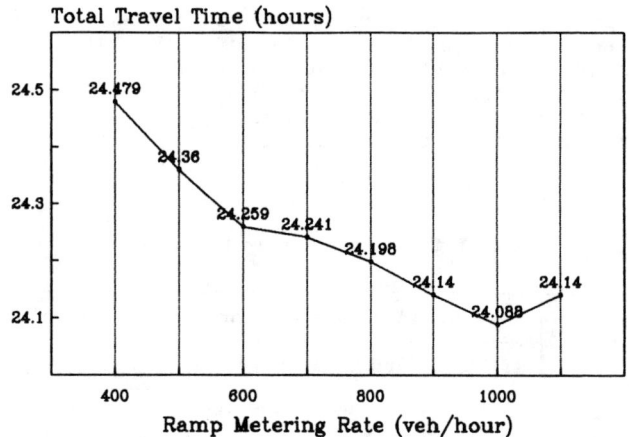

Figure 2. Simulated State Surface

REFERENCES

1. Wattleworth, J. and D. Berry (1965). Peak period control of a freeway system - Some theoretical investigations. Highway Res. Rec. 89, 1-25.

2. Papageorgiou, M. (1980). A new approach to time-of-day control based on a dynamic freeway traffic model. Transp. Res. 14B, 349-360.

3. Imada, T. and A. May (1985). FREQ8PE : A freeway corridor simulation and ramp metering optimization model. UCB-ITS-RR-85-10, U. Calif., Berkeley.

4. Kaya, A. (1970). Optimization of traffic flow for an urban transportation system. Ph.D. Thesis, U. Minn., Twin Cities.

5. Yuan, L. and J. Kreer (1971). Adjustment of freeway ramp metering rates to balance entrance ramp queues. Transp. Res. 5, 127-133.

6. Stephanedes, Y., E. Kwon and K. Chang (1991). Demand responsive ramp-metering control in freeway corridors. UM-CTS-5048-91, U. Minn., Twin Cities.

7. Lax, P. and B. Wendroff (1960). Systems of conservation laws. Commun. on Pure and Appl. Math. XIII, 217-237.

8. Chang, K. (1990). Optimal ramp-metering control for freeway corridors. M.S. thesis, U. Minn., Twin Cities.

APPLICATION OF CONTROL THEORY TO REGULATE TRAFFIC FLOW ON PERIURBAN RINGWAYS

Neïla BHOURI[1] and Markos PAPAGEORGIOU[2]

Abstract

Two on-ramp metering strategies are compared by a simulation study for the Boulevard Périphérique in Paris. The first one is a local strategy, based on linear feedback control theory. The second is a global strategy based on optimal control theory.

Introduction

Periurban ringways have been conceived as a common interconnection device for a high number of radial highways. If this central device is blocked or operated in degraded mode due to excessive congestion, its paralysing impact is imposed to a great part of the highway network, inside and outside the ringway, whence the necessity of improving the traffic situation on periurban ringways. One of the major technics aiming at preserving the freeway operational efficiency is on-ramp metering. Two on-ramp metering strategies are presented in this paper.

The first section of the paper gives the definition of macroscopic traffic flow variables. In the second there is a brief presentation of the local strategy. The third section is concerned with the optimal control strategy. It contains four parts: traffic flow model, system constraints, control objective and the definition of the optimal control problem resolved. The forth section presents the main characteristics of the Boulevard Périphérique in Paris (BP). The fifth gives results of the on-ramp metering strategies simulation tests on the BP, for two traffic flow scenaria. Some final conclusions are drawn in the last section.

I. Traffic flow macroscopic variables

Macroscopic description of traffic flow implies the definition of adequate flow variables expressing the average behaviour of the vehicles in a given freeway section i. For a space/time-discretized presentation, we define traffic density $\rho_i(k)$ as the number of cars in the section i at time $t = kT$ divided by the section length Δ_i where $k = 0,1,...$ is the discrete time index and T denotes the sample time interval. Similarly, we define mean speed $v_i(k)$ as the mean speed of the vehicles included in the section at time $t=kT$. Finally, $q_i(k)$ is the number of vehicles leaving the section during $kT \leq t < (k+1)T$, divided by T. On-ramp resp. off-ramp volumes $u_i(k)$ resp. $s_i(k)$ are defined in an analogous way.

II. ALINEA : a local feedack control law

Local strategies are characterized by the fact that on-ramp traffic volume is ordered taking into account the traffic conditions only in the on-ramp

---

[1] DART - INRETS - 94114 ARCUEIL - FRANCE
[2] Technische Universität München - P.O. Box 202420, MÜNCHEN - GERMANY

vicinity. Several local on-ramp metering strategies exist and are operating in different countries. A particularly efficient one based on a feedback concept is ALINEA. It was tested in real-life and proved superior to other local strategies [1]. ALINEA consists of the simple equation (1) obtained from a linear traffic flow model and an integral regulator [1].

$$u_1(k) = u_1(k-1) + K_R [\hat{\rho} - \rho_1(k)] \tag{1}$$

K_R is a positive constant parameter and $\hat{\rho}$ the desired density value. Eq. (1) means that ALINEA orders the on-ramp volume at each instant t=kT such that traffic density, $\rho_1(k)$ measured in the on-ramp vicinity, be close to the desired value $\hat{\rho}$.

III. COMET : optimal control startegy

Since in an optimal control problem, the system constraints and the performance index are expressed in terms of the model variables, it is the chosen model which mainly influences the efficiency and the computational effort required for the optimal control strategy. An available macroscopic traffic flow model called META is used as a basis for the development of the optimal control startegy and for the simulation tests. META has been tested on the basis of real traffic data and has given satisfactory results [2].

III.1. META : a freeway traffic flow model

Application of META requires a subdivision of the ringway into sections of some 500 m in length. In order to warrant the numeric stability of the model, the sample time internal should be in the order of 10 s. As presented in [2], META consists of the following equations to be applied to each section i.

1°) Conservation of vehicles equation:

$$\rho_1(k+1) = \rho_1(k) + \frac{T}{\Delta_1} [(1-\gamma_1) q_{1-1}(k) - q_1(k) + u_1(k)] \tag{2}$$

γ_1 expresses the off-ramp volume rate ($s_i(k) = \gamma_i \, q_{i-1}(k)$), which is supposed to be constant and known.

2°) A volume-density-mean speed relation, which is a consequence of the variables definitions:

$$q_1(k) = \rho_1(k) \, v_1(k) \tag{3}$$

3°) dynamic mean speed -density relationship:

$$v_1(k+1) = v_1(k) + \frac{T}{\tau} \left(V(\rho_1(k)) - v_1(k) \right) + \frac{T}{\Delta_1} v_1(k) \left(v_{1-1}(k) - v_1(k) \right)$$

$$- \frac{\nu}{\tau} \frac{T}{\Delta_1} \frac{[\rho_{1+1}(k) - \rho_1(k)]}{[\rho_1(k)/\lambda_1 + \kappa]} - \frac{\delta T}{\Delta_1} \frac{u_1(k) \, v_1(k)}{\rho_1(k) + \kappa} \tag{4}$$

$$- \phi \, \frac{T}{\Delta_1} \left(\frac{\lambda_1 + \lambda_{1+1}}{\lambda_1} \right) \left(\frac{\rho_1(k)}{\rho_{1,cr}} \right) v_1^2(k)$$

Where $V(\rho_1(k))$ is the steady-state homogeneous speed-density characteristic which is a monotonically decreasing function of the density ρ_1 ; $\rho_{1,cr}$ is the critical density and λ_1 the number of lanes in section i. τ, ν, δ, ϕ, and κ are constant parameters.

Eqs (2), (3) and (4) permit to describe the dynamic evolution of traffic flow on the mainstream. In order to reproduce the traffic behaviour on on-ramps, META uses the conservation of vehicles for each ramp queue :

$$l_i(k+1) = l_i(k) + T\,[d_i(k)-u_i(k)] \qquad (5)$$

where $l_i(k)$ is the queue length (veh) on on-ramp i at time $t = kT$, and $d_i(k)$ is the corresponding on-ramp demand (veh/h) which is assumed to be known.

III.2. SYSTEM CONSTRAINTS

Optimal specification of on-ramp volumes must be performed subject to some constraints:

$$0 < u_{i,min} \le u_i(k) \le u_{i,max} \qquad (6)$$

On-ramp maximum value is due to the geometric characteristics of the entrance and minimal value is posed for not closing the ramp. Besides these input volume constraints which are constant for each on-ramp, there are some other constraints which depend upon the traffic state either on the mainstream or on the entrance ramp. First, the queue length cannot be negative : $l_1(k) \ge 0$. By substitution in eq.(5), we obtain:

$$u_i(k) \le u_{max}(l_i(k)) \qquad (7)$$

Second, the density at the main street section cannot be negative: $\rho_1(k) \ge 0$. By substitution in eq.(2) we obtain:

$$u_i(k) \ge u_{min}(\rho_i(k)) \qquad (8)$$

Summarizing equations (6), (7) and (8), the admissible control region for on-ramp volumes is given by :

$$\max\,[u_{i,min},u_{min}(\rho_i(k))] \le u_i(k) \le \min\,[u_{i,max},u_{max}(l_i(k))] \qquad (9)$$

Moreover, queue lengths and densities should not exceed maximum values:

$$l_i(k) \le l_{i,max}\,, \qquad \rho_i(k) \le \rho_{i,max} \qquad (10)$$

These last constraints (eqs 10) will be considered in a smooth way in order to guarantee the existance of an admissible region for the optimisation problem.

III.3. CONTROL OBJECTIVE

Minimization of the total time spent by all drivers on the ringway system (Ts) is the most suitable control objective. It includes the total travel time on the ringway itself and the total waiting time at the on-ramps. However, using only Ts as a cost function may lead to discontinuous (bang bang) optimal control trajectories which cannot be accepted for on-ramp traffic volumes. Therefore, a term which penalizes the temporal variations of the input variables is added to the cost function. On the other hand, we have used the square of Ts in order to accelerate the algorithm convergence, and add penalty fuctions to consider state variable constraints (eq. 10). The cost function used is:

$$J = \left(T \sum_{k=0}^{K} \sum_{i=1}^{N} [\rho_i(k).\Delta_i + l_i(k)]\right)^2 + \beta_1 \sum_{k=0}^{K} \sum_{i=1}^{N} [(u_i(k)-u_i(k-1))]^2$$
$$+ \sum_{k=0}^{K} \sum_{i=1}^{N} \left\{ \beta_2.[\psi[(l_{i,max}-l_i(k))]^2 + \beta_3.[\psi[(\rho_{i,max}-\rho_i(k))]^2 \right\} \quad (11)$$

where K is the fixed optimization horizon and N the number of the ringway sections. β_1, β_2 and β_3 are weighting parameters. $\psi_1(\eta)=\min(0,\eta_1)$ are the components of the penalty function ψ.

III.4. CONTROL PROBLEM FORMULATION

The non-linear optimization problem solved is formulated as following :

(P): *Given an initial condition of state variables $\rho(0)$, $v(0)$, $l(0)$ and predicted trajectories $d(k)$, $k = 0,... K-1$. Find the optimal on-ramp trajectories $u(k)$, $k = 0,...K-1$ which minimize the performance index (11), subject to the model equations (2), (3), (4) and (5) and subject to the inequality constraints (9).*

The numerical solution of (P) is performed by a powerfull algorithm based on the discrete maximum principle, using conjugate gradient technics [3].

IV. Description of the Boulevard Périphérique of Paris (BP)

The BP consists of two closed ringways with opposit directions, each one being 35 km long. The BP is a crucial device in France traffic, it serves daily about 1,5 million vehicles, which offen leads to degraded traffic conditions. Only the inner part of the BP is considered in this paper. It includes 35 on-ramps and 38 off-ramps, out of which 5 on-ramps and 6 off-ramps are freeways. The greatest part of it has 4 lanes except for the southern stretch (7 km) which has 3 and partially even 2 lanes, and the merging area of some freeway entrances which is widenned to 5 or 6 lanes (Fig.1). Although it is similar to normal freeways, the BP has particular features like the high density of on-ramps and off-ramps, the priority of vehicles entering from on-ramps over mainstream vehicles, and the speed limitation of 80 km/h.

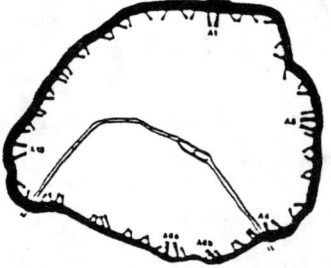

Fig.1 : sketch of the BPP

To apply META, the BP is subdivided into 70 sections of some 500 m in length. In order to obtain only one control variable and homogenous state variables for each section, 4 on-ramps, which are very closed to each other, were aggregated into two. Hence, 173 state variables and 33 control variables are used to represent the BP traffic flow in the optimal control problem.

V. RESULTS.

Only main results obtained from simulation tests for two traffic flow scenaria: without and with incident, are given here. These scenaria represent 45 min of the BP traffic flow, which is characterized by a high level of on-ramps demand. The only diffrence between them is in the initial conditions. For the first scenario, initial densities are equal to the critical ones (37.4 veh/km/lane), and initial queue lengths are equal to zero. The second scenario considers an incident which was cleared just before the simulation procedure. Initial densities reach 100 veh/km/lane on the nearest upstream sections of the incident and initial on-ramp queue lengths 57 vehicles.

	DENSE TRAFFIC	INCIDENT CASE	COMPUTING TIME
ALINEA	12.4%	10.3%	ms
COMET	13.6%	17.3%	1 min - 7 min

TOTAL TIME SPENT GAIN COMPARED TO THE NO-CONTROL CASE

For the first scenario, in spite of the severe congestions appearing in the no-control case, the local strategy ALINEA succeeds in eliminating the congestions. The optimal control strategy COMET provides almost the same degree of amelioration of the mainstream traffic flow and rather better traffic conditions on on-ramps (smaller total waiting time). For the incident case, ALINEA limits space and time propagation of congestions but doesn't reduce their gravity: while in the no-control case, congestion due to incident spreads over two thirds (23 km) of the BP and is present until the end of 45 min of simulation, with ALINEA congestion propagates over at most 7 km but with similar values of traffic density and speed (the speed is less than 10 km/h as in the no-control case). Finally, COMET succeeds to reduce significatly the congestion already in the first minutes of simulation period and eliminates them before the end of the 45 minutes. In the case of an incident, queue lengths with ALINEA respect imposed maximum values, this isn't the case for COMET, where these constraints are taken into account with penalty functions. Nevertheless the constraints violation is very limited in time and queue lengths are shorter than for the no-control case [4].

VI. CONCLUSION

For a normal traffic (without incident) and if maximum on-ramp queue length alloweded aren't very short, the local strategy ALINEA is sufficient to regulate traffic flow on the ringway. In other conditions, we should use COMET to improve the ringway traffic flow. This strategy requires more computing time but gives best results. In order to get an optimal use of these strategies, COMET can be used to provide desired values (set values) for ALINEA which applies them in real time.

REFERENCES :
1] PAPAGEORGIOU M., HADJ-SALEM H. BLOSSEVILLE J.M. : ALINEA: a local feedback control law for on-ramp metering. T.R.B 70th annual meeting. Jan. 13-17, 1991.
2] PAPAGEORGIOU M., BLOSSEVILLE J.M., HADJ-SALEM H. : Modelling and real-time control of traffic flow on the southern part of the Boul. Périph. in Paris. Transp. Res 24A 1990, pp.345-370.
3] BHOURI N., PAPAGEORGIOU M., BLOSSEVILLE J.M. : Optimal control of traffic flow on periurban ringways - Application to the Boul. Périph. IFAC Cong. Tallinn 1990, vol.10 pp. 236-243.
4] BHOURI N. : Commande d'un systeme de trafic autoroutier - Application au Boul. Périph de Paris. Doctoral dissertation. Paris-XI -Orsay. March 1991.

A DYNAMIC MODELLING FRAMEWORK OF REAL-TIME GUIDANCE SYSTEMS IN GENERAL URBAN TRAFFIC NETWORKS

R. Jayakrishnan and Hani S. Mahmassani[†]

Abstract

There is considerable interest around the world in developing and implementing real-time driver information and/or guidance systems to reduce congestion in the urban traffic networks. However most of these attempts have proceeded without much insight into the significant influence of many of the key elements and phenomena involved in a traffic system under information that could determine the effectiveness of alternative designs of such systems. These include: the context, extent and form of the information provided to the drivers, the response and compliance of the drivers to the information, the system-wide implications of particle equipping of the vehicle population, frequency of information update, etc. In this paper a simulation-based framework that incorporates these elements is presented along with illustrative results on its application to realistically sized networks. The traffic simulation is performed using macroscopic local speed-flow relations in discretized segments, while individual drivers' route choice decisions are modelled at the network nodes. The network path processing is based on k-shortest path tree-building that allows the drivers to select among several competing route. Efficient data structures are utilized for storing the network and the k-shortest paths are found using efficient binary heap-sorting procedures. The program can simulate networks with only a fraction of the drivers receiving information. The driver route switching behavior is based on candidate mechanisms of choosing to stay on a route unless an alternative route becomes sufficiently attractive. In addition, the model can also be used to evaluate the effectiveness of guidance instructions provided by a central controller. Simulations are carried out for different behavioral model parameters as well as for different fractions of drivers with information, the results of which are provided with a discussion of substantive conclusions and key insights. The simulations are carried out on a supercomputer, and the advantages of vector processing are also discussed.

Department of Civil Engineering, University of California at Irvine and Department of Civil Engineering, University of Texas at Austin. Austin, TX

Routing Based On Anticipated Travel Times

L. Rilett[1] and M. Van Aerde[2]

Abstract

The routing of drivers based on anticipated as well as instantaneous link travel times has presented route guidance system developers with an opportunity to achieve additional IVHS benefits beyond those that can be accrued by routing vehicles based strictly on current link travel times. This paper presents an analytical solution to this anticipatory routing problem for a simple 2 route traffic network and illustrates how this analytical solution can also be replicated and derived using the INTEGRATION simulation model.

1. Analytical Examination Of The Anticipatory Concept

While Kaufman and Smith(1990) presented a theoretical perspective on the anticipatory routing problem, it is useful to discuss the practical aspects of the concept of anticipatory routing and its implications using the example network that is illustrated in Figure 1. This network is representative of a commute from origin 1 to destination 2 along one of two routes. Route I (a-b-d-f) is identical to route II (a-c-e-f) in terms of length, speed, and capacity, as shown in Table 1. However, roughly half way (40%) along Route I origin 3 can contribute additional trips towards destination 2. The demand from node 3 to node 2 uses route III (d-f).

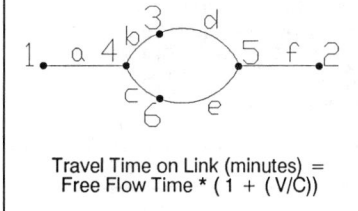

Travel Time on Link (minutes) = Free Flow Time * (1 + (V/C))

Figure 1: Sample Network Configuration

Link Name	Link Length (km)	Number of Lanes	Capacity (veh/hr)	Free Flow Travel Time (minutes)
a	4	2	4 000	4
b	8	1	2 000	8
c	12	1	2 000	12
d	8	1	2 0000	8
e	12	1	2 000	12

Table 1: Link Properties of Network

---
[1] Research Assistant and [2] Associate Professor, Department of Civil Engineering., Queen's University, Ontario, Kingston, Canada, K7L 3N5

Prior to examining any dynamic or anticipatory effects, it is convenient to introduce the equilibrium properties of the network for the 3 steady state traffic cases which are illustrated in Table 2. For example, if the demand from 1 to 2 is considered to be constant for a long period at a steady rate of 2000 veh/hr, and there is no demand from 3 to 2, it is clear that in view of the network symmetry the traffic volumes along Route I and II should be equal. Shown as case 1A in Table 2, a flow of 1000 veh/hr along each route results in a user equilibrium traffic assignment as the travel time along both routes would be equal to 42 minutes. As a second example one can consider a long term demand of 2000 veh/hr from 1 to 2, and of 1000 veh/hr from 3 to 2. Because of a lack of symmetry, this problem can not be easily solved by inspection but the application of a standard equilibrium assignment algorithm yields user equilibrium flows of 700, 1300, and 1700 veh/hr for links b,c and d, respectively (case 1B in Table 2). These flows represent a user equilibrium condition as the travel time along both routes is equal to 46 mins. In practice, however, static conditions do not prevail forever, and consequently a demand pattern as indicated as case 1C in Table 2 is somewhat more realistic. An initial estimate of the equilibrium routings for this dynamic traffic pattern is a composite average of the routing rates of case 1A for the first 2 hours, the routings of case 1B for the second 2 hour period, and finally the routings for case 1A for the fifth and the sixth hour. The traffic flows and the route times listed in Table 2 as Case 1C are estimates which were derived as weighted averages of the results of Cases 1A and 1B. However, such simple weighted averages are neither correct or optimum as indicated below.

2. Graphical Analysis of the Dynamic Equilibrium for Case 1C

If this composite routing of case 1C was in effect, and drivers were allocated to their routes at their trip origin, the link flows that are illustrated as case 2A in Figure 2a would likely be observed. It can be noted that the inflow into link b at the start of the first hour would be offset in time from the inflow into link a by an amount equal to the travel time along link a. Similarly, the inflow into link d would be offset by an amount of time equal to the combined travel times of links a and b. This same lag also exists 2 hours into the simulation when the routings change from case 1A to case1B. In this instance, vehicles, which departed just prior to the switch to the new routing rates, will continue to arrive at the start of link d at a rate of 1000 vph until approximately 18 minutes after the flow from origin c has started. Consequently, if it is assumed that in order to achieve the desired routing rates vehicles are given their routing instructions at the start of their trip, and they are not allowed to reroute, an excess flow of 300 veh/hr would be sent

Case	O-D Pair	Hourly Demand (vehicles per hour)			Average Flow On Route (vehicles per hour)			Average Route Travel Time (minutes)		
		0-120 minutes	120-240 minutes	240-360 minutes	I (a-b-d-f)	II (a-c-e-f)	III (d-f)	I	II	III
1A	1-2	2 000	2 000	2 000	1 000	1 000		42.0	42.0	24.0
	3-2	0	0	0						
1B	1-2	1 000	2 000	2 000	700	1 300	1 000	46.0	46.0	29.2
	3-2	1 000	1 000	1 000						
1C	1-2	2 000	2 000	2 000	900	1 100	667	43.3	43.3	25.73
	3-2	0	1 000	0						

Table 2: Steady State Route Flow and Times for Cases 1A, 1B, and 1C

along route I at the start of case 1B, as some of these vehicles should have been rerouted to route II. Similarly, up to the conclusion of the 4th hour, a reduced flow rate of only 700 veh/hr is sent down route I, such that in reality there is a flow of only 700 veh/hr on link d from the conclusion of the fourth hour until 18 minutes into the 5th hour. Clearly, the 50:50 split between routes I and II should have been invoked 18 minutes earlier in order to maintain a dynamic user equilibrium in which route times are kept equal.

In practice, one can also consider the routings for a full real-time route guidance system, as illustrated in case 2B in Figure 2b. As shown, even without an anticipatory function one could still improve on the case 2A routings, as one could take departures prior to the conclusion of the 2nd hour and divert some of them to route II as soon as the flows from origin 3 were observed on the network. This would mean that vehicles which left within the last 6 minutes prior to the conclusion of the second hour could still be re-routed at node 4, as they would still be on link "a". Unfortunately, any vehicles which had already passed node 4, and were on link b, would already be committed to their routings, and consequently a flow of 2000 veh/hr would still exist on link d for an amount of time equal to the travel time on b (12 minutes). At the conclusion of the 4th hour, the reverse situation would take place as some re-routing of traffic, which entered the network just prior to the end of the fourth hour (and therefore would still be on link a), would still be possible. However, for a period of time equal to the travel time along link b, the flow on link c would still drop from its previous equilibrium of of 1700 veh/hr to a disequilibrium rate of 700 veh/hr., before the new equilibrium flow rate of 1000 veh/hr becomes active.

It can be readily seen that the anticipatory routing should ideally start avoiding link b an amount of time equal to the combined travel time on links a and b, if routings were assigned at the trip origin and could not be reset. The resulting flows on a few representative links for this case 2c are illustrated in Figure 2c. They indicate that for a large portion of the analysis period the flows rates should be identical to the static component equilibrium solutions, which make up the entire dynamic demand time series. However, during the transitions from one static solution to another appropriate time shifts would need to be introduced. The difficulty of implementing the above anticipatory routing derives from the fact that the above time shift solution by inspection is not possible for actual networks. Furthermore, it should also be noted that even when the additional flow was introduced at 3, not all of the vehicles which travelled from 1 to 2 had to shift from route I to route II, but only a fraction did. Consequently, the derivation of practical anticipatory routings in the field is not an issue of simply finding a faster path into the future, but it is a matter of producing a new user equilibrium split of traffic in view of the future anticipated travel times.

3. Modelling Pre-specified Anticipatory Routings

In order to further quantify the potential impacts of anticipatory routing, the above network was coded for simulation using the INTEGRATION (Van Aerde and Yagar, 1990) simulation model. This model represents the routing behaviour of individual vehicles from the time they enter the network until these vehicles have traversed the network and have completed their trips. Figures 3 a, b and c illustrate the results for 3 simulation runs (Cases 3A, 3B and 3C) where the flows and routings for the entire 6 hour runs where made equivalent to those for Cases 2A, 2B, and 2C. Specifically, for Case 3A, all vehicles which departed during the first 2 hours and the last 2 hours of the simulation were given an equal probability of selecting either route I or route II, while the vehicles which departed during the 3rd and 4th hour were assigned only a 35 % probability of selecting route I, with the remainder being assigned to route II.

Case 3B respresents a similar simulation run in which vehicles were split with probabillities of selecting route I of 50%, 35% and 50%, as before, but in this case the new routings were invoked as vehicles reached node 4, rather than at the start of the trip at node 1. Consequently, vehicles which departed during the concluding minutes of the 2nd hour, and which reached node 4 after the start of the third hour, would still be re-routed. However, those vehicles which just passed node 4 prior to the conclusion of the second hour,

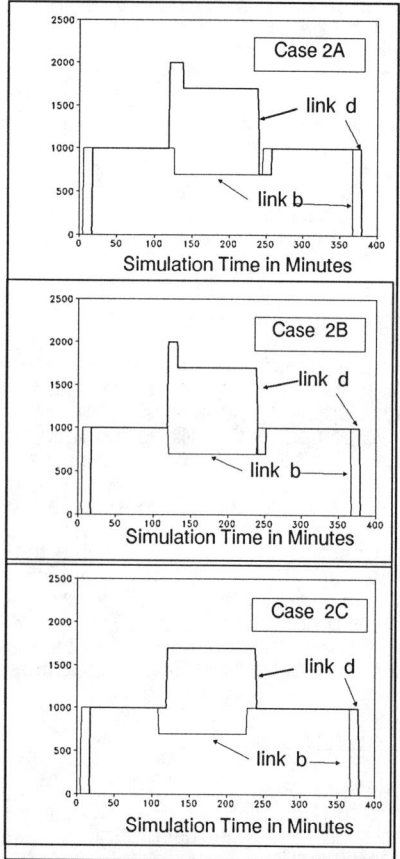

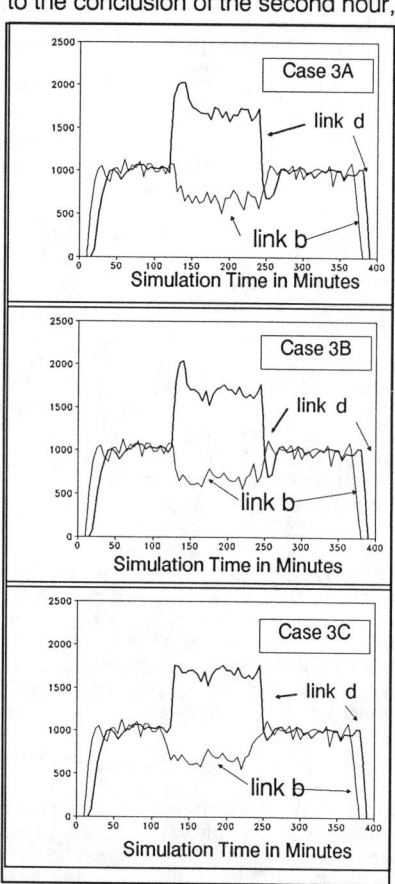

Figure 2: Flow Rates on Links b and d for Cases 2A, B and C in vehicles per hour

Figure 3: Flow Rates on Links b and d for Cases 3A, B and C in vehicles per hour

and which did not reach node 3 prior to the start of the 3rd hour, were still experiencing additional congestion. Finally, case 3C represents a simulation in which the new 35% : 65% split of traffic is invoked prior to the conclusion of the 2nd hour at node 4 in order to ensure that the combined flows on routes 1 and 3 never exceeds 1700 veh/hr and never drops below 1000 veh/hr.

Figures 3 a, b and c are shown to exhibit very similar patterns as compared to Figures 2 a,b and c. The only difference is that in the simulation the desired 50:50 and 35:65 splits of traffic routing rates are specified as probabilities such that the actual link flows are the result of the aggregate probabilistic decisions of individual vehicles. They are therefore subject to some predictable level of randomness. However, despite this randomness, it is clear that the vehicle flow patterns illustrated in Figure 2 are also replicated in Figure 3. The total system travel times for cases 3A, B and C were 572840, 572384 and 571733 veh.-mins., respectively, illustrating quantitatively the relative size of the benefits of improved anticipatory vehicle routings.

4. Automatic Generation Of Anticipatory Routings

Figures 4 illustrates the comparable total system travel times for a simulation run in which increasing percentages of a newly developed anticipatory vehicle type are introduced. This new vehicle type not only computes the probabilities of selecting routes I and II dynamically in view of anticipated future traffic conditions, but it also attempts to simultaneously approximate an equilibrium split about this best anticipated path in order to still maintain the whole network in multi-path equilibrium.

The above results indicate that such routings can be successfully implemented for a small percentage of participating vehicles. However, the real challenge derives from keeping these anticipatory routings also multipath in character. This will be required in order to maintain the anticipatory routings in a user equilibrium for larger percentages of participating vehicles and to avoid the increases in travel time which are indicated in Figure 4.

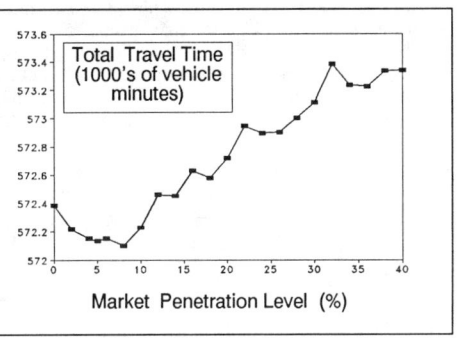

Figure 4: Total Network Travel Time for Different Levels of Market Penetration.

References

Van Aerde, M. and Yagar, S. (1990), Combining Traffic Management and Driver Information in Integrated Traffic Networks, Third International Conference on Road Traffic Control, IEE Conf. Pub. Num. 320, London.

Kaufman, D. and Smith, R.L. (1990), Minimum Travel Time Paths in Dynamic Networks with Application to Intelligent Vehicle-Highway Systems, IVHS Technical Report 90-11, University of Michigan, Ann Arbor, Michigan.

Geographic Database for IVHS Management

T.A. Yang†‡   S. Shekhar†‡   P.A. Hancock‡

ABSTRACT

A IVHS (Intelligent Vehicle Highway System) information management system obtains information from road sensors, city maps and event schedules, and generates information to drivers, traffic controllers and researchers. We extend the relational database to model traffic information in a relational database by abstract data types, and triggers.

1. Introduction

We are designing a traffic information base for Intelligent Vehicle Highway System (IVHS) application to create a shared resource, efficient disk based computation and integrity of data. As shown in Fig 1, the information stored in the IVHS database will be used by transportation system designers for traffic modeling and control. The same information will be used by human factors researchers to simulate driving conditions in intelligent cars with headsup displays and on-board computer. Efficient disk based computation to detect collisions and traffic incidents are needed. Spatial access methods, rules and triggers, provided by the database system will be used for efficient disk computation for incident-detection. One of the important benefits a database may offer is the integrity of stored data. Changes (deletion or addition) of maintenance work on highways, for example, shall not leave any management holes under the database's integrity constraints.

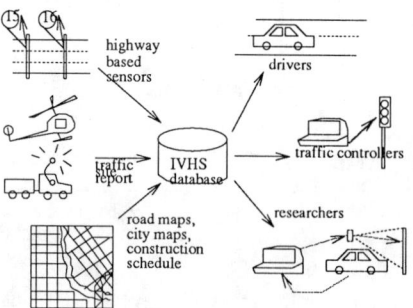

Fig. 1. Data Sources and Clients of a IVHS Database

† Computer Science Department, University of Minnesota, Minneapolis, MN 55455
‡ Human Factors Research Laboratory, University of Minnesota, Minneapolis, MN 55455

Database technology evolved over last two decades in response to the needs of commercial data processing. These databases are not able to provide reasonable performance in today's newer applications. Consider the example where a spatial object (e.g. road intersection) is stored in a traditional database by using the boundary representation approach. In this approach, a two dimensional region is represented as a collection of edges, and edges are represented by their endpoints. Three relations (regions, edges, and points) may be used to model the data. A relational database may be designed with following relations:

region: a pair (R_k, e_j) identifies a region R_k and one of its edge e_j.

edges: a triple (e_j, p_m, p_n) identifies an edge e_j and its two end points p_m and p_n.

points: a tuple (p_n, x, y) identifies a point and its coordinates x, y.

The representation smashes as simple an object as a square into parts spread over different relations and therefore over the storage medium. The question whether a region intersects given line L is answered by intersecting each of the edges of the region with the line L. If the tuple (R_k, e_j) in the relation region contains the equation of the edge, the intersection between edge e_j and L can be computed without accessing other relations. But to determine whether the intersection point lies outside or inside the edge e_j requires accessing the relations edge and points, i.e. accessing different blocks of storage, resulting in many more disk accesses than the problem requires. Efficiency requires, at least, retrieving as a unit all the data that defines a basic region such as square.

Two semantic domains that are essential to dealing with IVHS concepts, space and time, are provided by the formalism. Without excluding alternate views of space and time, we provide a kernel set of spatial and temporal concepts and operators. Based on spatial and temporal logic, they allow one to state that a fact is true at some point in time and in a particular place.

Five key characteristics distinguish IVHS data management from the other applications:

1. Data from individual sensors represent a stream of values, ordered by time of sensing. The values are accessed by their ordering in time. Data values associated with current and recent time is used more often than older data values. The average value over time intervals are often computed.
2. Many objects (e.g sensor, buildings, roads) are embedded in k-dimensional Euclidean space. For example, the record may be considered to be a point in attribute space. However, attribute space is not Euclidean space since the distance between two points (e.g., two names) may not be meaningful or satisfy the triangle inequality.
3. The objects are often accessed through their location in space. For example, partial match and orthogonal range queries are common in traditional applications. In contrast, queries for overlapping between two regions are more popular in IVHS databases.
4. A typical spatial object may be fairly complex. Even though the records contain a lot of data in traditional database, for search purposes it resembles a point in attribute space. A typical spatial object on the other hand may be a region of complex shape, and we may need to reduce them to predefined primitives such as points, edges, triangles.
5. IVHS data processing system needs to store and process information traditionally represented via maps[4] Example data include Landsat image data bank [9] , census data[5] , and city map of roads and buildings. Another important set of data includes the schedule of important events, such as road maintenance schedule, event schedule for major traffic sources including stadiums, shopping centers etc.

Our formalization views the world as a collection of entity instances. Information about individual objects are captured via the set of attributes defining its properties. For example, EE/CSci building is characterized by its location, office hour, capacity and other traffic relevant attributes. Related individuals are grouped into entities represented as a table with one column for each attribute and one row for each individual. All buildings are grouped in one table. All roads may be grouped in another table.

Spatial and temporal aspects of data are modeled via a set of abstract data types specifying the spatial and temporal attributes and operations. The space is modeled via a rectangular coordinate system. Objects are modeled approximately by a collection of primitive objects. Primitive objects include rectangles and rectangular solids. The embedding of objects to space is modeled by the coordinates of the center of object. Translation, rotation operation are supported and modeled by altering the values of relevant attributes of the objects representing the new embedding. Proximity relationship is preserved via the MoBiLe mapping function, which determines the disk address of an object from its space coordinates. The mapping function is monotonic and continuous to preserve the proximity relationships. The boundary traversal and other algorithms are supported efficiently by the mapping.

An important constraint in defining the formalism has been the desire to implement it in an extensible database such as Postgres. Postgres provides a template data base which can accept user defined data types and operators to model IVHS applications. We are implementing the formalism on Postgres version 2.0 in Unix environment on Sun Sparc machines using C and Lisp. Graphic interface for the map data is provided from the Xfig and Pic tools. The purpose of the experiment is to identify ways to overcome the limitations of traditional database systems in areas of efficiency, modeling and user interface.

2. Database Schema: Representing The Entities In The Traffic World

A database schema models application domain as a collection of entities with attributes and the relationships among the entities. To represent a domain efficiently, several data models have been proposed. Among them are the network model, the hierarchical model, the relational model, the entity-relationship model, the functional model, the semantic model and the object-oriented model. In addition, variants and extensions of the above models also exist. The extensible relational model [1,6] , for example, has been a main research area since the emergence of the relational model.

The first step of creating an extensible database application is to model the entities in the application domain as abstract data types (ADTs). An ADT contains attributes and operators specific for the type. A circle, for example, contains a center and a radius. A circle can be modeled as an ADT containing two attributes: a center coordinate (centerX, centerY) and a radius. In addition, operations such as 'get_circle_center (circle)', 'get_circle_radius (circle)', 'circle_equal (circle1, circle2)' may be defined for efficient manipulation of circles. An operation is usually defined as a function taking one or more parameters. The 'circle_equal' operation, for example, is a function taking two circles and returning a boolean value (equal/unequal or True/False).

We model the example IVHS data with the following entities: vehicle, building, traffic_sign, traffic_area, traffic_location, sensor, road, bridge, congestion, collision, and event. Each entity is represented as a relation table. Each table contains several columns representing attributes of the entity. Each column has a type, which is offered when building an entity into the database. A type specifies the domain of a column. The *building* entity, for example, includes the following attributes: *name, box, business, use-hrs*.

The semantic domain of a traffic information system can be classified into two classes: the spatial (geographical) domain and the temporal domain. The spatial objects are modeled with entities such as points, line segments, paths, boxes. The space is modeled by a coordinate system to embed objects in space. The temporal domain is modeled by time point (absolute time), time interval, periodic time (e.g., every Wednesday, every day at 9am, etc.), and schedule. The spatial and temporal classes are modeled as types and used together with the primitive types to specify the type domain of columns in a relation.

Each data type has a set of operations defined for easy and efficient use of the data type. The **box** data type, for example, has the following operations: *box_overlap(), inside(), box_center(), passesVia(), near(), enclosure(),* and *adjacent()*. The same operation may take different parameters. The operation *near()*, for example, has four different parameter pairs: point/box, box/box, lseg/box, and path/box. Although with the same name, each individual operation is actually defined differently, but the user does not need to worry about the details. This is one of the strong features of abstract data types.

3. Query Language of The IVHS Database

Once the application database is created, users can use the types and functions together with the data access commands to retrieve information from or change information inside the database.

The query language includes commands defining new types and functions not available in traditional database language like SQL or QUEL. These commands include **define type, define function, define rule, define index,** and **define operator**. Once a function is defined, it can be used in the queries. The database run time system will automatically load the corresponding function code when it processes the queries. A function can also be bound to an operator for improving the syntax of a query. The *inside(box,box)* function, for example, can be bound to the operator '<=' using the **define operator** command. The user of the database can then issue a more natural expression such as 'box1 <= box2' rather than 'inside (box1, box2)'.

Queries may be classified as relevant to drivers, experimenters, or traffic-controllers. The following are two example queries and their corresponding Postquel commands.

Query: *"Find adjacent camera pair on common lane with high difference (50% = 0.5) in average lane occupancy for last 5 minutes."*

    retrieve (c1.id, c2.id) from c1, c2 in sensor
    where c1.class = "camera" and c2.class = "camera" and c1 != c2
    and floatmi(
        average_overt(c1.occupancy, [timenow(), timemi(timenow(), @5 minute)]),
        average_overt(c2.occupancy, [timenow(), timemi(timenow(), @5 minute)])
) >= 0.5.

Query: *"Find all the objects existing within the safety envelope of the self vehicle."*

    define POSTQUEL function safety_check (vehicle) returns string is
        retrieve (building.name) where box_overlap(self.safety_envelope, building.box)
        retrieve (vehicle.name) where box_overlap(self.safety_envelope, vehicle.box)
        retrieve (traffic_sign.name) where inside(traffic_sign.point, self.safety_envelope)
    retrieve (cd_list = safety_check(self))

Queries can also be classified using a different criteria based on the data types involved. A query may be an instance of point queries, range queries, aggregate queries, join queries, transitive closure or boundary queries. Most of the above queries can be implemented efficiently. Point and range queries are implemented with the help of spatial and temporal access methods. Aggregate, join, and transitive closure queries, however, are expensive in terms of implementation cost and space and time cost. Transitive closure query, in particular, is very expensive. We are investigating the method of formulating path planning as a function. MoBiLe Files may be used in a rule-based route computation method to reduce the search space of transitive closure queries.

4. Event Detection and Database Triggers

Once built into the database, rules function like a demon and constantly monitors the database state for matching the condition part of the rule. If the state satisfies the condition, the action part is fired to conduct predefined actions (e.g., report collision). It is apparent that triggers are useful in a IVHS database application, where incidence detection and warning system for drivers are important features.

To integrate knowledge into a database system, two approaches have been proposed: the loose-coupling approach and the tight-coupling approach[7]. The tight coupling approach integrates rules into the database management system. The integration of rules into a DBMS has the capability of dealing with dynamic environment, where data are updated frequently. Another advantage of tight coupling is that it can be used with a database application which is not partitionable. One of the limitations of the loose coupling approach is that it needs a partionable database to make efficient inferences. Since a IVHS domain is dynamic, the tight coupling approach is more realistic.

In a IVHS database domain, events such as traffic accidents, traffic congestion, safety envelope violation need to be persistently monitored. Since the occurrence of events are not predictable, it is more appropriate to model them as triggers rather than as queries. Triggers are more like persistent queries or demons existing in the database. Triggers are usually implemented as rules. A rule contains a condition part and an action part. The rule "If collision with the self occurs, then report it and store all the colliding objects.'', for example, can be defined as follows.

```
define rule collision_report is
on Collision_Detection (self.box)
do report_collision()
```

5. Access Methods for Geographic Search and Motion

Access methods are data structures used to organize the data file on disk for efficient query processing. The access methods help in retrieving the data records of files in various sorted order to help answer interval queries efficiently. Recently a number of spatial data access methods, (for example, R-tree[2] and Grid Files [3]), have been proposed to retrieve objects which are n-dimensional points or solids. The spatial access methods optimize queries to retrieve all points or solids enclosed in or overlapping with a given search region. However, these access methods were designed with the assumption of static world with no moving objects. The update rates were assumed to be much smaller than the search query rate.

Motion distinguishes the IVHS database from the traditional geographic database. In presence of high update rates due to moving objects, traditional methods incur very high overhead of maintaining the tree structured indices. We have proposed a new access method called MoBiLe Files (MOnotonic Bounded mapping In LEotard Files) to address these concerns. We utilize the population distribution function and time-average traffic density function to determine the disk space needed to store the geographic data. A contiguous area on disk (t consecutive tracks and identical s sectors/track) is chosen to store the object. The shape of disk region corresponds to the population distribution in the real world to preserve the geographic neighbourhood relationship on the disk. The consecutive updates due to motion typically affect a common disk block unless the vehicle crosses over the block boundaries. Thus several update operations can be handled by updating a single disk block. The collision detection query is answered by examining the block containing the area occupied by the vehicle in question. For a detailed description of MoBiLe Files, refer to [8]

6. Conclusion and Future Work

We have provided a representation of transportation data which is semantically rich in capturing the needs for the driver, traffic controller and researchers. Abstract data types and rules are created using Postgres' type definition and rule definition facilities.

7. References

1. E.F. Codd, Extending the Database Relational Model to Capture More Meaning, *International Conf. on Management of Data*, ACM, (1979). Boston, Mass.
2. A. Guttman, R-Trees: A Dynamic Index Structure for Spatial Searching, *Proc. SIGMOD International Conf on Management of Data*, pp. 47-57 ACM, (1984).
3. K. Hinrichs and J. Nievergelt, *The Grid File: A Data Structure Designed to Support Proximity Queries on Spatial Objects*, Institut fur Informatik, Eidgenossische Technische Hochschule, Zurich (July 1983).
4. G. Nagy and S. Wagle, Geographic Data Processing, *Computing Surveys* 11(1) pp. 139-181 (June 1979).
5. J. Silver, The GBF/DIME system: Development, design and use, *U.S. Bureau of census*, (1977). Washington, DC
6. M. Stonebraker, Inclusion of New Types in Relational Data Base Systems, *Proceedings of Data Engineering,*, pp. 262-269. IEEE, (1986).
7. M. Stonebraker, *Readings in Database Systems,*, Mongan Kaufmann Publishers, Inc., San Mateo, CA. (1988).
8. T.A. Yang and S. Shekhar, Motion in a Geographical Database, *Working paper*, (1991).
9. A. L. Zobrist and G. Nagy, Pictorial Information Processing of Landsat Data for Geographical Analysis, *IEEE Computer Magazine* 14(11) pp. 34-41 (Nov. 1981).

Maglev Technology:
A Look at Guideway Construction and Maintenance Concerns

R. Scott Phelan[1], Student Member, ASCE and
Joseph M. Sussman[2], Member, ASCE

Abstract

Magnetically levitated vehicle technology, *maglev*, serves as a potential solution to ever increasing transportation demands. Though interest in maglev appears to be increasing, an appropriate level of concern directed towards guideway construction and maintenance issues has not been reflected. This paper presents an overview of maglev technology including a brief discussion of technical alternatives currently available. The paper places particular emphasis on guideway construction and maintenance concerns involved with high speed maglev systems and demonstrates these costs to be dominant. Suggestions of design methods capable of reducing guideway costs conclude the paper.

Maglev Overview

Maglev technology refers to magnetic levitation and propulsion of vehicles so that, during operation, no physical contact occurs between the vehicle and guideway. Maglev vehicles are levitated on magnetic fields either by an "electromagnetic suspension", *EMS*, or an "electrodynamic suspension", *EDS*, method as shown in Figure 1.[1]

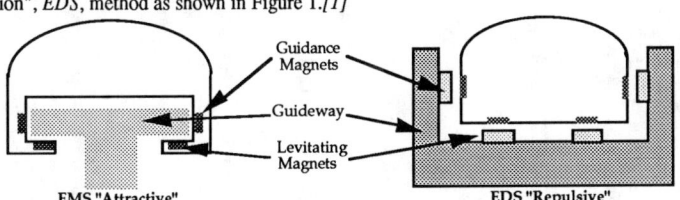

Figure 1. EMS and EDS Maglev Systems

An EMS *"attractive"* system uses conventional electromagnets attached to the lower portion of the vehicle and, due to the need for the vehicle to "wrap around" the guideway, requires extremely small air gaps between the vehicle and guideway. In contrast, an EDS *"repulsive"* maglev system operates when vehicle magnetic coils align with oppositely charged guideway magnets. Resulting repulsive magnetic forces levitate vehicles up to 8 inches (20 cm) from the guideway.[2] Due to the higher potential air gaps, EDS systems are felt to be more favorable than EMS systems for low cost guideway design.

State of the Art

The German "Transrapid" EMS system is the "super" speed maglev closest to commercial implementation. Current plans in the U.S. include a connection from the Orlando Airport to a vicinity near Disneyworld in Florida and a connection between Ontario, California and Las Vegas, Nevada. The Transrapid is limited to an air gap of 0.3 - 0.4 in (8-10 mm) as shown in Table 1, and therefore successful operation of the system hinges on precise guideway alignment. The only EDS system near commercial operation is the Japanese MLU system. Though technically proven, the MLU concept may not be commercially feasible until advances are made in magnetic shielding technology. The unshielded

---
[1] Research Assistant, Dept. of Civil Eng., Massachusetts Institute of Technology, Cambridge, MA 02139
[2] Professor of Civil Engineering and Director of Center for Transportation Studies, Massachusetts Institute of Technology, Cambridge, MA 02139.

Table 1. Characteristics of HSGT Systems

System	Vehicle Type	Suspension System	Speed Classification	Operating Speed (mph)	Number of Passengers/Car	Empty Car Weight (tons)	Health Impact	Energy Usage (BTU/seat·mile)	Air gap Tolerance
Birmingham	LIM	EMS	Low	37	40	5	none		15 mm
M-Bahn	LSM	EMS	Low-Intermediate	25-50	80	9.5	none		14 mm
HSST (HSST-05)	LIM	EMS	Low-Very High	25-200 [a]	160	40	none	275 @ 186 mph	9 mm
Transrapid (TR07)	LSM	EMS	Super	225-250	200	90	none	300 @ 186 mph	8-10 mm
								500 @ 260 mph	
JR (MLU-002)	LSM	EDS	Super	260	44	27		840 @ 260 mph	100 mm
Japanese Shinkansen	hsr	conventional	High	130-140		16 [b]			
TGV Atlantique	hsr	conventional	Very High	186	38-67	18.7 [b]		9400 @ 186 mph	
German ICE	hsr	conventional	Very High	186		11.4 [b]			
Italian ETR-500	hsr	passive-tilt	Very High	171	46-78				
Swedish X2	hsr	active-tilt	High	130-150	29-76			250 @ 150 mph	

System	Maximum Grade	Minimum Radius (ft)	Superelevation	Noise Level (dB @ 82 ft)		Capital Cost ($Million/mile)	Operating Cost (cent/PM)
Birmingham	10%	150	none used			[lower] [c]	[much lower] [c]
M-Bahn	15%	330	5.7 deg max	61 @ 25 mph		16-27	[much lower] [c]
HSST (HSST-05)	6%	8200 @ 186 mph				[somewhat lower] [c]	[lower] [c]
Transrapid (TR07)	10% allowable	1500 @ 60 mph	12 deg max	84 @ 186 mph		9-17	4-19
	3.5 % suggested	7500 @ 125 mph		92 @ 250 mph			
		13,100 @ 248 mph					
		21,400 @ 310 mph					
JR (MLU-002)	10% allowable	19,700 @ 310 mph			200 Gs field	19-31	[much lower] [c]
	6% continuous						
Japanese Shinkansen	2% suggested	8,200		95 @ 186 mph		[higher] [d]	[higher] [d]
French TGV Atlantique	2% suggested	13,127 @ 186 mph				5-10	6-29
	3.5% allowable						
German ICE	5% allowable	16,700		93 @ 186 mph		10-12	[somewhat less] [d]
Italian ETR-500	1.8% outside	12,100 @ 155 mph				9	[somewhat higher] [d]
	1.5% tunnels	17,800 @ 186 mph					
Swedish X2	5% allowable			70 @ 150 mph		[much lower] [d]	[much higher] [d]

Sources: [8], [9], [10], [11], [12], [13]

[a] 200 mph operation was proven with the HSST-04 prototype. Component changes would be required on the HSST-05.
[b] Maximum Axle Weight (tons).
[c] The subjective cost estimates for maglev systems is made with respect to the Transrapid cost values.
[d] The subjective cost estimates for hsr systems is made with respect to the TGV cost values.

MLU-002 passenger compartment currently experiences a magnetic flux density of 220 gauss[1].[3] The strong EDS magnets potentially will limit the use of reinforcing steel in guideway structural members and therefore will likely necessitate "magnetically inert" design and construction procedures.[2]

Implementation Cost Analysis

A cost model is used to determine and analyze broad overall costs for typical high speed ground transportation, HSGT, technology implementation scenarios as shown in Figure 2. The total cost is determined according to the following additive equation:

$$C_{\text{tot}} = C_{\text{acquis}} + C_{\text{const}} + C_{\text{oper}} + C_{\text{other}}$$

where C_{tot} is the total system implementation cost, C_{acquis} includes land, rolling stock and other special equipment acquisition costs, C_{const} represents construction costs including required material, labor and equipment expenses, C_{oper} represents operational and maintenance costs of the system, and C_{other} reflects other costs associated with system implementation including financial and legal arrangements, design and management charges and costs for feasibility studies.[3]

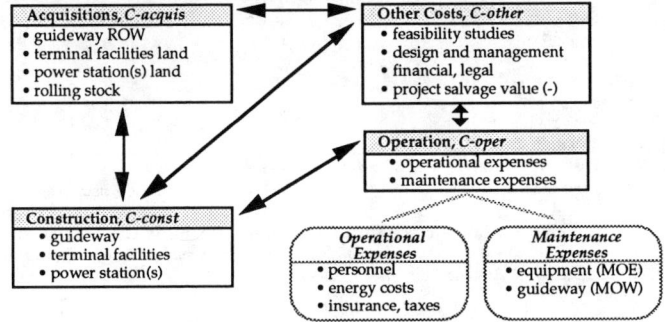

Figure 2. Interactive Nature of Cost Component Relationships

Applying cost data, taken from a 1984 Canadian Institute of Guided Ground Transportation, *CIGGT*, study of a 230 mile (370 km) Los Angeles to Las Vegas corridor[4], to the model reveals construction costs to be the most significant single cost component for both maglev and high speed rail, hsr. As shown in Table 2, construction costs represent 60% of the total cost for both technologies. A more detailed cost breakdown, shown in Table 3 indicates the dominant cost component for maglev, representing almost 40% of the total project capital costs, to be construction of the guideway superstructure.

Table 2. Cost Model Application Results, 230 mile LA-Las Vegas

Cost Category	Maglev, Transrapid ($1990 million)	% of C_{tot}	hsr, TGV ($1990 million)	%of C_{tot}
C_{oper}	560.1	17.44	628.1	21.20
C_{const}	**1926.8**	**60.01**	**1768.7**	**59.71**
C_{acquis}	428.8	13.35	305.0	10.30
C_{other}	<u>295.4</u>	9.20	<u>260.3</u>	8.78
C_{tot}	3211.1		2962.2	

Source: of cost data: 1984 CIGGT study, reference [4].

---

1 A "gauss" is a standard unit of measure for magnetic flux density. It is equal to one line of magnetic flux per square centimeter. *[1 tesla = 1 newton/(ampere•meter) = 10,000 gauss.]* Though the full extent of high magnetic field exposure on humans is not known, approximately 100 gauss is felt to be the human safety limit for extended periods of time. Current design criteria for passenger field exposure is approximately 0.5 Ga dc field and near 0.0 Ga ac field.
2 For additional discussion on both maglev and hsr systems, see references [1], [7], and [8].
3 A more detailed discussion of the cost model is found in reference [1].

Tables 2 and 3 demonstrate that for high speed maglev technology to be less costly, attention directed towards a more economical guideway design is needed. In addition to guideway construction costs, guideway maintenance costs are likely to be significant given the stringent alignment requirements. Maintenance cost projections used in the CIGGT study are shown in Table 4.

Table 3. HSGT Capital Cost Breakdown, 230 mile LA-Las Vegas

Cost Category	Transrapid ($1990 mil)	% Cap. Cost	TGV ($1990 mil)	% Cap. Cost
Guideway Superstructure*	**$ 894.8**	**37.98**	**193.6**	**9.34**
Rolling Stock	361.4	15.34	276.5	13.33
Power Station & Cables*	335.3	14.23	131.0	
Long Stator*	284.6	12.08		
Guidance Rails*	127.9	5.43		
Signals & Comm.*	106.3	4.51	106.5	5.13
Terminals*	69.4	2.95	90.8	4.38
Land	57.3	2.43	28.6	1.38
Fencing*	40.0	1.70		
Earthwork*	**39.4**	**1.67**	**1049.7**	**50.62**
Switches (Turnouts)*	23.2	0.98	8.0	0.38
Main. & Emer. Equip.	10.1	0.43		
Maint. Facilities*	5.9	0.25	87.9	4.23
Misc. Structures*			83.2	4.01
Buildings along route*			16.2	0.78
Trainset preparation centers*			1.9	0.09
Capital Cost Sub-total	**2355.6**	**1.00**	**2073.8**	**1.00**

* Indicates component of C-const   Source: of cost data: 1984 CIGGT study, reference [4].

Table 4. Maintenance Cost Calculations for Basic Guideways

System	Annual Cost ($1990 million)	Annual maint. costs	% of annual costs	% of all annual % Total System Cost
Transrapid	$ 2.660	3.83	**0.48**	0.74
TGV	$ 5.421	19.93	**7.00**	1.62

Source: of cost data: 1984 CIGGT study, reference [4].

The projected values in Table 4 appear to be much too low. For comparison, Table 5 shows projected maintenance requirements for the TGV system used in the CIGGT study with actual data from the Japanese Shinkansen high speed "bullet" train.[5] The values shown in Table 5 cast doubt on the optimistic TGV maintenance estimates. Likewise, considering the stringent alignment requirements for the maglev system, the Transrapid maintenance requirements shown in Table 4 appear to be unrealistic.

Table 5. Maintenance Comparison between TGV and Shinkansen

	TGV (projected) : (ballast)	Shinkansen: (ballast)	(ballast & slab)
Required Maintenance staff per track mile	**0.20**	5.63	4.21
Projected) Annual Cost ($1990 million/mile)	0.023	0.649*	0.486*

* Shinkansen cost values have been calculated based on labor requirements per mile of track. The TGV projects less than 0.2 maintenance men per track mile for its ballasted track, while more than 4 men per mile are required for the Shinkansen.

Need for Improved Guideway Design

With maglev guideway construction costs projected to dominate system implementation costs, reduction of the guideway cost through improved design is an immediate concern. A number of innovative guideway designs are possible including (a) narrow beam design, (b) segmental construction, (c) assembly-line type erection procedures and (d) real time alignment monitoring and control.

A narrow beam allows guideway optimization for lowest possible cost since the beam element is likely to be the most costly structural element for any moderately spanned guideway system.[6] Modularization techniques such as segmental bridge construction can be used for guideway column, footing and girder installation. Segmental construction techniques currently are used for a number of heavy civil engineering construction projects including bridges, buildings and tunnels. In addition, assembly-line type erection procedures are possible, where columns and footings are placed initially and girders are positioned and attached one by one starting from a single location. The outermost attached girder becomes a platform for transporting and placing the next girder into position. The method increases potential span lengths as restrictions imposed by highway right of way limitations (for transport to the construction site) are no longer valid. Finally, real-time monitoring and control of the guideway reduces the labor component of procedures even further as it allows the use of relatively inexpensive construction labor since precast components can be placed initially with relatively minor attention given to precision. Once the guideway is in place, real time monitoring and alignment is engaged. Such a "dynamically-adjustable" guideway system reduces both initial construction costs and long term routine maintenance costs.

Summary and Conclusions

Guideway construction costs dominate all other high speed maglev implementation costs. In addition, guideway maintenance costs can be expected to be significant given the stringent guideway alignment requirements for current maglev designs. With the implementation potential of high speed maglev technology linked to overall implementation costs, guideway cost reduction is an immediate concern. The following innovative approaches are proposed for improved guideway design:
- magnetically inert design and construction procedures,
- optimized, narrow beam guideway design,
- segmental guideway design,
- automated erection procedures, and
- real time, "dynamically-adjustable" alignment monitoring and control technologies.

References

[1] Phelan, R.S., "Construction and Maintenance Concerns for High Speed Maglev Transportation Systems", Masters Thesis, *MIT Department of Civil Engineering*, September 1990, Chapters 2 and 3.
[2] Johnson, L.R., et. al., "Maglev Vehicles and Superconductor Technology: Integration of High-Speed Ground Transportation into the Air Travel System", *Energy and Environmental Systems Division, Center for Transportation Research, Argonne National Laboratory*, Argonne, Illinois, April 1989.
[3] Hayes, W.F., "Magnetic Field Shielding for Electrodynamic Maglev Vehicles", *International Conference on Maglev and Linear Drives '87*, Las Vegas, Nevada, May 19-21, 1987, p. 54.
[4] CIGGT 9.2, *Maglev Technology Assessment, Task 9.2: Review, Validation and Revision of the Capital and Operating Costs for a Transrapid TR-06 Maglev System and for a TGV System in the Las Vegas-Souther California Corridor*, Super-Speed Ground Transportation System Las Vegas/Southern California Corridor Phase II, The Canadian Institute of Guided Ground Transport, July 28, 1989.
[5] Miyamoto, T., "History and Future Outlook of Track Maintenance of Shinkansen", *Japanese Railway Engineering*, Vol 19 No 2, 1979, p. 16-17
[6] Thornton, R.D., "Monorail Maglev", *Magnetic Levitation and Transportation Strategies*, Future Transportation Technology Conference and Display, San Diego, CA August 13-16, 1990, SAE Publication SP-834, Paper #90479, p. 61-67.
[7] Wyczalek, F.A., "Magnetic Levitation Transportation Strategy", *Magnetic Levitation Technology for Advanced Transit Systems*, Future Transportation Technology Conference and Exposition, Vancouver, BC, Canada, August 7-10, 1989, SAE Publication SP-792, p. 63-69.
[8] "Rail Technology", Texas Triangle High Speed Rail Study, *Texas Turnpike Authority*, February, 1989, p. 1-13.
[9] "Report of the Preliminary Feasibility Study of the Pittsburgh MAGLEV Project", *MAGLEV Working Group*, Pittsburgh, PA, Feb. 23, 1990, p. 5.
[10] Hayashi, A., and Ohishi, A., "HSST Maglev Train at Yokohama Expo '89", *Magnetic Levitation Technology for Advanced Transit Systems*, 1989, SAE Publication SP-792, p. 13-21.
[11] Dickhart, W.W., "The Transrapid Maglev System", *Magnetic Levitation Technology for Advanced Transit Systems*, 1989, SAE Publication SP-792, p. 13-21.
[12] Takeda, H., "Japanese Superconducting Maglev: Present State and Future Perspective", *Magnetic Levitation Technology for Advanced Transit Systems*, 1989, SAE Publication SP-792, p. 57-62.
[13] Welty, G., "High Speed Race Heats Up", *Railway Age*, Vol 191 No 5, May 1990, p. 70-80.

TWO APPROACHES FOR CHARACTERIZING IMAGES OF RAIL SURFACE FLAWS

Roemer Alfelor[1], Thomas Short[2], Behnam Motazed[3] and Sue McNeil[4], M ASCE

Abstract

Rail surface flaws can be removed by grinding to improve the life of rail and provide a safe, smooth running surface. To avoid time consuming and costly visual inspection in the determination of grinding strategies, high speed automated data collection has been explored. This paper describes and compares a frame-by-frame approach to the analysis of rail surface images obtained in the field using a high speed video camera with a pixel-by-pixel method.

1. Introduction

There has been increasing concern with rail surface condition as some railroads, for a variety of reasons, are experiencing more surface defects. If detected soon enough, surface flaws may be removed by grinding. If allowed to deteriorate they can have safety, speed, life cycle maintenance cost and energy cost impacts. Although the implications of surface flaws are known, knowing where, when and how much to grind requires comprehensive data on surface flaw location and severity. Collecting such data can be time consuming and expensive.

Systems for automated highway pavement surface defect data acquisition and processing have demonstrated the feasibility of automating this data collection process for rail head, as they are based on similar principles. A laboratory image processing system has been developed [McNeil 91a], and rail surface data has also been collected using a high speed video camera in the field at speeds of up to 50 mph [McNeil 91b]. To develop a production system, high speed image processing is required to classify surface flaws. This process is complicated for several reasons. First, data on surface flaws has not been routinely collected in the past and there is not a well defined, universally understood grammar to describe such defects. Second, image processing is computationally intensive in terms of time and memory. Finally, distortions of the recorded images add complexity to the image processing. These distortions arise from:

---

[1]Graduate Assistant, Department of Civil Engineering, Carnegie Mellon University, Pittsburgh, PA 15213

[2]Graduate Student, Department of Statistics, Carnegie Mellon University, Pittsburgh, PA 15213

[3]Research Scientist, Field Robotics Center, Carnegie Mellon University, Pittsburgh, PA 15213

[4]Associate Professor, M ASCE, Department of Civil Engineering, Carnegie Mellon University, Pittsburgh, PA 15213

- Distortion during the data acquisition in the form of image blurring and variable lighting.
- Artifacts on the rail such as oil spills and shadows.
- The nature of the rail surface, which may include stripes of discoloration.

Two approaches to processing rail surface images are explored to understand the characteristics, limitations and benefits of the two approaches. The first is based on a crude, aggregate classification of each image as defective or non-defective. It is based on well known image processing techniques such as edge detection and thresholding. This approach is referred to in this context as a global approach as it provides an aggregate classification of each frame. The second is based on statistical image processing algorithms. An iterative statistical procedure is used to reconstruct the image and classify defects on a pixel-by-pixel basis. Therefore, this approach is referred to as a local approach.

The following criteria are used to compare the two approaches:
- Usefulness of the classified images. Each method produces different data. The crude method produces aggregate data where the more complex statistical method produces disaggregate, detailed data. Each type of data has a role. The aggregate data is useful for maintenance planning at the strategic level, such as determining the average frequency of rail grinding, whereas the disaggregate data is used to plan maintenance at the tactical level, such as the selection of grinding patterns and depths.
- Speed of the processing. Image processing is computationally intensive. Although the crude method is significantly faster, the statistical image processing method demonstrates the potential to be relatively efficient.
- Accuracy of the procedure. No image processing procedure will correctly classify images all the time. The evaluation using this criteria also reviews the impact of improperly classified images, such as incorrect maintenance decisions.

2. Approaches to Image Processing

2.1. Global Approach: Thresholding and Edge Detection

For the purpose of identifying which sections of the railroad track have surface defects a procedure was developed which utilizes the overall characteristics of each frame or image. The objective is to define a signature or measure that can reasonably separate frames exhibiting surface defects to those that have no visible flaws. This procedure deviates from the conventional object recognition in that no specific objects are identified. Although rail surface defects appear as dark spots or visible streaks which can be identified by inspection, the irregularities in shape, size, location and orientation of these defects make pattern recognition difficult. An attempt to implement a more pattern-oriented approach using statistical techniques is presented in the next section.

Simple thresholding using pixel intensity values is not appropriate for defect recognition because of rail discoloration. Discoloration typically appears as dark bands running longitudinally on each side of the railhead. Because it shares the same intensity values as the defects, thresholding will result in misclassification of discolored frames. The width, location and intensity of the discoloration is not uniform, hence it cannot be subtracted from the image.

The global defect recognition approach adopts an edge detection algorithm. It was observed that the edges of the discoloration do not have well-defined edges in contrast with surface defects such as spalls. This means that an edge detection algorithm will result in bright pixels for edges of defects while the edges of the discoloration will not have very bright intensities. The edge detection used a Sobel filter [Gonzalez 77]. This algorithm is included in the image processing software Image-Pro II [Image Pro 88] and was chosen over other available edge detection operations because it is relatively insensitive to noise. A threshold based on the percentage of bright pixels was used to classify images as defective or non-defective.

These procedures were applied to 83 digitized images. Of these 33 were manually classified as non-defective and 50 defective. The images were further manually classified by severity of defects. For an arbitrarily defined range of 'bright' pixels (155-255), Figure 1 shows that the log of the percentage of bright pixels increases with increasing severity of defects.

For image classification using the above data, a threshold value needs to be determined. Two approaches for identifying this threshold was applied. Using discriminant analysis, the threshold value is set at 0.46% with an overall misclassification rate of 32%. Using an algorithm that minimizes the total number of samples misclassified (or the total cost of misclassification), the threshold value is set at 0.28% and the misclassification rate is 23%.

The global approach to image processing and classification simply uses edge detection and thresholding as preprocesses, and then determines the percentage of bright pixels as a signature that classifies images as defective or non-defective. The algorithms are fast and efficient.

2.2. Local Approach: Image Reconstruction and Segmentation

At the local, pixel-by-pixel level, image processing algorithms that are capable of **reconstructing** an estimate of the original image are required to account for the various forms of distortion encountered in rail surface images. These algorithms should also be flexible enough to handle different types and combinations of distortion. In addition, the algorithms must be capable of image **segmentation** to provide information about the defects present in the images. Segmentation involves the locating and measuring of objects in an image.

Existing statistical image reconstruction algorithms have been used as the foundation for extended algorithms that simultaneously reconstruct and segment the rail surface data. An estimate of the true image is calculated to maximize the joint probability distribution of the true, undistorted, pixel intensities and superimposed membership variables. The membership variables contain information about object membership for each pixel.

The algorithms proceed in an iterative pixel-by-pixel fashion, estimating the true intensity and membership variable values for each pixel based on the values of the pixel's neighbors. Eventually, the local information provided by each pixel and neighborhood accumulates to provide a reconstruction of an entire image, along with the geometric measurements of objects segmented.

The probability models that describe the relationships between pixels and their neighbors consist of two parts: a distortion component and a texture component. The distortion component models both the physical sources of distortion in the images and the distortion added during acquisition. It includes parameter values that can be adjusted as the physical context of each image changes along the length of the rail. The texture component allows us to model the behavior of pixels within the objects in the image. Functions that adjust the model according to the current values of the membership variable estimates are incorporated, and thus flexibility of the model is maintained even as the fundamental condition of the rail varies.

For investigative purposes, the algorithms produce reconstructed images that are compared to the observed video data as shown in Figure 2. Eventually, the output of the reconstruction and segmentation algorithms will be condensed to provide only the locations and geometric measurements of the objects detected in an image. When these summary data are fed into a classification routine, discoloration and other non-defective objects will be recognized and discarded. Objects that have geometric features similar to those of a particular defect category will be classified and reported. If objects are encountered that do not appear to fit any of the specified categories very well, they can also be reported and investigated further.

Ideally, the algorithms could proceed through large quantities of rail surface data, with little human intervention or constant monitoring. Maintenance decisions could be made by the railroad after review of the physical locations of the object and most probable classifications into defect categories.

3. Discussion and Future Work

The following table summarizes the attributes of the global and local approaches to image processing for rail surface defect data with respect to the criteria described in the introduction. As this evaluation is based on very limited data it is important to recognize that the results may not be broadly applicable. This exploration indicates that there are very clear trade-offs among quality and quantity of data, and speed of processing. Further exploration is required to understand the implications of misclassified data and the effort required to apply the techniques to large quantities of continuous data. Also, the properties of the methods such as convergence and accuracy, and the utility of the results must be pursued.

Criterion	Global Approach	Local Approach
Usefulness	Provides basic data for decision making	* Locates a wide variety of interesting objects, including those not previously encountered in the data. * Provides information needed to make classifications into specific defect categories
Speed	Relatively fast	* Basic algorithms are very slow. * Processing time increases when complex distortions are modeled. * Parameters of statistical models must be updated and monitored.
Accuracy	Poor, misclassification 23%	* Apparently good * Capable of locating individual defects.

4. References

[Gonzalez 77]   Gonzalez, R.C. and P.A. Wintz, *Digital Image Processing*, Addison-Wesley, 1977.

[Image Pro 88]   Media Cybernetics Inc., *Image Pro II - Image Processing System Version 2.0*, Media Cybernetics Inc., 1988.

[McNeil 91a] McNeil, S., B. Motazed, R. Alfelor and T. Short, *Automated Railhead Surface Flaw Inspection*, Gordon & Breach, 1991.

[McNeil 91b] McNeil, S., B. Motazed, R. Alfelor, T. Short, and J. Eshelby, "Automated Collection and Detection of Rail Surface Defects," *Heavy Haul Workshop*, Heavy Haul Association and RAC, June, 1991.

Figure 1: Defect Severity and Percentage of Bright Pixels

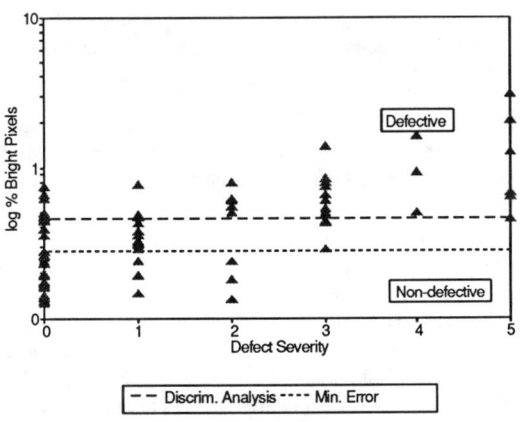

Figure 2: Local Approach: Reconstruction and Segmentation

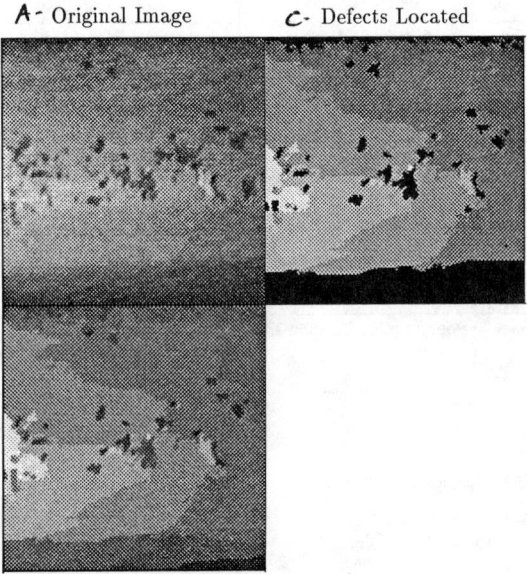

COMPUTER SIMULATION AND DISPOSITION IN RAILWAY

Leon Kos[1] - Werner Kretschmer[2]

ABSTRACT

The paper describes simulation package on high capacity personal computer. Sofware package PC-SIMU simulates extremely accurately the movement of trains within the region of the line suitable for central traffic disposition. The elements of simulation model and the field of its use are described. In the central supervision system the simulation subsystem serves as aid to the central dispatcher in discovering and solving traffic conflicts.

INTRODUCTION

Software package PC-SIMU serves for very accurate copying of the actual movement of trains on railway tracks to a mathematical logic model on personal computer. From here it is possible to make higly probable conclusions as to the positively expected traffic course.

The PC-SIMU simulation system provides very wide application range:
- evaluation of new track configuration
- evaluation of train schedule variants
- estimation of consequences of signalling system changes
- examining effects resulting from the changed traffic
- examining effects of operation disturbances (train delays, new trains)
- estimation of changes of dynamic train values, etc.

PC-SIMU is capable to run on personal computer under MS-DOS or SCO-XENIX operating system. The software package has been developed in FORTRAN.

---
[1] ISKRA, Stegne 21, YU - 61000 Ljubljana Yugoslovia
[2] HACON, Vahrenwalder Str. 7, D-3000 Hannover 1 Germany

SIMULATION MODEL

Constructional data (railway line):
- Transformation of track layout in conformity with EDP-form
- Description of signals, overlaps, release contacts, block sections, stop areas, approach routes, speed and gradient changes
- Description of the route directions and routes

Operating data (dynamics of train movements):
- Train model ranges
- Simulated train schedules: desired times, timing intervals, daily variation lines, distribution of delays.
- Determination of dispositive measures: priority of train classes, queuing up of the trains on the approach routes, destination tracks in the stations, waiting period regulations, connection of train and shunting movements.
- Simplified data for describing trains include: length of train, the highest speed of the train, medium acceleration, medium slowing down (Kretschmer, 1985).

There are different possibilities for disposition with PC-SIMU (Kos - Kretschmer - Dannenberg, 1989):
- disposition on access lines, keeping the smallest possible distance between trains and changing succession of trains
- disposition of track on the stations - one or more tracks are available for each train
- disposition of trains in accordance with priorities (train classes)
- reservation of longer or shorter part of train route
- taking into account waiting times for connections
- place and time connections among train and shunting movements.

COMPUTER SUPPORT FOR CENTRAL TRAFFIC CONTROL TASKS

Since the main lines, especially on European railways, are heavily loaded, the introduction of computer supported systems in traffic supervision centres has become indispensable. Railway Authority prepared special technical requirements for such computer systems and set up all relevant traffic and technical requirements. In this way the dispatcher gets all necessary information displayed on VDUs. The computer system for central supervision helps the dispatchers and gives them the urgently needed time and all information so that they can decide on the movement of trains.

By means of microcomputers based on up-to-date concepts we collect at a central place the data on position of points and state of signals. The trains are identified with their own number.A special microcomputer subsystem takes care of automatic following of train numbers in accordance with the actual movement of trains on the line. To this system we can add equipment which locally informs passengers in particular stations, as well as the system for automatic train route setting. These are especiallly useful, since they release the station master of the constant entering of individual commands for setting parts of the train route.

MODERN SYSTEM STRUCTURE

The control centre is based on subsystems connected into local computer network (LAN), ensuring fast exchange of messages among subsystems. Basic units of subsystems are personal computers PC-AT with Intel processors 80286 and 80386.

The communication subsystem provides exchange of message with remote stations. The system is doubled, and communication protocol ensures efficient and reliable transmission of data.

The main subsystem is also doubled and provides interactive work of the dispatcher. The dispatcher performs all manipulations on graphic colour video display with high resolution.

The track diagram subsystem consists of PC-AT computers without disk units and keyboards. Each computer has a graphic colour video display unit (VDU) with high resolution. Displays show track diagrams of the whole controlled area.

Subsystem for time-distance graph also consists of more VDUs and PC-AT computers, similar as in the track diagram subsystem (Kos - Kretschmer - Dannenberg, 1990).

The following displays are available to the dispatcher on the colour video display: system list, survey of alarms, standard running programmes, special running programmes, timetable, traffic diary, deviations from timetable, graphic timetable, data on traffic control, events protocol.

AUTOMATIC TRAIN ROUTE SETTING

The programme of running for a certain train is composed of a sequence of commands which fully determine the movement of the train from the beginning to the end station. The train route is set on the basis of train number. When the train occupies a corresponding section, the computer system sequentially sets the section of the train route in front of the train. The computer system recognizes a standard train route by the leading number (from 0 to 9), which is added to the train's own number and thus actuates setting of the train route according to the standard programme. The dispatcher can at any time allocate to any train a leading number. The computer automatic train route setting stops in case of a conflict between two or more trains. This conflict is resolved by the dispatcher's intervention into the system, at which he is assisted in finding the best solution by the subsystem for simulation of railway operation.

SUBSYSTEM FOR SIMULATION OF RAILWAY OPERATION

The man himself finds it difficult to anticipate in advance whether particular trains will obstruct each other and what will be the consequences of such obstructions for further running of traffic. In the first stage we plan computer assistance only in the form of advice to the central dispatcher, and in the second stage it will be possible to include the computer in the closed loop. The subsystem for simulation (disposition) will proceed its solutions of conflicts directly to the automatic train route setting. The central dispatcher still accepts the final decision upon the places for overtaking and crossing of trains, and the computer system on the basis of simulation only suggests one of the possible solutions. The computer becomes a handy tool for the dispatcher in his responsible work, i.e. solving of conflicts and searching for optimal movement of all trains.

The tasks of the computer supported system for disposition are:
- recognition of disturbances which will occur in the movement of trains (deviation from plan);

- ascertainning of effects of these disturbances on further movement of all trains (discovering of conflicts);
- effective informing of the dispatcher on the effects of these disturbances (graphic display on VDU);
- searching for solutions for non-conflicting movement of all trains (solving of conflicts).

The subsystem for simulation described in the previous paragraphs has to be connected with the central traffic control system. In this way the simulation subsystem receives constantly fresh data from the line about movement of trains on the basis of which it can perform, with simulation of train movement, its above mentioned functions in the future.

CONCLUSION

The foreseen computer automation releases the man - dispatcher of all routine activities, so that he can devote all his time to supervision of traffic flows. The computer can take over accurate calculations, simulation of train movements and prediction of conflicting states which are to occur in the near future. Use of high-capacity personal computers enables inexpensive implementation of this kind of complex software, resulting in real possibility of using theoretical results in practice.

REFERENCES

Kos, L., Kretschmer, W. and Dannenberg, H. (1989): Computer Simulation and Disposition in the Centralized Traffic Control System. IFAC Symposium: Control, Computers, Communications in Transportation pp. 239-246 Paris.

Kos, L., Kretschmer, W. and Dannenberg, H. (1990): Railway Operation Simulation in the Microcomputer-Based CTC System. Computer Applications in Railway Planning and Management. pp. 77-88. Computational Mechanics Publications. Southampton, Boston.

Kretschmer, W. (1985): Entwicklung eines DV-gestutzten Datenaufbereitungssystem fur eisenbahnbetriebliche Untersuchungen. D.Sc. Dissertation, Wissenschaftliche Arbeiten Nr. 21, Hanover.

RAIL REPLACEMENT PLANNING: A KNOWLEDGE SYSTEM APPROACH

Rabi G. Mishalani[1] and Carl D. Martland[1]

Abstract: REPLACER is a knowledge system that provides a consistent means for developing plans for rail replacement. It was developed, tested, and evaluated in cooperation with Burlington Northern Railroad. BN officials found that it has substantial potential for improving rail replacement planning. When compared to a network optimization decision model, REPLACER was found to produce essentially the same plans while using a more flexible and understandable methodology.

1. INTRODUCTION

REPLACER is one component of a larger system developed at MIT, referred to as REPOMAN, that assists railroad officials in making rail replacement decisions (Martland et al, 1989). The motivation behind the development of this rail management system is to ensure an acceptable rail quality for service provision, to avoid the inconsistencies of the existing decision process, and to avoid unnecessary maintenance expenditures. REPOMAN includes two other major components, the Categorization Model (CM) and the Model for Optimal Rail Relay Scheduling (MORRS). CM is an expert system that diagnoses each rail segment (a segment is a piece of rail with "homogeneous" attributes) of a region into one of four performance categories for each of the next five years based on the condition of the rail and the operating environment (Acharya et al, 1989). In decreasing order of severity, the categories are MUST, SHOULD, MAYBE, and OK. A MUST segment must be replaced to allow for the desired levels of operating performance. At the other extreme, OK segments are generally not candidates for replacement. REPLACER (Mishalani, 1989) and MORRS (Acharya et al, 1989) both take the five year categorization as input and produce a preliminary replacement plan. This paper focusses on REPLACER with some comparison with MORRS.

2. LITERATURE REVIEW

Several models have been developed to assist in rail replacement planning. The rail performance model (Wells, 1983) determines the optimal time to replace a single rail segment based on the number of defects predicted for the coming years, the expected costs associated with defects, and the expected costs of replacing rail. The model can be used to quantify the benefits of cascading older rail from a high density line to a low density line. This model has two main limitations. First, it is based on rail fatigue, which is no longer the primary reason behind rail replacement due to recently introduced maintenance techniques. Second, because the model is designed to deal only with one segment at a time,

---

[1]Massachusetts Institute of Technology, Department of Civil Engineering, Room 1-178 Cambridge MA, 02139

it does not deal with the network and system aspects of rail replacement.

TRACS (Hargrove and Martland, 1990) predicts rail wear and fatigue, taking into account the effects of grinding and lubrication. It determines the year in which each segment is expected to be replaced, along with the associated life cycle costs. TRACS can be used to test different replacement strategies under different scenarios and is ideal for strategic planning. However, like the rail performance model, it does not capture the network and system aspects of rail replacement and is not very helpful for determining a detailed rail replacement plan for the coming year.

REPOMAN is less concerned with deterioration modeling because BN has excellent information on rail condition that is suitable for planning rail replacements. The real problem is assembling a reasonable replacement plan given a multiplicity of options and constraints. Two approaches were pursued in parallel within REPOMAN: (1) MORRS, a network optimization decision model, and (2) REPLACER, a sequential selection approach designed within a knowledge system environment and based on economic analysis, heuristics representing engineering knowledge, and optimization principles. The motivation behind REPLACER's methodology is to investigate the possibility of developing an approach that is more flexible and more accessible to decision-makers than more traditional operations research approaches such as network optimization.

3. APPROACH

REPLACER selects segments for the coming year's replacement plan by going through a sequence of steps. At each step, more segments are selected until a complete plan is developed that satisfies operating and economic considerations. This plan identifies "stretches" of rail that are to be replaced, where a "stretch" consists of one or more contiguous segments. The structure of REPLACER is linked closely to the management framework of rail replacement. The **general goal** behind rail replacement is to strike a balance between maintaining desirable service levels and minimizing the life-cycle cost of the infrastructure. This general goal can be translated into more detailed **objectives**: (1) preventing the condition of the rails from "interfering" with service delivery or going beyond safety "standards", (2) finding ways to extend the life of rails, (3) installing rail (new or secondhand) that results in the lowest expected life-cycle cost for each particular location, (4) reducing the cost of replacement activities, and (5) utilizing steel gangs and equipment more efficiently.

Even though these objectives are more detailed than the general goal, they are not in an operational form. Therefore, even more detailed **criteria** are required for achieving the objectives. REPLACER arrives at a replacement plan by satisfying the criteria, thereby satisfying the various objectives and achieving the original goal of balancing costs and services. At each step of the sequence, one of the following criteria has to be satisfied before selecting or rejecting a rail segment:

Maintaining a desirable level of service ("poor performance"): The main idea behind this criterion is to replace rail before its condition reaches a level that does not allow an acceptable level of safety or service. This criterion reflects the level of service constraint (the first detailed objective listed above). Under this criterion, REPLACER selects a segment for replacement if its category in year 1 is a MUST. Even if the "rail performance" is satisfactory, premature rail replacement might still be economically justified. The remaining criteria deal with premature replacement.

Moving rail from high density locations to lower density locations ("cascading"): The economic rationale behind this criterion is to balance the demand for secondhand rail in low traffic density lines and the supply of secondhand rail from high traffic density lines. This criterion reflects the second and third objective mentioned earlier. Under this criterion, REPLACER selects segments based on the willingness-to-sell (WTS) or willingness-to-buy (WTB) economic indicators of a segment (McNeil and Martland, 1989) and the market price of secondhand rail, P. WTS is the minimum price of used rail that would offset any cost associated with premature replacement, and WTB is the maximum price of used rail that would offset any cost associated with using secondhand rather than new rail. These indicators are computed based on equating the net present values of the two cash flows with and without premature replacement. WTS is the indicator computed for high density lines that generate secondhand rail while WTB is the indicator computed for low density lines that accept secondhand rail. The premature replacement benefit (PRB) for a segment is defined as the difference between P and WTS for high density segments and the difference between WTB and P for low density segments. If PRB is positive, then the segment under consideration is selected for replacement under the "cascading" criterion. In the case where PRB is negative, it is referred to as premature replacement cost (PRC).

Coinciding rail replacement activities with other maintenance-of-way activities ("opportunistic scheduling"): The economic rationale behind this criterion is to save on the common costs among rail replacement and other activities such as tie replacement. This criterion basically makes use of economies of scope; it reflects the fourth and fifth objectives mentioned earlier. According to this criterion, REPLACER selects segments for replacement if the replacement benefits (characterized by the common costs avoided) exceed PRC.

Skipping rail replacement activities for a period of time ("skipping"): The economic rationale behind this criterion is to consolidate replacement activities for adjacent rail lines within one maintenance season in order to make better use of the fixed mobilization costs that will be incurred for labor and equipment irrespective of the miles of rail that are going to be replaced in a particular region. This criterion makes use of economies of scale; it reflects the fourth and fifth objectives mentioned earlier. According to this criterion, REPLACER selects a set of segments for replacement in order to skip one or more years of replacement activities, thus eliminating the fixed mobilization costs during the skipped years. The number of years to skip is determined such that the difference between the benefits and PRC are maximized, while avoiding long skipping periods for high density lines to avoid the problem of uncertainty in projecting the performance categories.

Replacing short gaps and Avoiding the replacement of short stretches: It is more economical to replace a few long stretches of rail than to replace many short stretches, simply because of the time and costs associated with starting, stopping, and moving a rail gang. REPLACER therefore combines nearby stretches to obtain longer stretches for the replacement plan and eliminates, where possible, extremely short replacement stretches.

Expanding replacement stretches to reflect replacement policies: At this stage, REPLACER selects additional segments in order to satisfy the two common replacement policies. First, if part of a curve is going to be replaced, then REPLACER selects the whole curve. Second, if rail on one side of a tangent track is going to be replaced, then REPLACER selects the corresponding rail on the other side (these considerations were incorporated after carrying out the case study presented in this paper).

These seven selection criteria and the information they require are represented in the form of rules and frames, respectively, within a knowledge system environment. As can be seen, the concepts are easy to follow and understand. Moreover, the simple sequential structure is very flexible for modification purposes and readily provides justification reasoning. In addition, the several economic indicators used at the different stages are essential for prioritization purposes at the system level.

4. CASE STUDY

REPLACER was used to analyze seven lines covering almost 600 miles of track (nearly 1,200 miles of rail) representing a cross section of traffic on the Burlington Northern Railroad network (a major North American railroad). This amounts to about 7% of BN's mainline tracks. The annual traffic densities ranged from below 10 Million Gross Tons per year to more than 50 MGT. The table below summarizes the results of CM and REPLACER (but without the consideration of "opportunistic scheduling" due to lack of data). REPLACER's results indicate that the major reasons for the recommended replacements were "poor performance" and "skipping", each of which accounted for about half of the plan. There were a few short gaps included in the program along with a few segments selected for "cascading", but together these totalled only a quarter of a mile and made up less than 3% of the plan. Also since there are at least 2 years skipped on each line, REPLACER recommends that no rail be replaced in years 2 and 3.

Line #	1	2	3	4	5	6	7
MGT/yr range	5-10	10-15	≤17	20-40	>50	>50	>40
cumulative MGT range	< 203	< 187	< 672	< 496	< 763	< 836	< 1355
rail-miles examined	367.82	211.18	202.32	211.03	26.38	46.35	88.30
"MUST" rail-miles in yr 1	0.62	0.40	0.90	0.09	0.44	0.87	1.15
"SHOULD" rail-miles in yr 1	1.02	0.76	0.67	0.20	0.02	0.86	1.01
"MAYBE" rail-miles in yr 1	21.26	2.04	0.42	0.85	0.34	0.36	3.50
"OK" rail-miles in yr 1	344.93	207.98	200.33	209.98	25.58	44.26	82.64
"poor performance" rail-miles	0.62	0.40	0.90	0.09	0.44	0.87	1.15
"cascading" rail-miles	0.06	0	0	0.05	0	0	0
"skipping" rail-miles	0.71	0.84	1.00	0.24	0.02	0.82	1.61
"short-gaps" rail-miles	0.07	0	0.03	0.02	0	0	0
"short-stretch" rail-miles	0	0	0	0	0	0	0
total rail-miles selected	1.46	1.24	1.93	0.4	0.46	1.69	2.76
skipping period (yrs)	3	3	3	3	2	2	2

The rail replacement plans produced by REPLACER were compared with those produced by MORRS. For year 1, there is more than 90% agreement between the plans for the nearly 1,200 rail miles compared. For year 2, the plans produced by each are identical, both suggesting no replacement. For year 3, the plans are almost the same. The differences in years 1 and 3 are related mostly to skipping. For example, REPLACER has an upper limit on the number of years to skip for high density lines in order to avoid the problem of uncertainty in predicting rail condition, whereas MORRS does not. On the other hand, MORRS examines the temporal and cumulative effects of multiple factors over a five-year period, whereas some of these effects are not captured in REPLACER.

REPLACER's plan was also compared to the actual replacement plan derived independently by BN for a line of around 110 rail miles. The total mileage selected by BN for replacement was 3.22 miles, 1.28 (non-MUSTs) of which was not selected by REPLACER. If the 1.28 miles are not replaced following REPLACER's recommendation, the savings from deferring replacement would amount to around $40,000 or around 12% of the cost of replacing the 3.22 miles according to the BN plan. Moreover, REPLACER selected 5.48 miles due to "poor performance" that BN did not select. By not replacing these segments, BN risks higher costs from unplanned replacements and train delays that might take place due to slow orders. The benefit of avoiding such delays, therefore, constitutes an additional advantage of REPLACER over the existing judgmental process. In any case, there appears to be inconsistencies between BN's existing decision process and their objectives that can be avoided by using REPLACER.

5. CONCLUSIONS

REPLACER produces effective rail replacement plans using a sequential selection process. This process has four main characteristics: (1) it is accessible to and understood by field engineers and managers, (2) it readily provides justification information, (3) it readily provides a prioritization mechanism to deal with system constraints, and (4) it is flexible for incremental development. All of these attributes give the knowledge system approach an advantage over the network optimization approach. The first two advantages are extremely important for the acceptance and usability of the system by railroad field engineers. REPLACER produced essentially the same replacement plans produced by the network optimization approach and it demonstrated an improvement over BN's existing decision process. Indeed, BN officials adopted REPLACER over MORRS for developing replacement plans for year 1. The main disadvantage of REPLACER is its slow processing speed. However, since this is mainly due to the computer environment in which it resides rather than its structure, REPLACER can be transferred to a data base environment or written using a structured language in order to circumvent the speed disadvantage. Given that REPLACER and MORRS produced essentially the same plans, both of which differed substantially from those produced by BN, the main issue is not to develop a procedure that finds "the optimal" solution, but rather to identify effective objectives and translate them consistently into operational plans using a procedure that will be understood and accepted by management.

REFERENCES

Acharya, D.R., "Using Expert Systems and Network Optimization Techniques for Rail Relay Scheduling," Proceedings of the 1st International Conference on Applications of Advanced Technologies in Transportation Engineering, San Diego, California, USA, ASCE, 1989.

Hargrove, M.B., and C.D. Martland, "TRACS: Total Right-of-Way Analysis and Costing Systems," Computer Applications In Railway Planning and Management, Vol. 1, Computational Mechanics Inc., Billerica, Massachusetts, 1990.

Martland, C.D., S. McNeil, D. Acharya, R.G. Mishalani, and J. Eshelby, "Applications of Expert Systems in Railroad Maintenance: Scheduling Rail Relays," Transportation Research-A, Vol. 24 A, No.1, pp. 39-52, 1990.

McNeil, S. and C.D. Martland, "The Economic Value of Secondhand Rail," Working Paper, Department of Civil Engineering, MIT, 1989.

Mishalani, R.G., "Planning Rail Replacements: Satisfying Multiple Criteria Within a Knowledge System Environment," M.S Thesis, Department of Civil Engineering, MIT, 1989.

Wells, T. R., "Rail Performance Model: Evaluating Rail," American Railway Engineering Association, 1983 Annual Convention, 1983.

POTENTIAL FREEWAY CAPACITY EFFECTS OF ADVANCED VEHICLE CONTROL SYSTEMS

Steven E. Shladover[1]

1. Introduction

Intelligent Vehicle/Highway Systems (IVHS) are intended to improve the productivity (capacity) and safety of our road transportation system. Indeed, those improvements are cited as the primary benefits to justify the costs of IVHS. However, considerable research is needed before the magnitude of those benefits can be estimated with much confidence. Much of the current work on IVHS is actually aimed at developing the knowledge base upon which benefit quantifications can rest.

The PATH program has placed a stronger emphasis on the Advanced Vehicle Control Systems (AVCS) aspect of IVHS than other current programs have, based on the potential that AVCS appear to offer for very significant improvements in capacity and safety. This paper explains how that potential for improvement has been estimated, and illustrates with some sample estimates of the lane capacity that could be achieved by operating vehicles in fully automated platoons.

AVCS was subdivided into three evolutionary stages of development by Mobility 2000 [1]. AVCS I systems will provide driver warning and assistance, to help drivers avoid potential accident situations, while the drivers retain control of their vehicles. AVCS II systems will provide for full automation of vehicle steering and engine and braking functions when the vehicles are operated on special restricted-access links in the highway network, while AVCS III will extend the AVCS II type of automation to complete networks, with automatic routing and scheduling of trips in the automated network.

2. Safety Criteria

The capacity of a roadway system cannot be treated in isolation from its safety. Generally, strategies that seek to squeeze more traffic into the same amount of space reduce safety by forcing vehicles closer together, where they are more likely to collide with each other. However, this is not necessarily the case, as the development of the Interstate Highway system proved that with a large enough improvement in design standards both capacity and safety could increase. AVCS II and III technologies appear to offer another opportunity for such a win-win situation.

The direct technical antecedents of AVCS were the automated guideway transit (AGT) systems that were studied intensively in the 1970s, but of which very few were

---

[1]Technical Director, Program on Advanced Technology for the Highway, University of California at Berkeley, Richmond Field Station, 1301 S. 46th Street, Bldg. 452, Richmond, CA 94804.

implemented. The single most controversial subject in the extensive AGT literature was the choice of the longitudinal spacing safety criterion to apply. Different investigators made radically different assumptions about acceptable safety levels and derived radically different capacity estimates. The differences in safety assumptions generally revolve around different answers to the following questions:

(a) How rapidly does a failed vehicle decelerate?
(b) How long does it take a vehicle to detect a failure of its predecessor?
(c) How rapidly can a following vehicle decelerate using its emergency brakes to avoid a collision?
(d) Are low-impact-velocity collisions tolerable or not?
(e) Should the nominal spacing between vehicles be based on safe accommodation of a single failure or must it accommodate combinations of multiple failures (such as an overspeed or brake failure of the following vehicle when the leading vehicle is decelerating)?

Conservative answers to all of the above questions will produce much larger spacings between vehicles than optimistic answers, with a consequent capacity reduction. Standard railroad practice, which was followed for all AGT systems placed in public service, assumes that failed vehicles decelerate instantaneously (i.e., hit a "brick wall"), even though that is known to be unrealistically conservative.

When we drive our cars, we implicitly make some very optimistic assumptions about the answers to the above questions, based on how closely we follow behind other cars. Either the average driver has an unrealistic perception of his reaction and braking ability or he is willing to accept the risk of the collision, or both. As designers of automated systems, we are forced to be more cautious than this.

Figure 1 shows the capacity of a standard freeway lane, based on the Highway Capacity Manual [2], along with the capacities that would be derived using several different safety criteria. The curve labelled "brick wall" is based on the assumption of the instantaneous stop of the failed vehicle, with the following vehicle needing 0.3 seconds to detect the failure and then applying its brakes at a 0.3 g rate (modest for dry pavement, achievable on wet pavement, but overly optimistic on snow or ice). The other curves show the effects of replacing the brick wall stop with a 1 g stop, and then seek to bracket the results by applying extremely optimistic and extremely pessimistic assumptions about performance, as indicated on the Figure. Note that unless we make extremely optimistic assumptions for the automated vehicle performance, it is difficult to match the capacity we already achieve with present-day drivers.

The AVCS I devices, which are designed to warn or assist drivers, may actually reduce effective highway capacity if they are designed to operate conservatively (that is, to give warnings well in advance of potentially hazardous conditions). On the other hand, if they are designed to operate less conservatively but are effective at significantly reducing the time to respond to a hazard (for example, by automatically applying the brakes), they may make it possible for vehicles to travel more closely together than at present, while still increasing safety. Until these devices are tested extensively, in use by drivers, it will not be possible to determine what their capacity impacts will be.

3. Effects of Failures

We must assume from the start that vehicle failures will occur, and that they will take a variety of unpredictable forms. Given the range of possible failure modes and their effects, it is impossible to configure a practical system that will ensure that collisions never occur. The system design challenge, then, is to decide how to manage the emergency

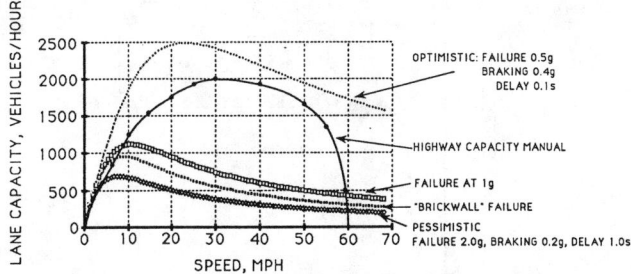

Figure 1. Effect of Safety Criterion on Lane Capacity for Single Vehicles

conditions to minimize, first of all, the deaths and serious injuries that will occur, and secondly the property damage. The deaths and serious injuries result from collisions that occur at high relative speeds, but not from "fender benders". The primary objective, then, should be to avoid the high impact speed collisions.

Although it is not obvious at first glance, a little careful study, which can be borne out by analysis, shows that high impact speed collisions can be avoided if the vehicles are either very close to each other or very far apart when the failure occurs, as shown in Figure 2. The far apart case is easy enough to understand, because it means that the following vehicle is able to stop before it hits the failed vehicle. The relative safety of very close spacings derives from the fact that the collision impact occurs almost immediately after the failure, before the failed vehicle has had a chance to lose much speed. In the extreme case of mechanically coupled vehicles, there is no impact because they simply decelerate as a single unit. Figure 2 shows that the impact speeds can be quite modest if the vehicles are only one meter apart when the failure occurs, but can be quite severe at the intermediate spacings. The relationships embodied in this figure were originally reported in [3], and the analysis approach is documented in [4].

The results shown in Figure 2, representing only one of many possible combinations of assumed parameter values, indicate the desirability of operating vehicles either very close together or very far apart, but not at the intermediate spacings. Figure 1 already showed that capacity could not be increased by operating individual vehicles at safe spacings, but Figure 2 provides the basis for grouping vehicles in platoons at very close spacings, while maintaining large spacings between the platoons.

4. Lane Capacity Achievable with Platoon Operations

Figures 3 and 4 show estimates of the lane capacities that could be achievable by operating vehicles in platoons, with one meter spacings between successive vehicles within each platoon. Figure 3 is based on rather conservative assumptions about failure conditions and vehicle performance, while Figure 4 is based on more optimistic assumptions. Each figure shows how lane capacity would vary with operating speed and with the average number of vehicles per platoon. It is apparent that increasing platoon length increases capacity and also increases the velocity at which the capacity reaches its peak.

The capacity estimates of Figures 3 and 4 are based on a simple kinematic analysis to ensure that the spacing between platoons corresponds to the difference between the stopping distances of the failed vehicle (platoon) and the follower, plus the distance that would be traveled by the follower during the time needed to detect the emergency condition. The capacity is derated by a further 20% to allow extra gaps for lane changing

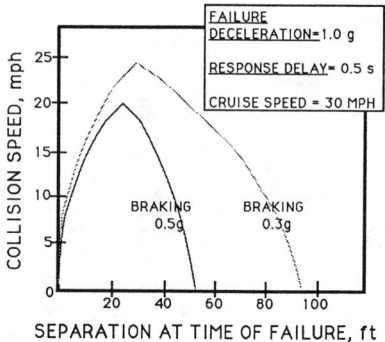

Figure 2. Dependence of Collision Speed on Spacing Between Vehicles

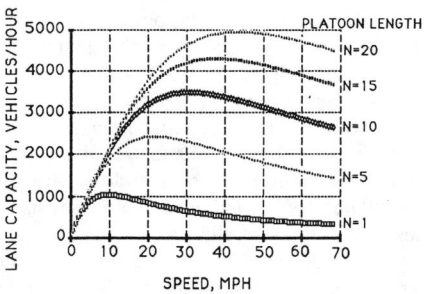

Figure 3. Lane Capacity of Platoon System Under Conservative Assumptions (2g Failure, 0.3g Emergency Braking, 0.3s Reaction Delay)

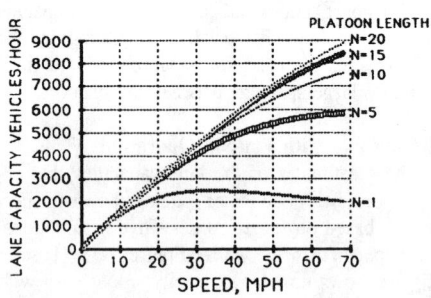

Figure 4. Lane Capacity of Platoon System Under Optimistic Assumptions (0.5g Failure, 0.4g Emergency Braking, 0.1s Reaction Delay)

and merging operations, based on the results of the detailed simulations of simple merges in [5]. However, much more elaborate simulations will be needed to determine whether that is a valid derating factor to apply in a complex network environment.

The important result to take away from this analysis is that even with the more conservative safety assumptions of Figure 3, the lane capacity with automated platoons could easily be at least double the present-day freeway lane capacity. The implications of that level of capacity increase for the operation of the entire road system could be profound. The potential productivity increase is so significant that it appears to clearly justify the investment of research resources to evaluate the feasibility of accomplishing the close-formation platoon automation.

The eventual capacity increase from full automation of a multi-lane freeway could be larger than that implied by the single-lane analysis. The additional capacity gain could potentially arise if automatic lateral (steering) control is accurate enough to make it possible to reduce lane widths below their present levels and thereby fit more lanes on the same rights of way and structures that we use today [6]. Research on lateral control presently underway will help to shed light on how much of a lane width reduction (if any) could be expected.

5. Conclusions

Some very simple analyses have been performed to produce initial estimates of the potential capacity increases that could be gained by use of AVCS II or III technology, with vehicles operated in automated platoons. These potential increases are sufficiently large to indicate the need for concerted research to seek to prove or disprove the estimates by developing experimental test data and higher fidelity analyses. The underlying assumptions about parameter values and operating protocols need to be examined carefully and adjusted as necessary to produce more credible results, which can then be used to develop policy recommendations.

6. Acknowledgement

This work was performed as part of the Program on Advanced Technology for the Highway (PATH) of the University of California, in cooperation with the State of California, Business and Transportation Agency, Department of Transportation, and the United States Department of Transportation, Federal Highway Administration.

7. References

1. Mobility 2000. Proceedings of a National Workshop on IVHS Sponsored by Mobility 2000, Dallas, Texas, March 19-21, 1990, hosted by Texas Transportation Institute.
2. Transportation Research Board, Highway Capacity Manual, TRB Special Report No. 209, 1985.
3. Shladover, Steven E., "Dynamic Entrainment of Automated Guideway Transit Vehicles", High Speed Ground Transportation Journal, Vol. 12, No. 3, 1978, pp. 87-113.
4. Shladover, Steven E., "Operation of Automated Guideway Transit Vehicles in Dynamically Reconfigured Trains and Platoons", Volume II, Report UMTA-MA-0085-79-3, July 1979, 220 pp.
5. Shladover, Steven E., "Operation of Merge Junctions in a Dynamically Entrained Automated Guideway Transit System", Transportation Research, Vol. 14A, 1980, pp. 85-112.
6. Shladover, Steven E., "Roadway Electrification and Automation Technologies", Journal of Transportation Engineering, Vol. 116, No.4, July/August 1990, pp. 417-425.

A Preliminary Systems-Level Evaluation of Automated Urban Freeways

Robert A. Johnston[1] and Dorriah L. Page[2]

ABSTRACT

Transportation planners predict that traffic congestion will increase dramatically in the future. With 63% of all congestion on urban freeways, much of it in high-density areas, traditional solutions such as lane additions, often may not be feasible. Automation is a proposed solution to this problem, as it increases freeway capacity without building new lanes. The technological capabilities for vehicle and highway automation are advancing, and with these advances the need for regional transportation systems modelling grows. This paper examines the impacts of urban freeway automation on regional travel behavior in the Sacramento, California region. An urban transportation systems model (MINUTP) is utilized to project the effects of automation on VMT, lane-miles of congestion, hours of delay, transit trips, average trip length, and auto occupancy rates. Four scenarios are examined: A 1989 Base Case, 2010 No-Build, 2010 Freeway Automation, 2010 New HOV Lanes, and 2010 Light Rail Expansion.

INTRODUCTION

Congestion is measured in terms of the reduction in average speed relative to that under free-flow conditions (Altshuler, 1979, p. 327). Congestion to the individual traveler is perceived primarily in terms of increased travel costs (Chen and Ervin, 1989). Automation technology serves to decrease travel time by increasing lane capacities and speeds, and decreasing vehicle headways. In this study, full automation of all existing freeway lanes in the central Sacramento urban region was modeled.

We define full automation as consisting of an urban freeway system with dual-mode vehicles operating at

---
[1] Prof., Div. of Envir. Studies, and Inst. of Transp. Studies., Univ. of California, Davis, CA 95616.
[2] Graduate Researcher, Div. of Envir. Studies, Univ. of California, Davis, CA 95616.

speeds of 80 mph at headways of 0.5 seconds under onboard and roadside computer control (Johnston, et al., 1989). Earlier studies by the authors (Johnston, et al., 1989; Johnston and Page, 1991) raised questions as to systemwide effects on the network. For example, CBD offramps and arterials could be overloaded. This is an important issue, because offramps and arterials are located adjacent to high-priced land and buildings in many CBDs. Widening many ramps and arterials could be very expensive and politically difficult.

Even more problematic is the likelihood that automation will induce more auto travel (VMT). The 1990 amendments to the Federal Clean Air Act require many large urban areas to control mobile emissions so as to "fully offset" any growth in VMT and trips. The California Clean Air Act goes farther and requires a "substantial" reduction in the rate of growth of VMT and an average vehicle occupancy of 1.5 during commute hours in the largest urban areas. In this paper, we report only model results for travel behavior; yet, viewed in the context of these air quality laws, the tradeoff between reducing delay and inducing more travel is of great interest to transportation engineers.

METHODS

MINUTP is an urban transportation systems model run on a microcomputer. MINUTP creates the zonal production and attraction data by means of a linear regression model developed in trip generation and distributes trips to all zones using a standard gravity model. A module is used to convert the production-attraction trip matrix to an origin-destination matrix, and the minimum impedance paths from zone to zone are developed. Mode split is developed by a logit model, the trip matrix is read and trips are assigned to the network by congestion-constrained iteration.

For this study, the Sacramento Regional Transit Model was used. The travel behavior models were developed by DKS Associates, using 1967 Sacramento O-D data and Seattle area equations, based on recent surveys. Zone and network files were created from 1989 data (Sacramento Systems Study, 1991). A 1989 Base-Year case was validated, and several future scenarios were run, including: 2010 No-Build, 2010 New HOV Lanes, and 2010 Extended Light Rail Transit (LRT).

We designed a 2010 Freeway Automation scenario. Only those freeway links at level of service (LOS) E or worse in the 2010 No-Build scenario were automated (see Figure 1).

Link changes were made that increased characteristics of the automated freeway lanes were altered to reflect changes in the volume-to-capacity ratio. The appropriate delay was calculated from the curve based on the V/C ratio and added to the congested impedance.

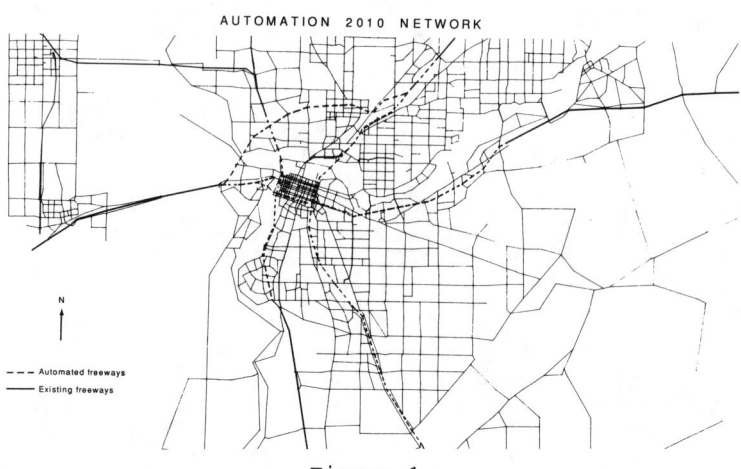

Figure 1

RESULTS

Compared to the 2010 No-Build case, the HOV scenario reduced VMT and trip length and slightly increased vehicle occupancy (Table 1). The HOV alternative also dramatically decreased hours of delay on freeways and on other roads (expressways, ramps, arterials and collectors), more so than the other scenarios. The LRT scenario slightly reduced VMT and had little or no effect on trip length and vehicle occupancy. Automation increased VMT and trip length, compared to the other scenarios, and reduced vehicle occupancy, as well as reducing hours of delay on freeways.

TABLE 1
SUMMARY TRAVEL CHARACTERISTICS (DAILY)
(Changes from 1989 Base Year)

Characteristics	ALTERNATIVES			
	No-Build	Automation	HOV Lanes	LRT
VMT	+ 77%	+ 88%	+ 73%	+ 75%
Average Trip Length (mi.)	+ 13%	+ 19%	+ 11%	+ 13%
Vehicle-Hours of Delay				
On Freeways	+ 986%	+ 824%	+ 456%	+ 865%
On Others	+ 426%	+ 385%	+ 321%	+ 410%
Average Vehicle				
Occupancy	+ 1.5%	- 0.8%	+ 2.0%	+ 0.7%
Transit Ridership Increase	+ 24%	+ 18%	+ 98%	+ 125%
Lane-Miles of Congestion (LOS E/F)				
On Freeways	+ 1818%	+ 1193%	+ 869%	+ 1751%
On Other Roads	+ 491%	+ 777%	+ 347%	+ 486%

Automation did not reduce delays on freeways as much as the HOV scenario, however, and actually increased delays on other roads. Automation resulted in fewer transit riders than the No-Build scenario, whereas the HOV and LRT scenarios substantially increased transit ridership.

Compared to the No-Build scenario, lane-miles of freeway congestion are reduced in the Automation and LRT scenarios, and to a much greater degree in the HOV scenario (Tables 1 and 2). This is because the HOV scenario reduces VMT whereas the Automation scenario does not. Automation also does not perform as well as the HOV scenario because of the increased congestion levels on the freeways beyond the ends of the automated lanes. Automation increased congestion on other roads because of overloaded ramps and nearby arterials.

TABLE 2
TRANSPORTATION SYSTEM CHARACTERISTICS (DAILY)

Characteristics	1989 Base Year	ALTERNATIVES			
		No-Build	Automation	HOV Lanes	LRT
Vehicle-Miles of Travel	31.5 M	55.9 M	59.2 M	54.6 M	55.3 M
Average Trip Length (miles)	9.3	10.5	11.1	10.3	10.5
Vehicle-Hours of Delay					
On Freeways	23,000	249,905	212,749	128,000	222,000
On Other Roads	80,000	421,000	388,312	337,000	408,000
Average Vehicle Occupancy	1.29	1.31	1.28	1.32	1.30
Linked Transit Trips	58,300	72,455	68,896	115,588	131,467
Lane-Miles of Congestion (LOS E/F)					
On Freeways	33	633	427	320	611
On Other Roads	106	627	930	474	622
Total	139	1260	1357	794	1233

DISCUSSION AND CONCLUSIONS

In previous research we deduced that the automation of urban freeways could increase VMT and cause overloading of offramps and adjacent arterials. That work led us to model the systemwide impacts of automation on the Sacramento network. We find that in a standard simulation, urban freeway automation slightly increases total roadway congestion, increases average trip length, and decreases transit ridership compared to the future No-Build case. Automation, however, also decreases average vehicle occupancy and increases VMT. These latter effects may make air quality planning more difficult in many urban areas in the U.S. and seem to be contrary to the law in California.

Outside of California, automation could be implemented if net mobile emissions were reduced, that is if the higher automated freeway speeds reduced emissions more than the other freeway and road congestion and higher VMT increased emissions.

In the future, we will design an improved freeway automation scenario, in which we include partial freeway automation (2 lanes in each direction) for a few miles past the automated segments and widen ramps and nearby arterials, where feasible. That should reduce congestion and emissions per vehicle-mile. We will also incorporate the effects of road congestion on auto ownership and on residential location. Impacts on vehicle emissions will be projected with a microcomputer version of DTIM. We will also analyze average trip times for all trips, by purpose, to see if the time-cost savings due to automation exceed the likely costs of automation.

REFERENCES

Altshuler, A., The Urban Transportation System. Politics and Policy Innovation, MIT Press, Cambridge, Mass., 1979.

Chen, K. and Ervin, R.D., "Developing a Program in Intelligent Vehicle-Highway Systems in North America", Proceedings of the 20th International Symposium on Technology and Automation, Volume 2, Florence, Italy, May 29-June 2, 1989.

Johnston, R.A., DeLuchi, M.A., Sperling, D., and Craig, P.P., "Automating Urban Freeways: Policy Research Agenda," Journal of Transportation Engineering, Vol. 116, No. 4, 1990.

Johnston, R.A. and Page, D.L., "Automated Urban Freeways: A Financial Analysis of Costs and Benefits for User Groups," Report, University of California, Davis, CA, April 1, 1991.

Sacramento Systems Planning Study, "Task 4.4: Model Run Results,"Parsons, Brinckerhoff, Quade and Douglas, Inc., Feb., 1991. (Sacramento, CA).

This research was funded by the Caltrans PATH program and we gratefully acknowledge that support.

Institutional Barriers to IVHS Introduction

Michael L. Patten[1] and John M. Mason, Jr.,[2] M. ASCE

Abstract

The introduction and application of Intelligent Vehicle/Highway Systems (IVHS) will help to mitigate the adverse safety, environmental, and economic impacts of traffic congestion on selected roadways in the United States. Before these advanced transportation technologies are implemented, numerous nontechnical aspects of product development and introduction must be addressed. This paper identifies critical institutional barriers that need to be resolved before IVHS technologies can be implemented significantly.

Introduction

Intelligent Vehicle/Highway Systems, often called "smart cars" and "smart highways," involve the interaction of advanced communications technology, computers, electronic information displays, warning systems, and vehicle/traffic control systems (USDOT 1990). The implementation of IVHS can help to mitigate the adverse safety, environmental, and economic impacts of traffic congestion on selected roadways in the United States and provide for safer traffic flows (Mobility 2000 1990).

To date, most of the research on IVHS in the United States has focused on developing and demonstrating individual IVHS technologies. For IVHS technologies to be integrated into a truly national system and to ensure that

---

[1]Research Assistant, The Pennsylvania Transportation Institute, The Pennsylvania State University, Research Building B, University Park, PA 16802.

[2]Director, Transportation Operations Program, and Associate Professor of Civil Engineering, The Pennsylvania Transportation Institute, The Pennsylvania State University, Research Building B, University Park, PA 16802.

the introduction of IVHS proceeds smoothly, many nontechnical aspects of product development and introduction must be addressed. This paper identifies critical institutional barriers that need to be resolved before significant progress toward the implementation of IVHS technologies can be realized.

Protocols and Standards

At its highest level, IVHS will rely on fast, accurate communications between each element of the system. Vehicles from various manufacturers will have to communicate with each other as well as with the roadway. Therefore, protocols and standards must be developed for both hardware and software to ensure compatibility and safe operation. The resulting standards must be acceptable to both the highway system operators and the vehicle manufacturers.

The early institution of standards for new technologies is important because, "if standards come first, quickly, and are driven by industry and market need, there is a much greater chance of success. Products emerge more rapidly, and intercompatibility of equipment from multiple vendors is guaranteed" (Kirson 1991).

The TRB IVHS Task Force has reported that a major impediment to the widespread implementation of IVHS is the absence of standards. The TRB Committee on Communications unanimously recommended that work on IVHS communications standards be initiated immediately (TR News 1991). Examples of effective and timely standards can be found in the commercial aviation industry and in the operating standards for computers.

Proprietary Rights to Technology

A key aspect of the envisioned IVHS program is the substantial involvement of private companies during the research and development phase (Mobility 2000 1990). If private enterprise is expected to make these large-scale investments, it is only reasonable for them to desire a fair return on their investments. Likewise, the firms developing IVHS technologies will have to collaborate to produce compatible equipment and systems.

Before firms can be expected to participate in such large-scale cooperative ventures, a fair and equitable process must be established that will, on the one hand, encourage the sharing of innovations while at the same time protecting the proprietary rights of the developers. Without the assurance that they will retain the rights to new technology and gain a fair return on their investments, few firms will be willing to risk their competitive advantage by participating in joint ventures. If this potential barrier to cooperation can be overcome, firms will be able to learn from one another, share technology, and minimize duplication of effort (Norman 1990).

Product and Tort Liability

As IVHS technologies become more pervasive, there is the potential for a shift in liability for automobile accidents from primarily vehicle-driver negligence toward the equipment and vehicle manufacturers and the organizations operating the systems. The risks of increased product liability and the associated costs of adequate liability insurance may deter many firms and, in some cases, government entities from entering the IVHS field (Syverud 1990).

Some limits to liability risk for the equipment and system developers and operators must be instituted. These limits can take the form of narrower definitions of negligence, limitations of awards, limits on joint liability, and federally subsidized insurance plans (Norman 1990).

Syverud (1990) has pointed out that the experiences of other new, ground-breaking technologies provide successful examples of how the liability issues involving IVHS can be overcome. Immunity from specific claims (as in the case of nuclear power plant contractors), liability limits (nuclear power plant disasters and international commercial aviation), mandatory risk pooling (nuclear power plants), and indemnification by the federal government (nuclear power plants) are all approaches that should be studied as having potential to reduce liability for the manufacturers and operators of IVHS equipment and systems.

Antitrust

Cooperative research and development efforts, which will be necessary for successful development of IVHS, may be illegal under the current U.S. antitrust laws. Because of the interactive nature of the IVHS component systems, the manufacturers of the various subsystems should be encouraged to work together to ensure compatibility of the final products. A degree of flexibility is needed to allow IVHS developers to work together up to the point of individual product development without being in violation of antitrust law or regulations (Norman 1990).

Inter-Jurisdictional Relations

Full implementation of IVHS will require unprecedented cooperation among many different governmental entities. Federal, State, and local transportation system operators and enforcement agencies must work together to meet common goals. Questions that need answers include: Who is responsible for the many individual IVHS subsystems? Who will own and maintain the equipment? What procedures will be followed to clear incidents? Who will be the lead agency in setting policy for the overall system?

Financial Considerations

The Federal Highway Administration (FHWA) has estimated that it will cost in excess of $35 billion through the year 2010 to develop, test, and deploy a nationwide IVHS system (Saxton and Bridges 1991). In addition to capital investments, the costs of system operation and maintenance must be anticipated. The added installation and maintenance costs will be passed on to vehicle owners who wish, or are required, to have IVHS systems in their automobiles. Several issues must be resolved: who will pay for the research and development costs of IVHS; will IVHS services be provided "free" or will some type of "user fee" be charged; and the equity problems inherent in requiring IVHS equipment to be installed in all vehicles.

Other Issues

Several other issues of importance need attention as IVHS technologies are introduced. One involves the decision of whether or not to allocate specific broadcast frequencies to IVHS and, if so, what they will be. Because many IVHS technologies are designed to make use of the already crowded broadcast frequency spectrum, it is critical that this decision be made as soon as possible. If specific frequencies are set aside early, IVHS systems can be designed to take the best advantage of those frequencies.

Maintaining the privacy of individuals who use IVHS systems is also of crucial importance. The most advanced stages of these technologies incorporate real-time identification and position monitoring of individual vehicles. Procedures need to be developed that will ensure the confidentiality of individual vehicle movements while allowing a free flow of the information required for system operation.

To gain the full benefit available from the different
IVHS technologies, substantial amounts of accurate, up-to-
date traffic information are required. This information
needs to be collected, interpreted, stored, and communicated to all system users and operators. Furthermore, in
the advanced stages of IVHS, the acquisition and dissemination of the data must occur in "real-time." A network
compatible with and interconnected to all four IVHS
components is required to coordinate and facilitate the
free flow of real-time traffic information to all users.

Summary

This paper has identified critical institutional
barriers, for the public and private sectors, that must be
resolved before significant progress toward the implementation of IVHS technologies can be realized. Solutions to
the problems outlined are not always readily apparent.
Significant and ongoing research regarding these "soft-
side" barriers to successful deployment of IVHS, and the
development of pragmatic solutions to mitigate their
effects, should be given top priority when establishing a
comprehensive IVHS program.

Appendix 1: References

Kirson, A. M. (1991). "RF Data Communications Considerations in Advanced Driver Information Systems." ITEE
Transactions on Vehicular Technology 40 (February), 51-55.

Mobility 2000. (1990). Mobility 2000 Presents: Intelligent Vehicles and Highway Systems (1990 Summary).
Texas Transportation Institute, College Station, TX.

Norman, M. R. (1990). "Intelligent Vehicle/Highway
Systems in the United States--The Next Steps." ITE
Journal 60 (November), 34-38.

Saxton, L. G., and Bridges, G. S. (1991). "Intelligent
Vehicle-Highway Systems: A Vision and A Plan." TR News
152 (January-February), 2-7.

Syverud, Kent. (1990). "Liability and Insurance Implications of IVHS Technology." In Automated High-
way/Intelligent Vehicle Systems: Technology and Socio-
economic Aspects, SP-833. Society of Automotive Engineers, Warrendale, PA, 83-96.

TR News. (1991). "IVHS Activities Within TRB." TR News
152 (January-February), 6-7.

U.S. Department of Transportation. (1990). Report to
Congress on Intelligent Vehicle-Highway Systems. Report
No. DOT-P-37-90-1. U.S. Government Printing Office,
Washington, DC.

CADD - Database Application for Facility Inspections

Axel J. Pollak
Rosalind Pierce-Spring
Sverdrup Corporation
New York, New York

Development:

In 1989, Sverdrup Corporation was hired to perform the Comprehensive Inspection of the Queens-Midtown and Brooklyn-Battery Tunnels in New York City. The scope of this project included the structural, mechanical, plumbing, and electrical inspections of approximately 30,000 linear feet of vehicular tunnel, seven ventilation/service buildings, and two parking garages.

From the onset of the project it was clear that a very large number of records would be collected by a large, multi-disciplined field crew. It was imperative to design a unified data collection and retrieval system, which would allow for efficient study, evaluation, cost estimating, budgeting, and management. Unified collection was accomplished by the use of preprogrammed, hand held data collectors. The use of relational database technology provided the ideal vehicle for management of the system.

As development of the database program progressed, the demands for the system requirements grew to include the storage and retrieval of photographs and finally it became clear that the information that was being collected by the inspection crews could be linked to a CADD package to automatically produce graphical representations of the subjects of interest. This included the automatic production of plans and wall elevations of the interior tunnel surfaces, as well as the production of histograms depicting the results of queries pertaining to specific inspection findings.

The result of this development was the creation of a microcosm of the tunnel facility within the computer. This system allows the engineer to sit at the computer and virtually walk the facility, by pressing the cursor keys or by using a mouse. In doing so, the engineer has the ability to filter the database contents for material of interest to him.

Additionally, the system can be expanded to include a facility maintenance program, which automatically generates maintenance work orders and keeps track of the facility inventory.

System Requirements:

While the program for the CADD-linked Database program will run under any MS-DOS or PC-DOS system, the data sorting and graphics program requirements virtually dictate a high capacity fast machine. For the above referenced project the following equipment was used:

1. **Programmable Hand-Held Data Collectors** - featured 64K RAM, a removable battery pack, a full function alpha-numeric keypad, a large liquid crystal display and a bar code laser scanner.

2. **Computer** - the personal computer (80386 processor and 80387 math co-processor) operated at 25 MHz, with zero wait state, had 8 Mb of RAM and featured 64K high speed cache memory, a 100 MB hard drive, a high resolution graphics card and a 19" color monitor.

3. **Digitizing tablet** - for use in creating drawings and for the option selections of the CADD graphics portion of the program.

4. **Optical disk** - this high capacity storage device was principally used to store scanned inspection photos, scanned original drawings, scanned equipment specifications, and scanned catalog cut data.

The Intergraph^R MicroStation software package was chosen as the engine to drive the graphics portion of the program. This choice dictated dBase file compatibility, and required the database program to be written in Clipper and the graphics linkage to be written in Microsoft^R C. The hand held data collectors were programmed using the manufacturer's own programming language.

CADD - Linked Database Program:

The following information was collected and stored as a separate database record by inspectors in the field, using the pre-programmed data collectors:

1. Building or Tunnel Identification
2. Building Floor or Tunnel Station
3. Building Floor Section or Tunnel Surface

4. Reference Code (code for item being inspected, i.e. WT-Wall Tile, EP-Exhaust Port, CS-Ceiling, slabs, etc.)
5. Deficiency Code (code for deficiency, i.e. SP-Spall CR-Crack, etc.)
6. Vertical and Horizontal Location of Deficiency
7. Estimated Quantity of Deficiency
8. Brief Notes on Condition
9. Recommended Repair
10. Recommended Repair Action (I-Immediate, P-Priority, R-Routine, N-None)
11. Photograph frame and roll numbers with a caption for each photo.

In order to ensure uniform data collection among the large field force, the data collectors were programmed to prompt the inspectors for the above information. The inspector's initials, the date and time were stored automatically with each database record. The information could be input into the data collectors by keying or by scanning bar code labels from a menu tablet. This technique proved to be far superior to manually recorded (hand written) inspection forms, because the information could be recorded far quicker and required no separate inputing into the computer. At the end of the daily inspection, the information in the data collectors was uploaded directly into the computer using an RS-232 serial communications connection.

The program allows for manipulation of the relational database to produce output of information in any desired manner, i.e. by location, by deficiency, by action, etc. This ability proved to be especially beneficial in the preparation of the Inspection Report and would not have been feasible using hand-written field notes. Further, this method of storage and retrieval made it easy to produce inventory records for the client. For example, one could sort all FVN (fire valve niche) in the Brooklyn-Battery Tunnel. A list of these could be produced, together with other pertinent information, such as location by stationing, etc.

All the photographs taken during the course of the inspection were scanned and are stored on the optical disk. The color photos can be viewed in either the Database or the Graphics portion of the program, providing a life-like view of the existing tunnel conditions.

The graphics portion of the program, since it makes use of the database, can be manipulated to show data in any desired manner. Graphics, for screen or plotted output, are generated, so to speak, on the fly, by superimposing, for a particular location, the graphical representations of each of the database items in the proper relation to each other. For the tunnel inspection the rapid generation of graphics was accomplished by indexing database items by fifty foot tunnel sections.

Capabilities of the CADD-Linked Database Program:

The program has many powerful features for sorting and presenting the inspection data as discussed below.

Queries

Queries may be established in order to retrieve a certain range of data. For example, in the Brooklyn-Battery Tunnel, it was necessary to locate all spalled and delaminated ceiling tile on the roadway ceiling requiring immediate or priority action. The computer input screen which facilitates such a query prompts the user to pick from multiple choice listings as follows: "All CT (reference code for ceiling tile) with SP or DEL (deficiency code for spall or delamination) and I or P (action on an Immediate or Priority basis). From this, all records which match the criteria are listed, along with the other information contained in their records, such as location, quantity, etc. The queries or any other type of data listing may be presented in various types of custom reports.

Reports

The data listings, queried or otherwise, can be presented in various types of report formats. The most frequently used report formats were included in the program as menu options. One frequently used report format had been devised by the client to be included in the inspection report. This format omitted certain fields of information, such as the date of the inspection and the inspectors initials, and included only the items that the client felt were pertinent for their use. Another report format, used most often by those writing the inspection report, sorted the data by action. This format facilitated the writing of the inspection report, when all items requiring the same action had to be discussed in the same report text section.

Estimates

Another feature of the program is the ability to assist in the rapid generation of cost estimates for a particular maintenance contract. The procedure used is to match the particular unit costs of typical repairs to the existing field conditions.

Histograms

Perhaps the most interesting aspect of the program is the ability to produce histograms of selected inspection data. Plotted along the tunnel alignment (See *Figure 1*) these histograms can visually enhance information, which is otherwise not obvious. The histograms facilitate the ability to see trends and patterns in certain deficiencies and permit an analysis of the causes of certain types of deterioration.

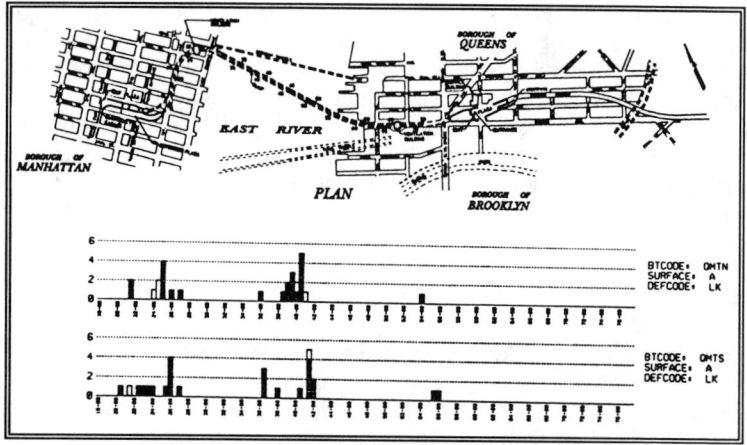

Figure 1

Further Applications

While this particular application was used for tunnel facilities, the techniques presented herein can be applied to any facility, be it major bridges, airports, industrial facilities or complete highway systems. The future will demand that such database systems be put in place at the design stage and that contract drawings and specifications be structured for inclusion in the database.

Microsoft is a registered trademark of Microsoft Corporation.
Intergraph is a registered trademark of Intergraph Corporation.

AUTOMATED ANALYSIS OF PAVEMENT DISTRESS DATA

Haris N. Koutsopoulos[1], Rabi G. Mishalani[1], Allen B. Downey[1]

Abstract: A system model that can be used for the analysis of pavement distress data collected by various visual devices (e.g. photographic film) is presented. The models operate at two levels: a) the microscopic level for processing of the images and classification, and b) the macroscopic level for reduction and aggregation of the information (based on spatial relationships among distresses) so that it can be used for pavement management purposes.

1. INTRODUCTION

The collection and analysis of pavement distress data is a primary component of any pavement management system. Currently pavements are usually manually inspected for collection of surface distress data. This form of inspection is slow, labor intensive and expensive with low sampling rate. It is also subjective with low consistency between surveys and poor repeatability. These drawbacks result in inaccurate pavement condition assessment and subsequently wrong maintenance decisions. To eliminate the drawbacks of manual inspection, automation of the process has been suggested. Various systems exist or are under development to record the surface of the pavement on video tape or photographic film and subsequently analyze it either manually in a laboratory or automatically using image processing and pattern recognition methods.

Automation of the data collection process provides, in most cases, a large amount of data. Data at this "microscopic level" are useful for developing detailed "distress mapping" of a particular highway segment. Such detailed mapping is a necessary input to (robotic) systems that may automate certain maintenance activities (e.g. crack filling). However, the main use of the collected data is as input to other components of the infrastructure management process, namely pavement evaluation, deterioration modeling, future condition prediction, and appropriate maintenance and rehabilitation strategy selection. This creates the need for methods to aggregate the information so that it can be used more effectively by the agencies. Consequently, in addition to automatically processing the individual images collected by the various technologies, another important and practical need is for methods to aggregate individual images in a systematic way, without loss of information, so that it can be further used in the later stages of the infrastructure management process.

In this context the objective of this paper is to present a prototype system for **automated** analysis of pavement data that has the potential to provide a unified platform to various existing technologies and bridge the gap between the microscopic data collected and the macroscopic data needed for decision making. Such a system requires models at two levels:

---
[1]Massachusetts Institute of Technology, Department of Civil Engineering, Room 1-179 Cambridge MA, 02139

the **microscopic** level, which is concerned with the processing of individual images; and the **macroscopic** level which, based on the results obtained from the microscopic level, models spatial relationships among distresses and summarizes the information in a useful way.

Figure 1 presents the necessary components of an automated inspection system and their relation to other elements of the pavement management process mentioned above. The bold blocks represent models (developed or under development) directly related to this paper. At the microscopic level, such models include **image enhancement** (improvement of digital images and preparation), **segmentation** (extraction of objects of interest from the background), **classification** to a distress class and **quantification** (measurement of the extent and severity of the distresses). The macroscopic level includes models for the identification and development of spatial relationships among distresses.

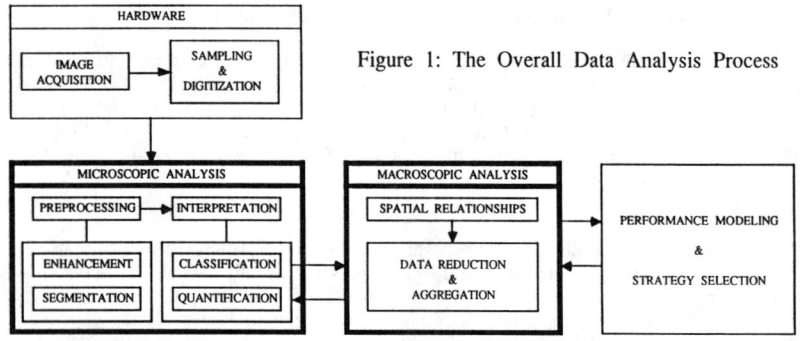

Figure 1: The Overall Data Analysis Process

2. MICROSCOPIC LEVEL MODELS

Methods and algorithms that have been developed for preprocessing and classification of pavement image analysis are summarized below:

Preprocessing: A method based on background subtraction has been developed for image enhancement. Subsequently segmentation on the enhanced images is performed. Segmentation is a very important and difficult preprocessing step. The effectiveness of segmentation greatly depends on the quality of the acquired images, which is affected by the hardware configuration employed for image acquisition. Various approaches for segmentation of pavement images have been examined in detail (Koutsopoulos et al). The methods that have demonstrated potential and gave promising results are a) a modification of the general relaxation method; b) a thresholding method which relates the threshold for a given image to statistics of the corresponding histogram through a regression equation; and c) a simple segmentation method that assigns one of four values (0-3) to each pixel based on its probability of being an object pixel.

Interpretation: In general classification methods use a representation of the object to be classified by a **feature vector**. The choice of the descriptors included in the feature vector is based on their discriminatory power, i.e. their ability to differentiate among the various classes. The descriptors suggested for the feature vector of pavement distresses include global characteristics, such as area, inertia ratio, angle of inclination and aspect ratio and building blocks (primitives), such as line segments, junction points etc, of which the various distress classes are comprised. Therefore the basis of our approach for global classification is that the presence of a particular primitive contributes to the likelihood of the distress belonging to a particular type. Furthermore, the use of global descriptors such as area, inertia ratio, aspect ratio and angle facilitates the classification. The design of the classification strategy aims at addressing the most difficult task --distinguishing between alligator and block cracking. It is based on the observation that the primary distinction between alligator and block cracking is one of scale and texture. Block cracks tend to be patterns of distinct linear segments outlining large square blocks, and they are often dominated by a traverse or longitudinal crack. Alligator cracks are typically an intricate pattern of small segments with many diagonal elements forming small polygon shapes. The primitives used in the feature vector describe textural and structural characteristics of the various distresses and capture the above differences between alligator and block cracking.

The determination of global descriptors is a quite straightforward task. To identify primitives the images are subdivided into a n by n array of regions. Each region is classified according to the primitive element it contains. The primitives used in the analysis include linear segments, joints and plain regions.

The method that has been used for classification is based on the logit model (Ben-Akiva and Lerman, 1985). The various classes constitute the alternatives and the systematic utility of each class is defined using the descriptors in the feature set as alternative specific variables. The probability that an object with feature vector x belongs to class i is given by:
$Prob(i) = e^{V_i} / \sum_m e^{V_m}$ where V_i is the systematic utility of class i defined as: $V_i = \sum_{j=1} a_{ij} x_j$

Application: The above methods have been applied to a set of 79 images (39 undistressed, 21 transverse, 6 longitudinal, 6 alligator and 7 block cracks), provided by PCES. Figure 2 demonstrates the image resulting from the segmentation of a block crack and the classification of the regions to primitives: linear (-, |, \, /), joint (*), plain.

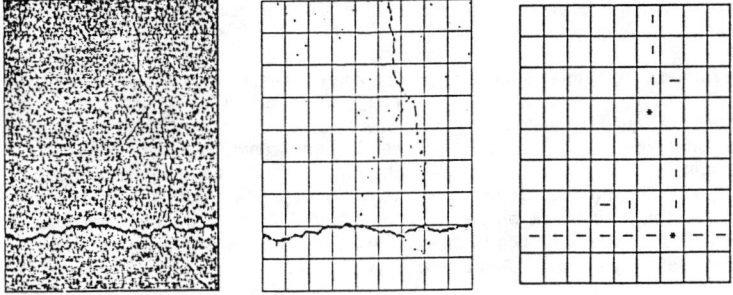

Figure 2. Image of a block cracking, segmentation and primitive classification

The table that follows summarizes the results of the classification of the 79 images:

Distress Type	Total Number	Classified as				
		Alligator	Block	Longitudinal	Transverse	Plain
Alligator	6	5 (83%)	1	0	0	0
Block	7	0	5 (72%)	0	2	0
Longitudinal	6	0	0	6 (100%)	0	0
Transverse	21	0	1	0	19 (90%)	1
Plain	39	0	0	0	0	39 (100%)

The results are very promising. The system correctly distinguishes between alligator and block cracking. The few misclassifications that occurred are not serious. For example, a faint transverse crack with a small patch of alligator cracking was classified as an alligator crack, a classification which, for all practical purposes, is acceptable.

3. MACROSCOPIC LEVEL MODELS

The data collected and the information extracted via interpretation at the microscopic level is very local, whereas performance and impact modeling and decisions on maintenance and rehabilitation activities are made on facilities of significant length. Therefore, there is a need to identify, and model the long range spatial patterns exhibited by distress data. Three main applications with potential need for a macroscopic analysis of distresses have been identified and are discussed below:

Improving performance definition: Due to the level of detail at the microscopic level, the microscopic level data needs to be aggregated in order to provide relatively large "homogeneous" sections whose sizes are practical for strategy implementation purposes without loosing valuable spatial variation information.

Enhancing the accuracy of image interpretation: The locally isolated examination at the microscopic level overlooks the value of information about the identity of neighboring locations and its potential to provide additional input to the classification and measurement task. Spatial patterns have the potential to quantify and provide this additional information.

Reducing data processing and collection cost: Modeling the spatial distress relationships may permit processing a subset of the collected images or limit data collection needs and thus reducing the respective costs. Furthermore, this will also allow for determining the optimal subset to process or collect.

In the infrastructure literature there is very limited work related to this problem. There are methods, though, that aim at identifying homogeneous sections. One approach (Acharya) examines only condition data and aggregates it based on differences in the condition level. The other approach (Alfelor and McNeil) considers the condition, the maintenance policy, and maintenance implementation constraints and is based on optimization principles. Although existing methods deal with spatial relationships of infrastructure distresses in the sense that similar adjacent observations are grouped together to form "homogeneous" segments, they are very limited in their scope. They do not examine spatial relationships explicitly and they provide little theoretical justification.

The spatial environment can be characterized by the **scale** of the problem at hand and the **causality** behind distress spatial occurrence. In relation to the causality aspect there is no well accepted theory that explains why a particular distress occurred in a particular location, and how it propagates. This is mainly due to the large number of influencing factors involved (e.g. traffic, soil conditions, weather, construction quality, and maintenance history, etc). This situation, therefore, renders the development of causal spatial models difficult.

Given that one of the main objectives of developing spatial relationships is to bridge the microscopic level and the macroscopic level, it is important to consider the **detail** associated with the microscopic level and the **coverage** of the macroscopic level. To reduce the complexity of the problem two control parameters are used to build a conceptual representation: **resolution** (the level of detail captured) and **coverage** (the size of the area examined). Resolution controls the definition of events. Coverage controls the boundaries of the area under examination. The combination of different values of these parameters determines the level of analysis of spatial relationships. For example, at one end of the spectrum there is the HRSC (high resolution small coverage) abstraction that considers each distress feature as a single event (i.e. high resolution) but examines only a short section (i.e. low coverage) instead of an entire highway link. At the other end there is the LRLC (low resolution large coverage) abstraction that considers spatially adjacent distress features as a single event (i.e. low resolution) but examines the entire facility (i.e. high coverage).

For developing models that capture the different abstractions discussed above, ideas from stochastic spatial processes are used since they have the flexibility and potential to capture the important characteristics of the problem. In this framework the events of interest (distresses) are classified as follows: point (potholes), lineal discrete (e.g. longitudinal), areal discrete (e.g. alligator), longitudinally continuous transversally discrete (rutting), and strictly continuous (roughness). They are characterized by two attributes: spatial location and severity. Consequently, two distinct approaches are possible. One is a hierarchical approach where the two elements are separated; first, the spatial location attribute is characterized and then the severity element is treated conditional on the location characterization. Alternatively, the two elements can be characterized simultaneously in a single model. Both approaches are currently under investigation. Moreover, alternative estimation techniques that explicitly capture the characteristics of the problem are also under development.

Two interesting extensions to the models and methods described above are worthwhile to mention: (1) modeling different distress types simultaneously in order to make use of and capture the potentially existing cross-distress type correlations; (2) incorporation of additional information. This second extension deserves further explanation. Despite the complexity of the causal mechanism there is potential for improving the spatial modeling of surface distresses by taking into consideration some proxy causal variables. There exist for example (radar) technologies that continuously measure the thickness of the layers of the pavement structure (Maser). This additional information therefore can be used as an added characteristic of the events to be modeled.

REFERENCES

Acharya, D.R., "Using Expert Systems and Network Optimization Techniques for Rail Scheduling," Ph.D. Dissertation, Department of Civil Engineering, M.I.T., 1990.

Alfelor, R.M., and S. McNeil, "Definition of Homogeneous Rail Segments Based on Condition Data," Working Paper, Dec. 1990.

Ben-Akiva, M., and S.R. Lerman, Discrete Choice Analysis: Theory and Application to Travel Demand, MIT Press, Cambridge, MA, 1985.

Koutsopoulos, H.N., I. M. El Sanhouri, and A. B. Downey, "Analysis of Segmentation Algorithms for Pavement Distress Images," submitted for publication to *Transportation Engineering*, February 1991.

Maser, K., personal communication, February 1991.

PAVEMENT IMAGE PROCESSING USING NEURAL NETWORKS

Mohamed S. Kaseko[1]
Stephen G. Ritchie[2]

ABSTRACT

This paper discusses the potential for employing neural network models in the pavement image interpretation and classification stage of an automated pavement condition evaluation system. It discusses the choice of a suitable model for application, the effects of various input parameters on the training of the model and performance of the model in terms of the accuracy of correct classification of pavement images. Preliminary results are presented.

INTRODUCTION

An essential element of any pavement management system is the collection and evaluation of pavement surface distress data. The data include the types, severity and extent of pavement cracking. Although the data collection process is still largely a manual process involving visual inspection by field personnel, considerable progress has been made in developing automated systems utilizing computer vision and image processing technology (Mendelsohn, 1987, Ritchie, 1990, Ritchie, et.al, 1991). However, none of the systems currently available does provide a complete classification and quantification of the major pavement distresses.

The objective of this research is to develop a neural network-based methodology for classification of pavement images by type, severity and extent of distress. The paper discusses the implementation of a neural network model known as the "multi-layer perceptron".

---

[1] Graduate Research Assistant,
[2] Associate Professor, Institute of Transportation Studies and Department of Civil Engineering, University of California, Irvine, CA 92717

NEURAL NETWORK APPLICATION

Neural networks are information processing structures that consist of many simple processing elements (PE's or "neurons") with dense parallel interconnections. Each neuron can receive weighted inputs from many other neurons, and can communicate its outputs, if any, to many other neurons. Information is thus represented in a distributed fashion, across massive weighted interconnections. To implement a neural network model for pattern recognition, a set of patterns is repeatedly presented to the network during a "training session", and the system is supposed to "learn" to which class each of the input patterns belongs, so that later, when a related pattern is presented, the system will classify it properly.

The Multi-layer Perceptron (MLP)

Of the several neural network models available, the "multi-layer perceptron (MLP)" is probably the most studied model. The MLP consists of an input layer, an output layer and one or more layers of neurons in between, with each neuron in a layer connected to all neurons in the preceding and/or following layers through weighted interconnections (Figure 1). The output of each neuron is a function of the sum of the weighted outputs of the neurons in the immediate preceding layer. When the MLP is used as a pattern classifier, a vector to be classified is presented in the input layer and the computed vector at the output layer corresponds to the class to which the input pattern belongs.

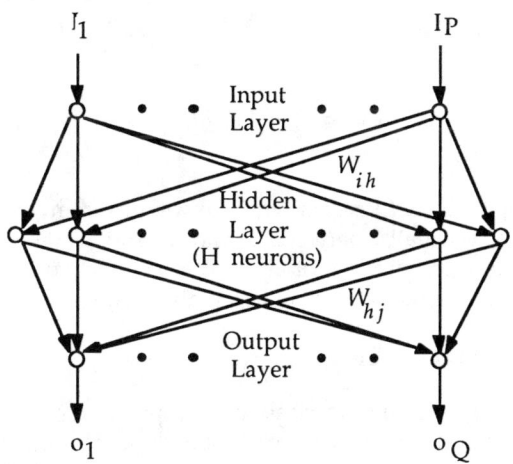

Figure 1: A 3-Layer Perceptron

Implementation of the MLP for processing of pavement images was presented in Ritchie, et.al, (1991). In the paper, a training algorithm for the MLP based on the back-propagation algorithm (Rumelhart, et.al 1986) was presented. The weights of the perceptron are updated after each iteration as summarized in the equation below:

$$W_{hj}(t+1) = W_{hj}(t) + \Delta W_{hj}(t+1)$$

where

$W_{hj}(t+1)$ = the interconnection weight between hidden neuron h and output neuron j at time $(t+1)$;
$\Delta W_{hj}(t+1) = \eta e_j o_h o_j(1 - o_j) + \alpha \Delta W_{hj}(t)$ if a sigmoid activation function is used, i.e.,
$o_j(t+1) = 1/\{1 + \exp(-net_j(t+1))\}$
$net_j(t+1) = \sum_{h=1}^{H} o_h(t) W_{hj}(t) - \theta_j$

e_j is error in output of neuron j in the output layer at time $(t+1)$, given as the difference between the desired and the actual outputs; o_h is the output of hidden neuron h at time$(t+1)$; o_j is the output of output neuron j at time $(t+1)$; θ_j is the threshold value for output neuron j; η is the training rate, normally $0 < \eta < 1$; α is the momentum term, also $0 < \alpha < 1$; and H is the number of neurons in the hidden layer. Similarly, for connection weights between the input and hidden layers,

$$\Delta W_{ih}(t+1) = \eta e_h o_i o_h(1 - o_h) + \alpha \Delta W_{ih}(t)$$

where

$$e_h = \sum_j e_j W_{hj}$$

Training of the MLP starts with the weights being assigned small random values, normally between 0 and 1, and is terminated when either the maximum number of iterations is reached or the sum of squares error (SSE) is reduced to an acceptable value.

Implementation of the MLP

The MLP was trained to classify, by type of cracking, 32x32 pixel sub-images extracted from raw 512x512 gray-scale pavement images and the effects of different parameters on the performance of the MLP were evaluated. Implementation involved (1) pre-processing the raw images for "noise" removal and image enhancement; (2) extraction of the 32x32 pixel-sized sub-images or "tiles" from the processed images; (3) generation of features vectors from the tiles for use in (4) training and testing of the MLP for classification of the tiles by type of cracking.

For pre-processing of raw images, *thresholding* was used. In this technique, the brightness value of each pixel is compared to a *threshold* value and pixels with values lower than the threshold are identified as distressed pixels, while those with higher values are identified as background pixels. In training and testing of the MLP, each tile was reduced to a vector whose elements are parameters representative of the orientation and thickness of cracking in the tile. These parameters were based on summary statistics of the row and column sums of the number of distressed pixels in a tile. In the case study reported in Ritchie, et.al (1991) three parameters, the mean number of distressed pixel per row, the variance of the number of distressed pixels per row and the variance of the number of distressed pixels per column, were used to train the perceptron for classification of tiles by type of cracking. In this case study two additional parameters are included, namely, the variances in the two diagonal directions, and the perceptron is trained to classify the tiles by severity of cracking present in the tile. Three severity levels were identified, *zero* for no cracking, *high* for average crack width > 1/4" and *low* for average width <= 1/4". Training and testing of the perceptron was simulated on a Sun SPARCstation.

Analysis of Results

Table 1 summarizes the performance of the perceptron in terms of its accuracy in correctly classifying tiles in the test set. Each row in the table presents a breakdown of how the tiles of a given cracking severity were classified by the perceptron into the three different classes. Almost all mis-classifications were on those tiles with a lot of noise, which means performance of the perceptron would increase significantly if the images are better preprocessed.

Table 1: Classification of Test Tiles by the Perceptron

	Classification by the Perceptron				
	Zero	High	Low	Total	Accuracy
Zero	4256	21	91	4368	97.4%
High	2	149	9	160	93.1%
Low	48	0	288	336	85.7%

The above results are based on the perceptron trained for 10,000 iterations using 256 training samples and three neurons in the hidden layer. However, training was repeated several times with different combinations of the number of hidden neurons and the size of the training set. It was observed that, although the final SSE generally decreased with increased number of hidden neurons, indicating better perceptron training with more hidden neurons, there was no corresponding consistent increase in the accuracy of classification as a function of the number of hidden neurons.

CONCLUSIONS

This study has further demonstrated the potential for using the MLP to process pavement images for distress classification and quantification. The accuracy with which the perceptron was able to classify the test set images is quite impressive, since most of the misclassifications were associated with marginal cases and "noisy" images. Therefore, with better pre-processing of images, the performance of the perceptron is likely to be dramatically improved.

However, the results indicate that the determination of the appropriate number of hidden neurons must be based on accuracy of correct recall of a test set, rather than a lower SSE during the training session.

REFERENCES

Mendelsohn, D.H. (1987). "*Automated Pavement Crack Detection: An Assessment of Leading Technologies*" Proceedings, Second North American Conference on Managing Pavements, Toronto, Canada.

Ritchie, S.G. (1990). "*Digital Imaging Concepts and Applications in Pavement Management.*" ASCE, Journal of Transportation Engineering, Volume 116, No. 3.

Ritchie, S.G, Kaseko, M.S. and Bavarian, B. (1991). "*Development of an Intelligent System for Automated Pavement Evaluation*" Paper No. 910394 presented at the 70th Annual Transportation Research Board Meeting, Washington, D.C.

Rumelhart, D.E., Hinton, G.E., and Williams, R.J. (1986). "*Parallel Distributed Processing, Vol.1*", MIT Press.

Design Of HCM Signal Timings Using Signal Expert

Stewart, J, Allen [1] And Van Aerde, Michel [2]

Abstract

While computer implementations of the procedures for evaluating the operation of traffic signalized intersections are receiving widespread use, these implementations are still primarily analysis as opposed to design tools. Consequently, the user must apply them iteratively, while recoding the input data between iterations, in order to perform a search for optimized designs of new signal timing plans. This paper presents an illustration of the Signal Expert approach to the analysis and design of signal timing plans, which can fully optimize these plans during a single execution of the program from a single user input data file. While the model was initially designed for use in Canada, this paper demonstrates the very practical benefits of also applying this new model in conjunction with the U.S. Highway Capacity Manual.

1. Current Practices in North America

The Highway Capacity Manual (HCM) Chapter 9 Appendix II (TRB 1985) and the Canadian Capacity Guide for Signalized Intersections (CCGSI) (ITE 84) were developed to systemize a method for analyzing signalized intersections in the United States and Canada, respectively. In order to facilitate the application of these procedures, the Highway Capacity Software (HCS) (FHWA, 1987) reproduces in a computer format all of the procedures contained in the HCM, while the Canadian Micro-Sintral program (Teply et al 1989), similarly incorporates most of the procedures of the CCGSI.

Both of these computerized procedures are adequate for analyzing and designing signal timing plans for simple non-congested intersections. However, for more complicated conditions both methods have some basic limitations that make the design process not only very tedious and time consuming, but often result in the ultimate selection of less than optimum signal timing parameters. The limitations are encountered in three main areas; namely the need to pre-select a seed cycle time, phase split and phasing scheme prior to the start of the design process, the need to pre-determine unprotected opposed saturation flow rates for left or right turn lanes, and the need to determine saturation flows and traffic volume

---

[1] Captain and Assist. Prof. of Civil Eng., Royal Military College of Canada, Kingston, Ontario, K7K 5L0. (Ph.D. Candidate at Queen's Univ.). CANADA

[2] Assoc. Prof. of Civil Eng., Queen's Univ., Kingston, Ontario, K7L 3N6. CANADA

allocations for lanes that are both shared and opposed. All of these limitations are further compounded if some lanes discharge during multiple phases.

In order to deal with these problems both the HCM and CCGSI must be used in an iterative fashion. Specifically, an initial seed solution, in the form of a trial cycle length, green split and phasing scheme, are evaluated and modified until such time as the calculated optimum cycle time and green split become self-consistent with those that were utilized to calculate the saturation flow rates that produced these cycle time and green time estimates. Unfortunately, as the demand volumes for competing approaches reaches v/c ratios near 1.0, the number of required iterations increases very rapidly, and in some cases the iterative procedure may actually diverge.

2. Development and Purpose of Signal Expert Program

In order to overcome these limitations a new approach to the design of signal plans for pre-timed traffic signals, (which has been called Signal Expert) was developed. It attempts to design a signal timing plan that is both self consistent and minimizes average vehicle delay within a single model run and without the need for updates to the input data between iterations. The main difference between Signal Expert and the HCM and CCGSI is the way in which the optimum cycle time and green splits for each phasing schemes are determined. The main inputs that are required by Signal Expert are the number of lanes for each approach, the kinds of movements allowed on each lane and the basic unopposed saturation flow rate for each lane. This allows Signal Expert to determine on its own how the specified pedestrian volumes and vehicular approach demand volumes for each approach will result in the prevailing opposed saturation flow rates and lane volume allocations. Consequently, Signal Expert can then, for a user-specified range of minimum/maximum cycle and green times, evaluate all possible cycle time and green split combinations without any further user intervention. This allows the optimum plan to be selected as simply the one that is evaluated to have the minimum delay, L.O.S., probability of clearance, etc...

What sets Signal Expert apart from most other approaches to signal optimization is that no trial cycle time or green times needs to be assumed or specified, and that the magnitude of any opposing vehicular or pedestrian flow effects on the lane saturation flow rates need not be pre-specified by the user prior to the start of the optimization. While the typical 24 hour computational time that is required on a simple microcomputer may seem impractical in day-to-day applications, it should be noted that this time will be probably reduced for the forthcoming version for a dedicated RISC or SPARC workstation to about 30 minutes. Furthermore, in terms of the more important user time, the analysis requires only about 15 minutes to code.

3. A Simple Case Study Comparison of HCS with Signal Expert

In order to demonstrate the practical significance of the Signal Expert program, the results of several typical runs of the Signal Expert program and the HCS package for the same intersection are shown in Table 1. The intersection lay out and traffic parameters of this example, which are illustrated in Figure 1, are those used in Chapter 9 of the HCM to illustrate the operational analysis of a pre-timed intersection with an unknown cycle time and unknown green splits.

HCM SIGNAL TIMINGS

Column A-1 in Table 1 indicates whether the scenario was analyzed using Signal Expert(SE) or the Highway Capacity Software's Signalized Intersection Module (HCS). Column A-2 indicates the number of phases that was considered for each run, where number is only provided if it represented a preset constraint. When three phases were used they consisted of a protected eastbound phase followed by a permitted east/west phase, and ending with a northbound phase. When only two phases were permitted they consisted of a permitted east/west phase followed by a northbound phase. The term TBD denotes that the number of phases was left to determined internally by the program as part of the actual optimization process.

The columns A-3 and A-4 provide the durations of the cycle time and effective green times, respectively. Again, the term TBD indicates that these parameters were left to be determined internally by each model as part of the optimization process. For the HCS solutions, the cycle times and green splits were estimated using the methods described in the HCM and using its Equations II. 9-1 and II. 9-2. Column A-5 describes which traffic volumes were utilized for each model run, where the term "Default" indicates the HCM example volumes, while the term "45% LT" lable indicates that a scenario with

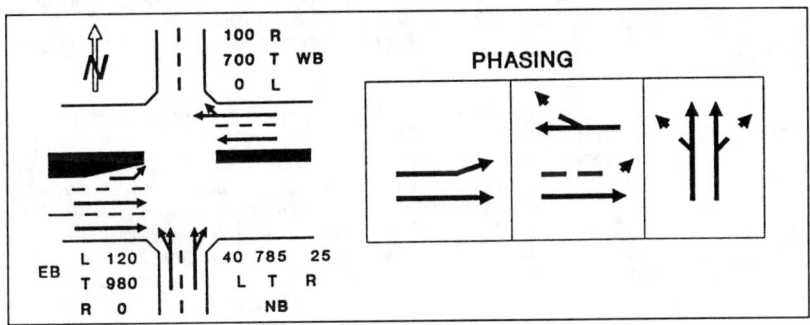

Figure 1: Characteristics of HCM Example Intersection Analysis

Case Number	Model (A-1)	# of Phases (A-2)	Input Cycle Time (A-3)	Input Green Time (A-4)	Eastbound Left-Turn Volume (A-5)	Output Cycle Time (B-1)	Output Green Time (B-2)	Average Delay (B-3)
1a	SE	3	TBD	TBD	Default	60	5, 23, 23	20.5
1b	HCS	3	50	5, 18, 20	Default	50	5, 18, 20	10.5
1c	HCS	3	60	5, 23, 23	Default	60	5, 23, 23	10.9
2a	SE	TBD	TBD	TBD	Default	40	15, 19	15.7
2b	HCS	2	40	15, 19	Default	40	15, 19	8.1
3a	SE	3	TBD	TBD	45% LT	50	5, 19, 17	29.0
3b	HCS	3	61	20, 17, 18	45% LT	61	20, 17, 18	27.8
3c	HCS	3	53	10, 19, 17	45% LT	52	10, 19, 17	21.3

Table 1 : Comparison of HCS and Signal Expert for the HCM Example.

a greater percentage of left-turns was analyzed, as will be discussed later. The basic saturation flows were derived from the values that are provided in Appendix IV of Chapter 9 for the HCM Calculation 2. The values in column B-1 and B-2 are a repeat of Columns A-3 and A-4, if the cycle times and green splits were external to the model, as was typically the case for the HCS. Otherwise, the numbers represent the optimum cycle time and green splits specified during the optimization. Finally, Column B-3 provides the estimated average vehicle delay for the intersection. It should be noted that in comparing the results of the two models, the absolute value of the delay estimates is not necessarily the same due to the slightly different delay equations that are utilized by the HCM as opposed to signal expert, and due to the manner in which left turn sneakers are modelled.

Case 1a in Table 1 provides the results of a Signal Expert optimization of the phase structure shown in Figure 2. Case 1b is an HCS analysis of the same problem using the cycle times and green splits found according to the procedures in the HCM. Case 1c is also an HCS analysis, however, the cycle times and green splits that are evaluated this time are those that were determined to be optimum by Signal Expert in Case 1a. In comparing Case 1b to Case 1c one can note that the HCS only finds 0.4 seconds of difference between the 50 second cycle length, which it found, and the 60 second cycle length, which was found by Signal Expert. This slight difference is due to the fact that Signal Expert performs its optimization based on a slightly different delay formula than the HCM.

However, when the full potential of Signal Expert was used to determine not only the optimum cycle time and green splits, but also the number and types of each phase, the real potential of the Signal Expert approach becomes more evident, as indicated as Case 2a in Table 1. It shows that when Signal Expert examined a set of 80 possible phasing schemes, the optimum phasing plan was found to be a simple two phase plan. Furthermore, Case 2b provides the results of an HCS analysis of this Case 2a timing plan found by Signal Expert and substantiates that this plan is an improvement over a 3 phase plan, as the HCS calculates that the new plan provides a decrease in average delay per vehicle from 10.9 to 8.1 seconds.

As Cases 1 and 2 represent relatively simple design problems, Case 3 was introduced to illustrate more clearly the benefits of the Signal Expert approach when a large number of opposed left turners need to be dealt with. Specifically, Case 3a shows the cycle time and green split found for the same phase scheme that was analyzed in Case 1, except that the number of left turners using the eastbound left turn bay was increased to represent 45% of total approach flow, while the through volume was decreased accordingly in order to maintain the same total approach demand volume. Signal Expert indicates that a 50 second cycle length should be utilized and that the protected left turn phase should still be relatively short with the majority of the green time split being allocated between the permitted east/west phase and the northbound phase.

In Case 3b the HCM method was used to determine the saturation flow rates necessary to calculate the new cycle time. This yielded a cycle time 61 seconds and a green split of 20, 17 and 18 secs, respectively. The HCS found the delay for this HCM phase plan to be 27.8 seconds. The Case 3a signal timing plan, as found by Signal Expert could not be analysed directly by HCS due to the manner in which the HCS allocates volumes between the permitted and protected phase. Specifically, as the v/c ratio for the protected phase was above 1.0 the delay for this lane group could not be calculated using the HCS delay equation.

However, Case 3c gives the results of a phase plan that could be considered by the HCS and which closely resembles the one found by Signal Expert. Specifically, a cycle time of 53 seconds and an allocation of green time equal to 21% of available green was evaluated using the HCS as Case 3c. The delay for the Case 3c timing plan turned out to be 28% less than for the Case 3b recommnedation by the HCM . This indicated that the new plan would result in a L.O.S. of B rather than a L.O.S. of C.

4. Summary Discussion

The above comparison illustrates two main findings. Firstly, eventhough Signal Expert implements the Canadian Capacity Guide procedures, it is still capable of determining in a single model run a signal timing plan, for a given phasing scheme, which is at least equivalent and often better than the one generated by the HCM optimizer, even when evaluated by the HCS evaluator. The second important finding is that Signal Expert can for the same user effort, and for the same input data file, also determine the best signal timings for all other phasing schemes that may be possible.

While the above findings are for a single sample intersection problem and have not been generally proven for all possible intersections, they are consistent with earlier findings which compared Signal Expert to the Canadian Micro-Sintral Program (Stewart and Van Aerde, 1991.). Furthermore, there appear to be no theoretical reasons as to why Signal Expert would not always provide equivalent or better signal timings that the HCS software and in less user time, especially if the formulas that are currently imbedded in Signal Expert were converted to be those of the HCM, rather than those of the Canadian Capacity Guide.

References

ITE, 1984, Canadian Capacity Guide for Signalized Intersections, First Edition, S. Teply, Editor, Inst. of Transp. Eng. - District 7, Canada

Federal Highway Administration, 1987, Highway Capacity Software User's Manual, U.S. Department of Transportation, Federal Highway Administration, Washington D.C.

Stewart, J. A. and Van Aerde, M., 1991, An Expert Assistant for Selecting Green Splits, Cycle Times and Phasing Schemes based on the Canadian Capacity Guide Procedures, 1991 Annual Conference of Canadian Institute of Transportation Engineers, Victoria, B.C.

Teply, S., Stephenson, B., and Kua, H. C., 1989, MICRO-SINTRAL Version 3.0 User Manual, Institute of Transportation Engineers District 7, University of Alberta, Edmonton, Canada.

Transportation Research Board, 1985, Highway Capacity Manual, Special Report 209, National Research Council, Washington, D.C.

DESIGN AND EVALUATION OF MULTI-BAND PROGRESSION SCHEMES

Nathan H. Gartner[1], Susan F. Assmann[2],
Fernando Lasaga[1], and Dennis L. Hou[1]

Abstract

MULTIBAND is a new arterial progression scheme that has the capability to assign each directional road section with an individually volume-weighted band. The method offers to the traffic engineer a much wider range of design options than do existing arterial progression methods. In this paper we apply the MULTIBAND method to three arterial streets and compare its performance with conventional bandwidth maximization.

Introduction

Arterial progression methods are widely used in the U.S., as well as in other countries. The conceptual basis for the progression design is that traffic signals tend to group vehicles into a "platoon" with more uniform headways than would otherwise occur. The platooning effect is accentuated on the major streets which have signalized intersections at frequent intervals. It seems desirable, in these circumstances, to encourage platooning so that continuous movement (or progression) of vehicle platoons through successive traffic lights can be maintained. The signal timings, in this case, are designed to maximize the width of continuous green bands in both directions along the artery at the expected speed of travel. In general, such signal systems operate best when the main-street flow is predominantly through traffic and when the number of vehicles turning onto the main street is small.

Advances in optimization techniques and computational capabilities have steadily increased the sophistication of arterial progression methods. Two of the most advanced and versatile of these methods today are PASSER-II (Chang et al, 1988) and MAXBAND (Little et al, 1981). Popular delay-based methods such as TRANSYT also use bandwidth optimization programs in a 'hybrid' manner to improve progression on the principal arteries (Cohen and Liu, 1986; Liu, 1988).

A basic limitation of existing bandwidth-based programs is that their progression design criterion does not depend on the actual traffic flows on the arterial links, and therefore is insensitive to variations in such flows. The total bandwidth that is obtained for the arterial can be allocated in any desired ratio among the two directions of travel. A common practice is to apportion it according to a single directional volume ratio k. Because of turn-in and turn-out traffic we do not, generally, have constant volumes along each direction of the arterial. Consequently, the idea of a uniform platoon moving through all the signals in one direction, which forms the conceptual basis for the bandwidth approach, does not always hold. Moreover, the ratio of volumes on opposing road sections between each pair of adjacent signals is also varying. It is, therefore, inconceivable that the single parameter k for the entire arterial can adequately reflect this diversity.

(1) Dept. of Civil Engineering, University of Lowell, Lowell, MA 01854.
(2) Department of Mathematics, Regis College, Weston, MA 02193.

The Multi-Band Approach

MULTIBAND is a new computer method that is designed to remedy the deficiencies mentioned above (Gartner et al, 1990). It uses mixed-integer linear programming for optimizing the design variables.. MULTIBAND places the arterial bandwidth optimization concept on a more solid foundation by incorporating into the calculation procedure a systematic traffic-dependent criterion. The volume on each link of the artery, together with other traffic parameters (such as capacity, speed, etc.), has an effect on the optimization outcome through suitably chosen link-specific weighting factors, as contrasted with a single weight in existing programs. The multi-band approach offers a wide range of design options that can be used by the traffic engineer to tailor the control strategy to each particular arterial street situation. In this way we can obtain improved traffic performance beyond what is possible with conventional progression schemes.

To explain the structure of the MULTIBAND model we compare it with the MAXBAND optimization model which is shown in the first box in Table 1. The green splits can be given or, alternatively, the user can provide traffic volume and capacity information for each intersection and the program will calculate the splits. An example of a MAXBAND design is shown in Figure 1.

In MULTIBAND we calculate a different bandwidth for each directional road section of the arterial. Each band section can be individually weighted in the objective function. Thus we obtain a method that is sensitive to varying traffic conditions and we can tailor the progression scheme to the different possible traffic flow patterns. The user can still choose uniform bandwidth weightings if he so desires, but this is now only one of many user options. The optimization model is shown in the second box of Table 1.

The most important departure compared to existing progression methods occurs in the definition of the objective function. Since in MULTIBAND the bands are link-specific, they can be weighted disaggregately to achieve desirable traffic objectives for each link. The new objective function has the following form:

$$\text{MAX } B = \frac{1}{n-1} \sum_{i=1}^{n-1} \left(a_i b_i + \bar{a}_i \bar{b}_i \right)$$

where $a_i(\bar{a}_i)$ are the link-specific weights in the two directions. There are a multitude of options available for choosing the weighting coefficients. We chose to investigate the following weighting options:

$$a_i = \left(\frac{V_i}{S_i}\right)^p \qquad \bar{a}_i = \left(\frac{\bar{V}_i}{\bar{S}_i}\right)^p$$

where, $V_i(\bar{V}_i)$ =   directional volume on section i, outbound (inbound); either the total volume or the through volume can be used. The latter is called the "platoon volume."

$S_i(\bar{S}_i)$ = saturation flow on section i, outbound (inbound); this is the capacity volume in vphg.

p = exponential power; the following values were used: $p = 0$ (unit coefficients), $p = 1$ (i.e., volume/capacity ratio), $p = 2$ (i.e., $(\text{vol/cap})^2$), $p = 4$ (i.e., $(\text{vol/cap})^4$).

Other weighting options can easily be specified. This provides considerable flexibility to the user. An example of a MULTIBAND design is given in Figure 2.

TABLE 1

MAXBAND		
	Given:	splits, queue clearances target ratio of bandwidths limits on: cycle time, link speeds, changes in speeds
	Find	cycle time, offsets, interferences bandwidths, b, $\bar{b}$ link progression speeds left-turn phase patterns
	To:	maximize $b + k\bar{b}$
	Subject to:	cycle time constraint bandwidth ratio constraint interference constraints loop integer constraints speed and speed-change constraints

MULTIBAND		
	Given:	splits, queue clearances target ratios of bandwidths for each section limits on: cycle time, link speeds, changes in speeds allowed left-turn phase patterns
	Find:	cycle time, offsets, interferences link-specific bandwidths $b_i, \bar{b}_i$ link progression speeds left-turn phase patterns
	To:	maximize $B = \dfrac{1}{n-1} \sum\limits_{i=1}^{n-1} \left(a_i b_i + \bar{a}_i \bar{b}_i \right)$
	Subject to:	cycle time constraint bandwidth ratio constraints interference constraints loop integer constraints speed and speed-change constraints

Evaluation Results

The multi-band approach has been evaluated for three different arterial systems. We simulated the signal settings that were calculated by the different MULTIBAND design options and the two available MAXBAND options. The NETSIM model was used for the simulation. Each setting was simulated five times and the results were statistically analyzed. Average delay values for all options considered are given in Table 2. For each street the lowest MULTIBAND delay was below the lowest MAXBAND delay. In two of the three cases that were analyzed, every MULTIBAND option that we tested produced delays that were markedly lower than the lowest MAXBAND delay. Similar improvements were also observed in the other measures-of-effectiveness, such as number of stops and average speed.

The results obtained so far indicate that MULTIBAND offers the opportunity for considerable improvements in traffic performance compared with existing practice.

References

1. CHANG, E.C.P., MESSER, C.J., and GARZA, R.U. (1988). Arterial signal timing optimization using PASSER II-87. *ITE Journal*, November 1988, 27-31.
2. COHEN, S.L. and LIU, C.C. (1986). The bandwidth-constrained TRANSYT signal-optimization program. *Transpn. Res. Rec.* **1057**, 1-7.
3. GARTNER, N.H., ASSMANN, S.F., LASAGA, F., and HOU, D.L. (1990) MULTIBAND: A variable bandwidth arterial progression scheme. *Transpn. Res. Rec.*
4. LITTLE, J.D.C., KELSON, M.D., and GARTNER, N.H. (1981). MAXBAND: A program for setting signals on arteries and triangular networks. *Transpn. Res. Rec.* **795**, 40-46.
5. LIU, C.C. (1988). Bandwidth-constrained delay optimization for signal systems. *ITE Journal*, **58** (12), 21-26.

TABLE 2

Progression Scheme		Canal Street		Main Street		Mass. Avenue	
		Avg. Delay	Difference	Avg. Delay	Difference	Avg. Delay	Difference
MAXBAND	Symmetric	29.69	2.94%	66.26	7.34%	26.94	-18.63%
	Proportional	28.84	0.00%	61.73	0.00%	33.11	0.00%
MULTIBAND	1.0	25.62	-11.17%	54.31	-12.02%	26.05	-21.33%
	V/C	25.20	-12.62%	52.84	-14.40%	26.10	-21.17%
	$(V/C)^2$	25.35	-12.10%	52.93	-14.26%	26.06	-21.29%
	$(V/C)^4$	24.11	-16.40%	51.71	-16.23%	38.45	16.13%
	P/C	25.08	-13.04%	56.03	-9.23%	27.00	-18.45%
	$(P/C)^2$	25.25	-12.44%	56.86	-7.89%	27.00	-18.45%
	$(P/C)^4$	25.25	-12.44%	53.81	-12.83%	27.00	-18.45%

TRANSPORTATION TECHNOLOGIES APPLICATIONS

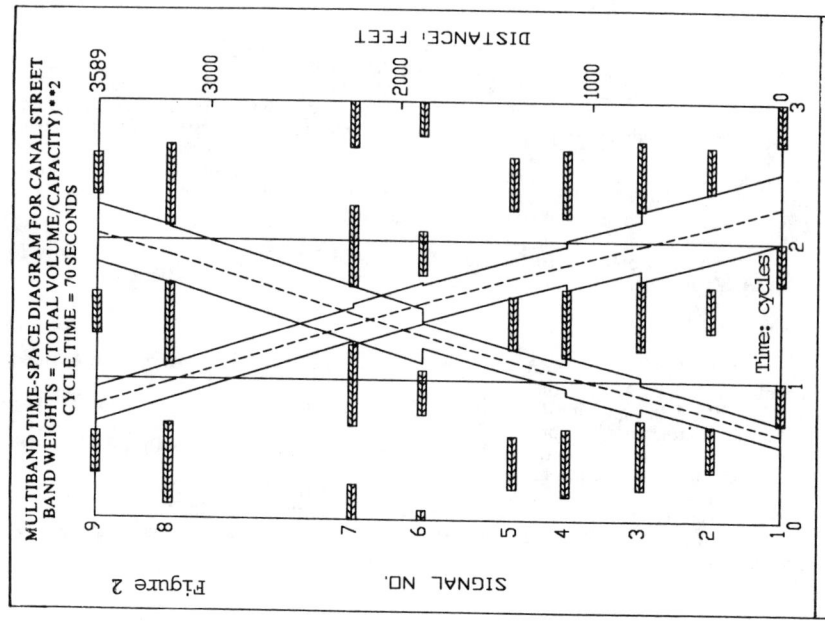

Figure 2. MULTIBAND TIME-SPACE DIAGRAM FOR CANAL STREET
BAND WEIGHTS = (TOTAL VOLUME/CAPACITY)**2
CYCLE TIME = 70 SECONDS

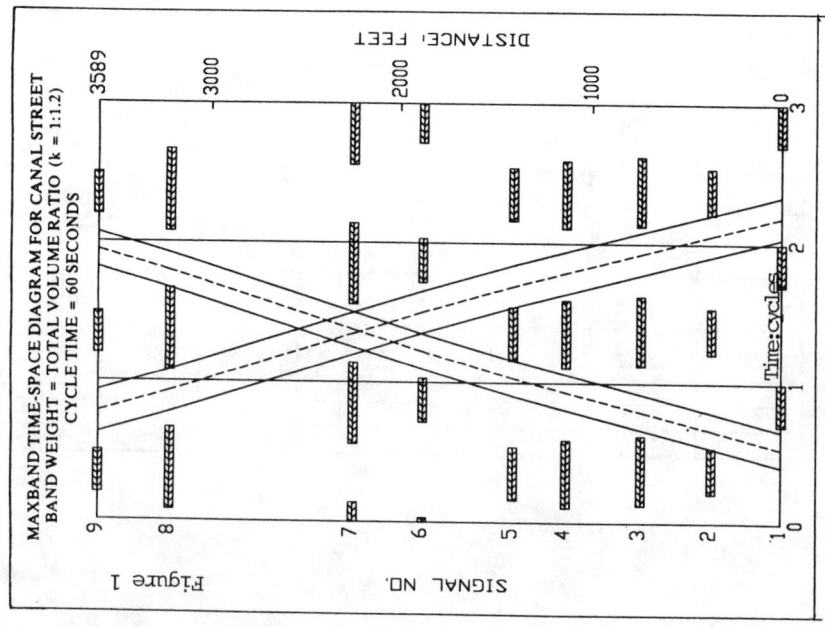

Figure 1. MAXBAND TIME-SPACE DIAGRAM FOR CANAL STREET
BAND WEIGHT = TOTAL VOLUME RATIO (k = 1:1.2)
CYCLE TIME = 60 SECONDS

An Intelligentual Real-time Traffic
Control System Suited To Chinese Cities

He Guoguang[1] Lu Baichuan[2] Liu Bao[3]

Abstract

Our new researched and designed real-time traffic control system has been installed in Tianjin City, in which we presented a machine-learning algorithm suited to the traffic characteristics in Chinese cities. This paper introduces its basic principle, hardware and software structure, ML-1 traffic controller and control effects.

Introduction

As cities grow so the traffic density for vehicles and pedestrians increases. For commuters to move efficiently and smoothly traffic system must control this flow. But the traffic flow in Chinese cities is much different from that in developed countries, such as, there are many bicycles; vehicles and bicycles drive on the same roads; traffic flows are less stable ect..

Now there are several typical area traffic control systems, such as TRANSYT, SCOOT, SCAT, whose principles are based on the traffic characteristics of developed countries' cities. ALthough some Chinese cities have installed TRANSYT, SCOOT or SCAT system, they didn't get satisfied control effects, for they are not suitable to the traffic characteristics of Chinese cities. In order to improve urban traffic, we have researched, designed An Intelligent Real-time Traffic Control System for Tianjin city (TICS). Tianjin has 4 million people and is the third largest city of China. Its traffic is very typical and reflects almost all characteristics in Chinese cities. Because it is difficult to accurately describe traffic

---

1. Professor, 2. Lecturer, 3. Professor, Institute of Systems Eng.,Tianjin University, Tianjin, P.R. China

characteristics by mathematic equations and prevent using uncorrect estimates of traffic states, here we apply machine-learning method in Artificial Intelligence to the system. The system will constantly improve control parameters in accordance with the changes of circumstance or other factors and store those which have the best control effects. So it is suitable to almost all kinds of traffic flow.

Principle of TICS

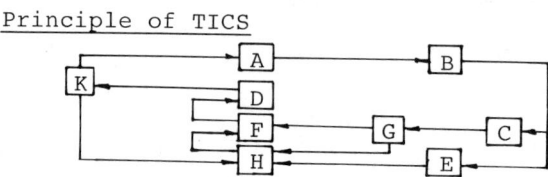

Fig. 1　principle block

　　TICS is a machine-learning area traffic control system. Its basic principle block diagram is shown as Fig.1. Each block represents a subsystem having its own function. A—Controlled process. B—Detecting subsystem. There are two kinds of detectors, vehicle detectors and bicycle detectors. Vehicle detectors receive the data of arriving vehicles and their change rules by which the traffic states and control effects, such as stops, delay, flow etc., can be calculated. Bicycle detectors collect the volume of bicycles, whcih greatly influence the urban traffic in China, especially during the peak hours. Here they are used as one of the factors to calculate traffic states. C—Traffic state distinguishing subsystem. D—Real-time control subsystem. This subsystem select the control parameters (C,O,S), C: cycle, O: offset, S: split. E—Control effect evalution subsystem. In general, the objective function $I=F(I_1,I_2,I_3)$. I_1: delay, I_2: stops, I_3: flow. The objective function is inverse proportion to the control effect. F—Knowledge-base of control parameters. G—Control parameter inference engine. There are some inferencing rules between traffic states and control parameters. H—Machine-learning subsystem. There are a set of learning rules to determine whether or how to renew the knowledge. K—Lamp driving subsystem.

　　　Fig.1 includes two main loops: real-time control loop　A—B—C—G—F—D—K—A, and learning loop A—B—E—H—F—D—K—A.

　　　Process of real-time control loop: All vehicles and bicycles are driving in the road network and signals are operating at intersections, i.e. A. B constantly collects data of vehicles and bicycles arriving or passing intersections. Using this data, C distinguishes the traffic states of every intersection and transmits them to G. Under certain inferencing rules, G generates inferencing results. Using these results and traffic states, F chooses the optimal control parameters. Then by transforming parameters into pulse, K drives traffic signal lamps at the intersections.

Process of learning loop: when the signals are driven by a group of control parameters (C,O,S), B calculates the performance indices $[I_1, I_2, I_3]$. E calculates objective function value I and compares it with I^* which is the best function value at same state before now. H decides how to change control parameters by using inferencing rules and $I-I^*$. The rest F—D—K—A is same as in the control loop. This loop will constantly compare objective function values and renew the control parameters to keep the knowledge improved.

Hardware structure of TICS

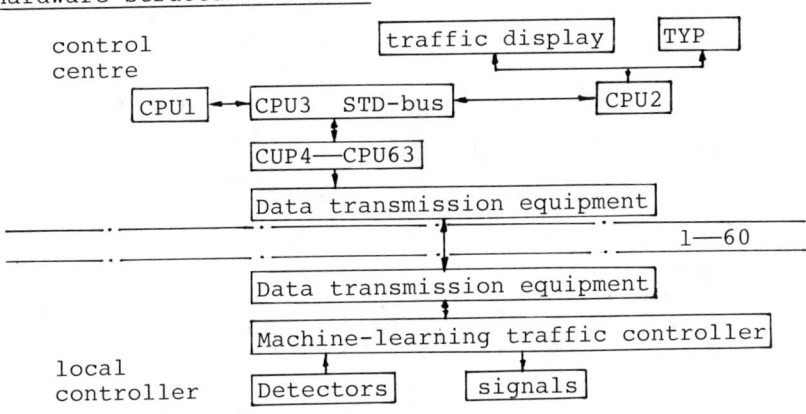

Fig.2 Hardware structure

TICS has capacity to control the traffic of an area about 60 intersections. Its hardware structure is shown in Fig.2. It consists of 1 control centre and 60 local controllers. Local controller controls traffic signals and receives data from detectors. Here is ML-1 (see later). Control centre controls and coordinates local controllers. It is a multi-CPU structure and has 63 CPUs. Each CPU has its own function. CPU1 is a management CPU, in which the monitor and management programs are operated and a MIS of traffic data is founded. CPU2 calculates coordinative and optimal control parameters. CPU3 is s STD-bus to manage and transmit the data of CPU1, CPU2 and CPU4-63. In general, local controllers are not near the control cnetre. We use long distance data transmission equipment with CPU4-63 to separately transmit data between control centre and local controller. TICS parallelly processes data transmission of 60 intersections, so it has better real-time control capacity.

Software structure of TICS

The software structure of TICS is designed to complete all functions. Here we only give that of CPU1.

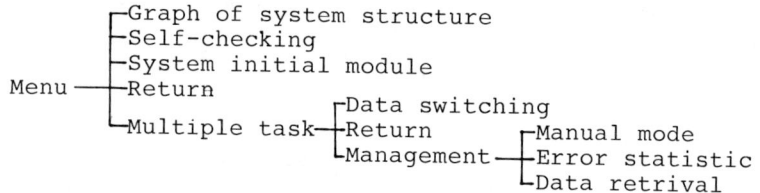

Fig.3 Software structure of CPU1

ML-1 traffic controller

The ML-1 is a microprocessor based traffic controller at intersections. Machine-learning control mode is its most wonderful characteristic. When fully equipped with communications, the ML-1 controller integrates perfectly into the TICS. It will upload data received from its detectors and accept control instructions directly from the central computer over a cable. In the case of communications failure or isolated, the ML-1 has 5 operation modes, especially machine-learning, to keep this intersection traffic controlled well.

A. Operating modes and basic principle

There are five standard modes of operation: (1). Mannual. (2). Multi-plan fixed time. (3). Flexilink. (4). Machine-learning. (5). Computer masterlink. Its basic principle is almost same as that of TICS. The difference is that when linked to TICS, It accepts cycle, split, and optimizes its own offset. When isolated, it optimizes its own cycle, split and offset.

B. Hardware structure

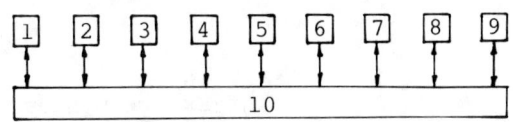

Fig.4 Hardware structure of ML-1

The hardware of ML-1 is built around the M8031 processor and shown as Fig.4. 1. Processor card. 2. State distinguishing card. 3. Real-time control card. 4. Clock. 5. Printer. 6. Protection card. 7. Output interface. 8. Input interface. 9. Long distance communication card. 10. STD-bus.

C. Control effects of ML-1

The ML-1 was installed at a intersection in Tianjin in October, 1989. Since then, it has been operating efficiently without faults. The control effects of two operating modes are shown in table 1. Machine-learning mode has better performance indices.

Table 1

operating modes	fixed time	machine-learning
average delay	0.160	0.127
average stop	47.4	44.7

Conclusion

The principle of TICS is based on the traffic characteristics of Chinese cities. Bicycle detectors are specially designed to determine the influences of bicycles. From the Table 1, we know that machine-learning mode in ML-1 has improved the traffic of single intersection. Machine-learning mode in the TICS has better simulation results. It will be installed to control an area by July, 1991. TICS is a multi-CPU structure, so it has better real-time control capacity. Because all functions can be realized by micro-computers, TICS has lower costs and is easily installed in other cities.

An Application of Expert Systems to Traffic Signal Control

S. Manzur Elahi[1], A. Essam Radwan[2], Member, ASCE, and K. Michael Goul[3]

ABSTRACT

Signal Control at Isolated Intersections (SCII) has been developed to emulate an adaptive controller using AI technology. It is an application of expert systems to the real-time traffic signal control. The first version of the prototype (SCII-1) was developed in 1987. Subsequent research works were conducted to develop the second version (SCII-2). SCII-2 can be applied to three different types of intersection geometries. It determines the appropriate cycle length, phasing pattern and split, and updates these based on the prevailing traffic demand. It is capable of switching the operation from pretimed to actuated mode and vice versa. Tests were performed to validate the prototype using field data as well as computer simulation. The field test verified the appropriateness of switching the signal operation from one mode to the other. The tests with simulations indicated the superiority of SCII-2 over pretimed and actuated controllers.

INTRODUCTION

The conventional pretimed and actuated controllers lose their performance quality under unanticipated traffic demands and saturated conditions. This research

---

[1] Traffic Engineer, Bureau of Traffic Services, District of Columbia Department of Public Works, 2000 14th Street, N.W., 7th Floor, Washington, DC 20009.

[2] Professor and Chair, Department of Civil and Environmental Engineering, University of Central Florida, Orlando, Florida 32816.

[3] Associate Professor, Department of Decision Information System, Arizona State University, Tempe, Arizona 85287.

study utilizes a knowledge-based expert system (KBES) orientation to this problem domain.

$\underline{S}$ignal $\underline{C}$ontrol at $\underline{I}$solated $\underline{I}$ntersections (SCII) is an expert system prototype developed for the real-time traffic signal control operations. The first version of the prototype (SCII-1) was developed in 1987 at Arizona State University (Goul, et al. 1987). The experiences from the project, especially during the verification stage, is documented by Radwan, et al. (1989). However, SCII-1 was limited with respect to the type of intersection geometry and signal phasing schemes it could accommodate.

Further research works were conducted to enhance this prototype to operate under a variety of conditions. The second version of the prototype (SCII-2) is the result of this effort.

BACKGROUND INFORMATION

The Highway Capacity Manual (1985) Chapter 9 Delay Model is used for calculating delays in the SCII-2 prototype.

Queue-length is another important measure of effectiveness used in SCII-2. Cronje (1983) conducted a research to assess the existing queue-length formulas. He identified Miller's queue equation (Miller 1968) as the most realistic model. This model is used to calculate the queue-length in the SCII-2 prototype for under-saturated conditions.

For near and over-saturated conditions, a deterministic queue-length model is considered (Newell 1982). This method assumes a constant rate of vehicle input and output.

STRUCTURE OF THE SCII-2 PROTOTYPE

SCII-2 is coded in LISP programming environment on a microcomputer. This prototype can be applied to three different intersection geometry types. It can analyze up to 8 different types of phases, out of which a maximum of 6 phases can occur in a single cycle.

At the top level of SCII-2, the user selects whether the mode of operation will be conventional (actuated and pretimed) or adaptive control (Figure 1). He can also select the option for simulation.

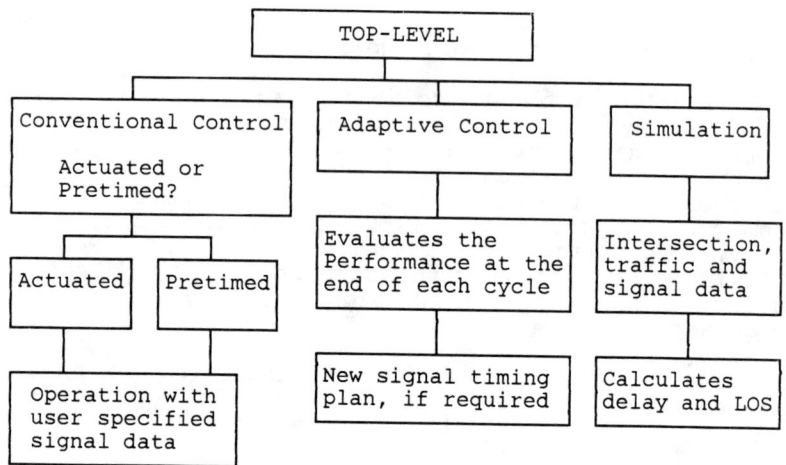

Figure 1. Structure of SCII-2

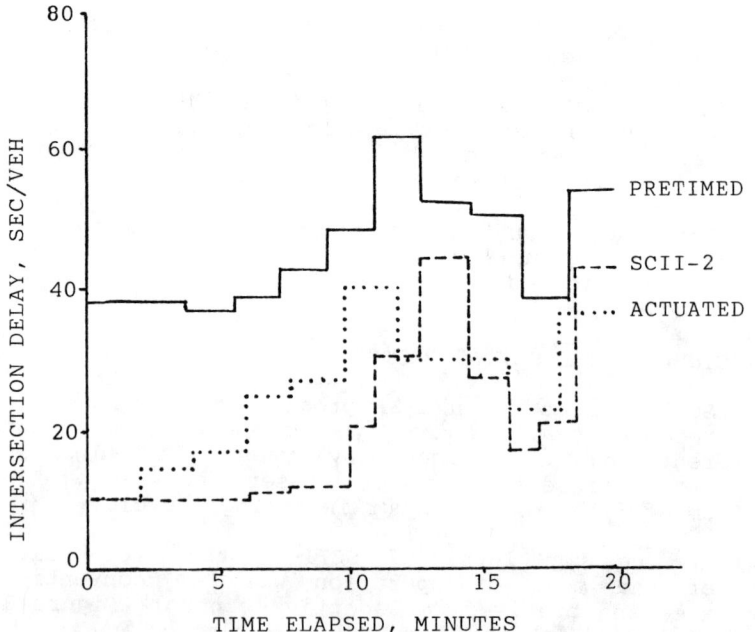

Figure 2. Comparison of Intersection Delays

Conventional Control

At this mode, SCII-2 evaluates the signal performance at the end of each cycle and determines the point where signal control needs to be switched from actuated to pretimed and vice versa. SCII-2 uses the methodology outlined in NCHRP Report 233 (Tarnoff and Parsonson 1981) to determine the appropriate mode of signal operation under specific traffic conditions.

Adaptive Control

The adaptive control operation strategy constitutes the major portion of SCII-2's computer code. SCII-2 evaluates the signal performance at the end of each cycle, and calculates a performance rating as a function of the stopped delay and the queue-length.

If the performance rating falls below some user-set threshold value, SCII-2 determines the parameters for a new cycle. To do that, it predicts the traffic volume for the subsequent signal cycle. Based on the predicted volume, it uses a table look-up procedure for determining the appropriate cycle length. Then, it calculates the phasing pattern and splits.

SCII-2 maintains a database of significant volume data of previous cycles. It checks whether the current traffic volume represent a new trend not reflected in the database. If it is so, it stores the data in the database.

Simulation

Using this mode, the user can calculate the approach and intersection delays and levels of service for a given set intersection geometry, traffic and signal conditions.

VALIDATION

The TEXAS microscopic simulation model was used for testing SCII-2. A 20 minute data set previously collected at a local intersection in Phoenix, Arizona was utilized for this purpose. The data represented the early period of the evening peak at which volumes rise to the peak and remain there.

The settings suggested by SCII-2 were simulated by the TEXAS model for the 20-minute time period. Five replications were made for each run and stopped delays were noted. The pretimed and actuated controller operations were also simulated with the same traffic data and the delays were noted. Figure 2 shows the comparison of the three types of operations. It indicates that SCII-2 reduces the delay substantially when compared to other controllers.

Field tests were also conducted to validate the switching operation of the conventional control mode of SCII-2. Statistical analyses of the field data supported the concept used.

CONCLUSIONS

SCII-2 represents a unique application of AI technology for traffic signal control. The simulation and field tests have produced encouraging results and indicated the potential of the SCII-2 prototype for future implementation.

APPENDIX. REFERENCES

1. Cronje, W.B. (1983). "Analysis of Existing Formulas for Delay, Overflow, and Stops." Transp. Res. Rec. 905, Transportation Research Board.
2. Goul, K.M., O'Leary, T.J., and Radwan, A.E.. (1987). "Expert Systems for Traffic Signal Control." Proceedings of the Microcomputer Applications in Transportation - II International Conference. Boston, Massachusetts. American Society of Civil Engineers, New York. 629-638.
3. "Highway Capacity Manual". (1985). Special Report 209, Transportation Research Board.
4. Miller, A.J. (1968). "The Capacity of Signalized Intersections in Australia." ARRB Bull. 3, Australian Road Research Board.
5. Newell, G.F. (1982). "Applications of Queueing Theory 2nd ed." Chapman and Hall, London, Great Britain, 287-300.
6. Radwan, A.E., Goul, K.M., O'Leary, T.J., and Moffitt, K.E. (1989). "A Verification Approach for Knowledge-Based Systems." Transp. Res., Vol. 23A, No. 4, Great Britain.
7. Tarnoff, P.J., and Parsonson, P.S. (1981). "Selecting Traffic Signal Control at Individual Intersections." NCHRP Report 233. Transportation Research Board.

Bus Pre-Emption: A Real Time Control Strategy For Privatized Transit Operation

By

Snehamay Khasnabis[1], M.ASCE and Bharat B. Chaudhry[2]

ABSTRACT

Bus pre-emption strategies are designed to provide priority to transit buses over passenger cars for urban travel. A comparative analysis of a proposed suburban bus service for regular operation versus prioritized operation is presented. The study shows that reduced travel time brought about by pre-emption, if properly implemented, may translate to a smaller fleet size and a lower cost of operation.

1. INTRODUCTION

Delay to buses at signalized intersections on urban arterials comprises approximately 20% of average bus trip time. Bus pre-emption strategies are designed to provide priority to transit buses over passenger cars by providing continuous green phases at successive signalized intersections, thereby reducing travel time. The technology involves the use of instrumented buses, transmitters, loop detectors and a real time control system for estimating arrival times at the intersection and for triggering signal pre-emption. Techniques of Green Extension, Red Truncation, and Red Interruption have been used in the past for pre-emption purposes.

Signal Pre-emption Technology

In the mid 1970's, experiments were conducted in a number of US cities to test various methods of

---

[1] Professor, Department of Civil Engineering, and Director, Urban Transportation Institute, Wayne State University, Detroit, MI

[2] Graduate Research Assistant, Department of Civil Engineering, Wayne State University, Detroit, MI

minimizing bus delays at intersections, under the auspices of the UTCS/BPS/program (1,4,5). The results of these experiments can best be categorized as 'mixed' successes. Although specialized signal controls are used widely in Europe today, a number of factors have, thus far, prevented their widespread application in the U.S. These include: the absence of a reliable technology to monitor the arrival of buses and to trigger pre-emption, lack of standards to determine pre-emption warrants and inordinate delays to motorists travelling along the cross street. With increased application of Intelligent Vehicle Highway Systems (IVHS) concepts, bus pre-emption is likely to re-emerge as a tool for alleviating urban congestion. In the past, standard loops reacted to the presence of any vehicle, thereby making the system incapable of distinguishing buses from passenger cars. However, today's technology makes instrumented buses automatically distinguishable and candidates for preferential treatment. As an example, the "Vetag" system devised by Philips is used in the Hague by light rail transit vehicles as well as buses (6).

Purpose of the Paper

A framework for evaluating the operating cost consequences of bus pre-emption for a proposed privatized transit operation has been presented in this paper. The data-base is derived from a recent study in which markets for transit privatization for suburban travel in the Detroit Metropolitan area were identified (2). The basic premise of the above study is that high quality transit service, whether private or public, has a high potential of penetrating the market that is currently dominated by the private automobile.

Background

In the aforementioned study, the authors developed a procedure for identifying markets for transit privatization for suburban travel in a large metropolitan area. First, a demand based approach was developed that identifies zone-pairs with high travel demand. A procedure for identifying potential markets from these high demand sectors was developed by considering other explanatory variables, e.g., travel time, congestion levels and land use density (3).

The methodology, when applied on the Detroit suburban area resulted in a total of 14 potential markets that were merged in various combinations into five sectors for transit privatization. Operating plans for these five sectors for express, non-stop service were developed based upon an assumed 'market capture' of five percent (of the total travel demand) by transit, from all available modes.

2. METHODOLOGY

The following formulations were used in developing the operating plans.

$Nv \geq (Dp \times C)/(Vc \times 60)$ (A)
$H = C/Nv$ (B), where;

Nv = Number of buses required (Fleet size)
Dp = Hourly Passenger Demand at the maximum loading section
C = Cycle Time (minutes) including: Driving Time; Boarding/Unboarding Time and Layover Time.
H = Headway (minutes)
Vc = Bus Capacity (number of passengers)

Equation A shows that for a given demand Dp and bus size Vc, the fleet size can be minimized by reducing the cycle time C. Signal pre-emption is designed to reduce delays (and hence, driving time between two terminal points), thus resulting in reduced cycle time and a reduced fleet size and hence in reduced operating cost. The procedure for signal pre-emption evaluation consisted of estimating the driving times based upon (assumed) higher speeds and recalculating the reduced fleet size and reduced operating costs, directly attributable to signal pre-emption. The data thus generated was compared with operating data for the base condition.

3. RESULTS

Two sets of analysis are presented. Table 1 shows the basic operating data along with annual operating cost and fare-box revenue for privatized operation for the five sectors described earlier (termed 'Base Condition'). Table 2 presents similar data for bus operation under signal pre-emption for the same five sectors.

A comparison of the two tables indicate that reduced cycle time, resulting from signal pre-emption, is instrumental in reduction in fleet size and in operating cost. In all the cases analyzed, a reduction in deficit has resulted because of the reduction in fleet size.

4. CONCLUSIONS

The study shows that pre-emption strategies may help reduce operating deficit. Since the services proposed are for express, non-stop operation, pre-emption is likely to result in a significant reduction in cycle time. Reduced cycle time translates to a smaller fleet size resulting in a lower cost of

TABLE 1

Basic Operating Data & Fare-Box Revenue for Base Condition

Sector	Fleet Size Peak/ Off Peak	Headway (H) (Minutes)	Cycle Time (C) (Minutes)	Annual Operating Cost x 10^6	Annual Fare-box Revenue x 10^6	% Profit (Deficit)
1	P - 16 O - 7	6 12	96 - 100 84	$2.420	$1.335	(44.8)
2	P - 11 O - 5	6 12	66 - 70 60	$1.692	$1.413	(16.5)
3	P - 5 O - 2	12 30	60 - 70 60	$0.728	$0.536	(26.4)
5	P - 4 O - 2	15 40	60 - 70 60	$0.642	$0.497	(22.7)
7	P - 20 O - 8	5 10	100 80 - 85	$2.656	$1.793	(32.5)

TABLE 2

Basic Operating Data & Fare-Box Revenue for Signal Pre-Emption Condition

Sector	Fleet Size Peak/ Off Peak	Headway (H) (Minutes)	Cycle Time (C) (Minutes)	Annual Operating Cost x 10^6	Annual Fare-box Revenue x 10^6	% Profit (Deficit)
1	P - 11 O - 6	6 12	65 75	$1.842	$1.335	(27.5)
2	P - 9 O - 5	6 12	50 60	$1.520	$1.412	(7.1)
3	P - 4 O - 2	12 30	40 60	$0.642	$0.536	(16.5)
5	P - 3 O - 2	15 40	45 60	$0.557	$0.497	(10.8)
7	P - 16 O - 8	5 10	80 80	$2.570	$1.793	(30.2)

operation. Further, reduced travel time, brought about by pre-emption, is likely to contribute to a larger market share of the travel demand and a higher fare - box revenue. No effort has been made in this paper to assess the fiscal consequences of larger market shares in this study. Also, the adverse consequences to motorists travelling along the cross street need to be determined to make a comprehensive evaluation of pre-emption strategies. Further research is recommended to address these questions.

5. ACKNOWLEDGEMENTS

The study from which this paper is developed was made possible through a grant provided by the Urban Mass Transportation Administration (UMTA). Matching support was also provided by the College of Urban, Labor and Metropolitan Affairs, Wayne State University. The authors would like to express their appreciation to the above agencies for their support. The opinions and comments are those of the authors and do not necessarily reflect the official policies of any one of the agencies mentioned above.

6. REFERENCES

1. "Evaluation of UTCS/BPS Control Strategies." Washington, D.C.: Federal Highway Administration. Prepared by JHK and Associates, March 1975.

2. Khasnabis, S., and Chaudhry, B., "Privatization of Transit Services Between Suburban Communities in the Detroit Metropolitan Area." Final Report, prepared for UMTA, at Wayne State University. Detroit, 1990.

3. Khasnabis, S., Chaudhry, B., Nahan, N., and Neithercut, M., "Developing Markets for Transit Privatization for Suburban Travel in Large Metropolitan Areas," To be published in a forthcoming Transportation Research Record, National Research Council, 1991.

4. Tarnoff, P.J., "The Result of Urban Traffic Control Research: An Interim Report" Traffic Engineering, Vol. 45, #4, April 1975.

5. "Urban Traffic Control and Bus Priority System-Volume 1, Design and Installation." Federal Highway Administration. Prepared by Sperry Systems Management Division. November 1972. (NTIS-PB 214 788)

6. Vuchic, V.R., Urban Public Transportation System and Technology, Prentice Hall, 1981.

Advanced Software Design and Standards for Traffic Control

by Darcy Bullock,[1] AM. ASCE and Chris Hendrickson,[2] M. ASCE

Abstract: Improved traffic traffic management and control systems are widely reported to be cost effective investments [Neudorff 88, Kessmann 85]. However, many hardware and software obstacles have impeded the actual implementation of advanced traffic management systems. This paper identifies several issues that should be addressed in order to improve the integration capabilities of existing hardware. In addition, cheaper and more powerful computers have made the software used in such systems increasingly important. New computing, communication and software engineering standards are recommended to facilitate the future development and integration of advanced traffic management systems.

1. Introduction

Introducing advanced software environments and standards may have several advantages. First, configuration of a system controller for a particular intersection could be simplified, thereby reducing implementation time and the required programming expertise. Second, communication options may be increased and made more flexible, thereby aiding corridor level control technologies. Third, traffic signal control might take advantage of the hardware production economies and experience in other markets, thereby reducing the cost of controllers. Finally, flexible software platforms would allow "standard" devices to be used in custom applications such as weigh-in-motion systems and reversible lane installations.

In the following sections, we first review the limitations of existing traffic control software environments. Requirements for a new generation of control systems are proposed and these requirements are compared with the current practices for real time control. A final section suggests an evolutionary process leading to improved traffic control systems.

2. Problems with Current Traffic Control Technology

The current technology for traffic signal control and configuration is based on standards and protocols that are over a decade old [Chase 89]. This technology has a number of associated problems:

- Existing standards are microscopic in nature. They detail electrical interfaces, standard signaling practices, physical dimensions, and rigid, inflexible microprocessor platforms.

- The interfaces to controllers are overly complex and difficult to manipulate. This problem has been recognized and many companies provide microcomputer based software for configuring controllers. This software allows better editing facilities, but preserves the underlying complex interface. Little has been done to improve the transition from conceptual design to actual implementation.

---

[1] Research Assistant, Department of Civil Engineering, Carnegie Mellon University, Pittsburgh, PA 15213; 412/268-3781; Internet: bullock@ce.cmu.edu

[2] Professor, Department of Civil Engineering, Carnegie Mellon University, Pittsburgh, PA 15213; 412/268-2948; Internet: cth@ce.cmu.edu

- The existing microcomputers are not powerful enough to accommodate the increased memory and processing demands required for more sophisticated control.
- Finally, budget constraints and low-bid selection often dominate decisions when evaluating any signalization project. In order to stage improvements to signalization systems, the hardware must not only be modular, but support common software and communication protocols.

The traffic control community is not isolated in its quest for modern, easy to use systems. Many of the same issues facing traffic engineers currently face engineers in manufacturing and industrial control. Organizations involved in these areas have sought to address system integration issues by adopting standards. In developing such standards, it is always difficult to balance general standards (such as specifying a communication port) and specific standards (such as specifying an RS-422 serial port supporting a specific protocol). If standards are precisely detailed, but leave room for flexible implementations, they benefit the user community by ensuring compatibility between vendors and promoting competitive bidding. However, if they are loosely detailed and do not provide an adequate foundation for applications without additional *proprietary* features, they leave room for vendors to craft custom features that are not uniform from vendor to vendor. These issues can be illustrated by examining two of the most common standards in traffic control: the National Electrical Manufacturers Association (NEMA) and the California Department of Transportation (Caltrans) Type 170 standard.

The NEMA standard specifies details for the mechanical and electrical nature of all the connectors on a controller unit.[3] In concept, the modular connectors attached to one controller conforming to the NEMA standard can be disconnected and another vendors controller (also conforming to the NEMA standard) can be reconnected. This is true if the controller is operating at an isolated intersection and the software configuration is disregarded. However, the NEMA standard is electrical in nature and does not specify how controllers are configured. The problems inherent in the existing NEMA standard are widely recognized. In particular, the NEMA standard has not adequately addressed system coordination, time base control, preemption, uniform code flash, communications, or diagnostics [TCT 90]. Consequently, each manufacturer has implemented custom features in a unique way on their controller. These proprietary features counteract the interchangeability spirit of the NEMA TS1 standard and as a result, technicians installing a new unit must understand new configuration software. Further, connecting a standard NEMA controller to detectors, conflict monitors, and other controllers requires the use of one conductor for each signal. If all the connections were used, the unit would have 322 conductors leading into the controller [TCT 90]. This type of networking is not cost effective in view of modern communication technologies and provides a multitude of potential failure points. This communication problem has been recognized by various vendors and resulted in additional proprietary communication and printer ports.

The Caltrans 170 controller specification addresses standardization from a portable software perspective by precisely detailing a 6800 based microcomputer. These Caltrans 170 controllers are general purpose computers with diverse applications ranging from pump control to networked traffic control systems. In this architecture, a general purpose rack mountable computer is fitted with traffic control firmware. In concept, this standard is very effective at providing a software migration path for controllers because of the flexibility afforded by a general microcomputer. However software is rigidly tied to the existing hardware architecture because no standard operating system or kernel is defined [Chase 89]. The reality of the Type 170 standard is that the application software produced for Caltrans 170 controllers must be crafted using low level code and cross compilers. Much of the application software developed for the Caltrans 170 controller is capable but difficult to configure. Minor modifications to accommodate nuances of unique intersections often must be contracted out to specialists or simply ignored. This often results in sub-par performance because minor changes can be too difficult or expensive to implement and maintain.

Reviewing the NEMA TS1 standard and the Caltrans 170 standard demonstrates the recognized

---

[3]In this comparison we will only consider the 1988 NEMA standard, designated TS1. The now proposed NEMA TS2 standard addresses some of the deficiencies in the TS1 standard, but as of this writing has not been officially released or adopted.

need for uniform electrical interfaces and flexible software platforms. However, neither standard provides an adequate framework for more sophisticated hardware platforms or software engineering methods. Development of software standards for configuring controllers, network installation, and system integration, should be examined in order to reduce software maintenance costs. In fact a recent paper reviewing current traffic control systems noted: *"Software used in today's control system architectures often plays a more crucial role than does the hardware"* [Chase 89].

3. Requirements for New Software Technology

Simply inventing a new generation of controllers with the latest and fastest processors, huge quantities of memory, and fast networking will not guarantee major traffic control improvements. In order to achieve effective traffic control, systems must be easy to assemble, configure, and most importantly maintain. Current notions on what constitutes maintainable hardware are very well established. However, attempting to define maintainable software is much more difficult and falls under the broad classification of *software engineering*. This fundamental difference between hardware and software maintenance can be illustrated by examining the failure rate of both hardware and software systems. Figure 1 illustrates a typical pattern associated with software systems. During initial startup a number of errors are detected and corrected. Latter, small changes designed to improve the operating efficiency or integrate new functions result in a sudden increase in problems. These deficiencies are corrected and the failure rate begins to settle down until the next change. Consequently, the idealized failure curve is never realized. A similar curve for hardware failure rates would be shaped like the idealized curve shown in Figure 1, except, it would begin to increase when the life expectancy was reached. The resulting curve would be somewhat 'U' shaped.

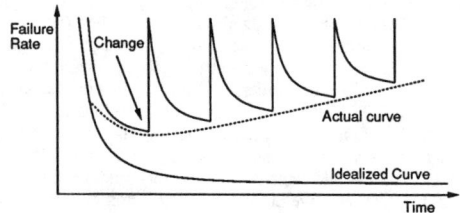

Figure 1: Software Failure Curve [Pressman 87]

By viewing traffic control systems as a software engineering project, the entire life cycle is considered. Application of software engineering concepts provide a structured framework for development, coding and documentation. This avoids the problems associated with "one-man-show" development and defines metrics for evaluating the software before it is developed. Imposition of a structured development does not eliminate the need for creativity in the overall software concept.

As one article discussing the software development for several real time systems noted, *"A sound problem-relevant philosophy is the key to achieving successful implementation of complex computer based systems. Software engineering methods and tools will naturally flow from this foundation."* [Lawson 90]. This leads us to present several issues which should be addressed by the implementation philosophy:

- Traffic controllers need to be configured by traffic engineers and a standard representation should be developed for configuring them. Currently, manufacturers' control strategies, symbols, and terms differ from those provided in ITE Recommended Practices [Marshal 90]. Integration and standardization of this representation with stochastic simulation and design programs will provide better tools for effective strategy design and timing.

- The configuration architecture needs to be open and extensible with provisions for incorporating interfaces to special purpose systems used in Advanced Traffic Management

Systems (ATMS) and Advanced Traffic Information Systems (ATIS). For example variable message signs, driver information systems, vision systems, sonar systems, ramp metering, and incident detection should be accommodated by a new architecture.

- Existing electro mechanical and solid state controllers cannot be discarded when new controllers with modern technology are introduced. Rather, older existing control zones must be capable of interfacing with new control zones. Proprietary circuits and custom microprocessor systems used to bridge incompatible equipment should be replaced by highly configurable, distributed controllers conforming to standard software metrics.
- Advances in communication technologies have made networks ranging from twisted-pair cabling to fiber optic strands superior alternatives to the electrical interconnection mechanism defined in the NEMA TS1 standard. Obviously, the varying costs, ranges, capabilities, and capacities of the various network alternatives cause different media to be selected for different applications. Software realizing the benefits of networking must also be abstract enough to migrate to more sophisticated media as needs change.
- System costs must remain competitive with existing units. Advances in hardware manufacturing have caused software costs (development, maintenance, enhancements) to account for a significant portion of a systems costs. Hence, systems will have to provide clear engineering benefits (reduced development, lower maintenance, flexible extensible architecture) to warrant the additional cost of hardware and software.

Recognizing the importance of software engineering in traffic control, it is important to develop quality software engineering techniques that reduce costs associated with steep learning curves and minimize software maintenance costs. No longer is it acceptable to purchase hardware and then try to design software to meet a regions' needs. Rather, the software must first be carefully selected and engineered to provide the maximum flexibility with the least amount of maintenance. This criteria can be interpreted very widely (from assembly language coding, to merely entering cycle, split and offset). Most likely, regional and local traffic control needs are best served by a compromise between these two extremes.

4. Discussion

Previous sections have outlined problems with current traffic controllers and cited requirements for new control system software. Keeping in mind the existing huge investment in traffic control systems and the fact that most of the systems do work, an evolutionary process towards new software environments and standards must be followed. This does not preclude introducing new controllers, new programming methods, or new networking concepts, as long as adequate provisions are made for interfacing to existing equipment. Following these guidelines, an outline detailing desirable software and complementary hardware is described below. The most important point of this system outline is that the software and hardware technologies described already exist and have been implemented in several other domains. The points below represent some required steps to take advantage of these technologies.

1. Intuitive Representation.
The single most important advance needed in the area of software engineering for configuring traffic control devices is an intuitive representation. Such a representation would allow the engineer to design control algorithms using graphical icons and data exchange with simple graphical connections. This *function block*, or object oriented approach was pioneered by the Foxboro company in the early 1970's and has evolved to a point where there is almost a one to one correspondence between Process and Instrumentation Diagrams (P&ID's) and the graphical implementation of the control system [Elwart 90]. The engineering tasks for porting this type of system to traffic engineering require the definition of suitable functions blocks, in combination with updating the ITE Recommended Practices.

2. Integrating Simulation and Control.
An approach often used by large system integrators in other industries involves staging and simulating an entire control system. The technology currently exists to construct entire computer simulations with displays indicating microscopic intersection performance and macroscopic network performance.

However, these simulations are often done using traffic control algorithms different from those implemented. If geometry and network topology are included in a standard representation, then stochastic modeling and simulation programs can be used directly with the defined control system to evaluate performance and potential impact of proposed improvements. Conceptual models could be constructed by the traffic engineer and adjusted to improve system performance. Confidence would be gained in the performance of the control system and economic viability could be justified before the system would be purchased and installed. Considering the conservative posture of most public works departments, this capability would be more important as systems become more complex and intimidating.

3. High Level Networking.
The traffic control community needs to develop an abstract network representation for defining communication channels. Industrial networking and protocols may provide an effective standard or starting point.

4. Integration with NEMA Standards.
While the NEMA TS1 standard has been widely criticized, the contributions it has made in standardizing electrical connections should be recognized. The proposed NEMA TS2 standard also make significant contributions by recognizing the need for high speed communication ports, reducing the number of hard connections, and recognizing the need for better user interfaces. Because the NEMA standard is predominantly an electrical standard and does not rigorously define the communication protocol to be used by the communication ports or the software interfaces for configuring the units, it must be supplemented with software and networking standards.

5. Integration with Industrial Hardware.
The past decade has seen tremendous growth in the power, quantity, and quality of industrial computation. Industrial platforms are becoming more powerful and highly configurable. Because industrial systems employ much if not all of the technology required for traffic control, it seems likely that traffic control manufacturers may benefit by entering into an Original Equipment Manufacturer (OEM) arrangement with controller companies and then tailoring software and electrical interfaces to produce a powerful traffic control device. The multitude of industrial VME bus systems appear a likely candidate at the current time.

5. References

[Chase 89]   Chase, M.J. and Hensen, R.J., "Traffic Control Systems - Past, Present and Future," *Applications of Advanced Technologies in Transportation Engineering*, ASCE, pp. 257-262, February, 1989.

[Elwart 90]   Elwart, S.P. and Martin, P.G., "New Software Structures Extend Control Capabilities," *Control Engineering*, Vol. II, June 1990.

[Kessmann 85]   Kessmann, R.W., C.S. Ku, and D.L. Cooper, *1.5 Generation Feasibility Study*, Technical Report, Kessman and Associates, Houston, TX, February 1985.

[Lawson 90]   Lawson, H.W., "Philosophies for Engineering Computer-Based Systems," *Computer*, Vol. 23, No. 12, pp. 52-62, December 1990.

[Marshal 90]   Marshal, P.S. and Berg, W.D., "Evaluation of Railroad Preemption Capabilities of Traffic Controllers," *Transportation Research Record*, No. 1254, pp. 44-49, 1990.

[Neudorff 88]   Neudorff, L.G. and Terry, D.C., *Communications in Traffic Control Systems, Volume II: Final Report*, Technical Report FHWA-RD-88-012, FHWA, Washington, DC, August 1988.

[Pressman 87]   Pressman, R.S., *Software Engineering: A Practitioner's Approach*, McGraw-Hill Book Company, 1987.

[TCT 90]   Engineering Department, *Migration Path For Controllers*, Technical Report, Traffic Control Technologies, Liverpool, New York, 1990.

Error rate measurements of RDS-FM transmissions Application to an RDS-beacon

by M. Heddebaut[1], M. Berbineau[1], M. Szelag[1]

ABSTRACT : The Radio Data System is now commonly implemented on many European VHF-FM broadcasting transmitters. RDS provides for the transmission of a silent data channel on existing VHF-FM radio stations. RDS achievements on open areas ie roads and motorways have been extensively assessed in Europe these last years. This paper will present the performances of an RDS retransmission using leaky feeders in road tunnels. Then starting from these results we propose a new open road application of leaky feeders dealing with the possibility to transmit very locally specific RDS informations without altering the received audio program by the car driver.

Introduction

DRIVE I is a precompetitive, prenormative R&D program involving collaboration between over 1,000 transportation-related experts from 300 European organizations. It has a conceptual goal of a fully integrated road transport environment (IRTE) in which intelligent vehicles are linked to an intelligent road network.

CERACS (Comparative Evaluation of the different RAdiating Cables and Systems technologies) is one of around seventy projects within DRIVE I. It is aiming to establish possible new applications of leaky feeders for road to vehicles communications.

Today a couple of almost available technologies for which the standards are already made, are available: the Radio Data System and the Pan European Cellular Mobile Telephone, the GSM.

The Radio Data System (RDS) is a facility, defined by European Broadcasting Union (EBU) specification (1, 2) that provides for the transmission of a silent data channel on existing VHF-FM radio stations. Its primary purposes are to identify radio broadcasters, and allow self tuning receivers to automatically select the strongest signal carrying a particular program. The Radio Data System Traffic Message Channel (RDS-TMC) will be introduced in Europe in the mid-1991s. TMC will provide interfaces linking it to other intelligent vehicle highway system technologies. As an example, it will become a valuable complement to the autonomous navigation devices, which are being developed and tested, based on dead reckoning and map matching.

RDS leaky feeder receiving quality measurement

From the initial work done into CERACS, useful cables for road applications correspond to the leaky feeders working in a frequency range situated between 100 and 900 MHz. Two categories of leaky feeders working into this frequency range can be distinguished : the so called "coupled modes cables" and the "radiated modes cables" (3). We distinguish in these cables various properties as the structure of their shield, the working frequency range and the variation of the field around the cable.

---
(1) INRETS-CRESTA 20, Rue Elisée Reclus F-59650 Villeneuve d'Ascq - FAX (33) 20 67 08 32
France

The "coupled modes cables" include the coaxial structure and the bifilar line. The shield of the coaxial structure is then drilled by small apertures spaced by intervals very much smaller than the wavelength. This structure gives a wide frequency bandwidth. The coupling mechanism between the inner and the outer part of the cable is governed by a coupled mode process and the electromagnetic field is focused in a virtual cylinder around the cable.

The "radiated modes cables" include only the coaxial structure in which the radiating elements are periodic inclined slots made on the shield. For this reason the electromagnetic coupling is governed by the radiation of an array of equivalent magnetic dipoles. The coupling with the outer part of the cable is good and the electromagnetic field is expanded around the cable. However this condition is satisfied in a frequency range which is directly related to the period of the slots. For mechanical reasons, into the VHF band, only the coupled modes leaky feeders can be easily built. Figure 1 shows an illustration of the electromagnetic radiation from a conventional antenna and from a leaky feeder.

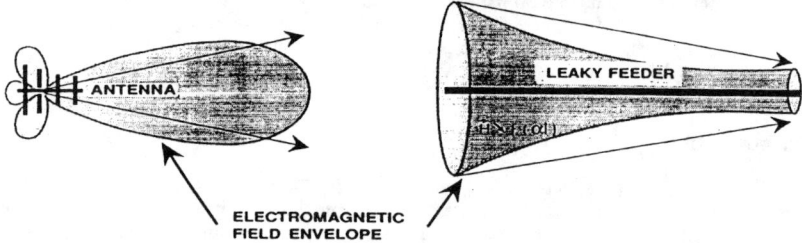

Figure 1: Comparative radiation of a conventional antenna and from a leaky feeder

Let us now consider a radio FM retransmission in tunnels using a leaky feeder. An RDS frame is obtained by the continuous transmission of groups of 104 bits. Each group is composed of 4 blocks of 26 bits. In each of these blocks, 15 bits transmit data and 11 bits are used for redundancy and forward correction. Each of these blocks are analyzed using a specialized code implemented on a personal computer. Results are presented on the following figure 2. We have represented on the x-axis the distance. On the y-axis we indicate firstly the number of errors recorded per block: 1, 2, 3, 4 or rejected block due to too many errors. On the y-axis we have also represented the amplitude of the received VHF signal inside the car. These results have been obtained in the St Cloud tunnel near Paris.

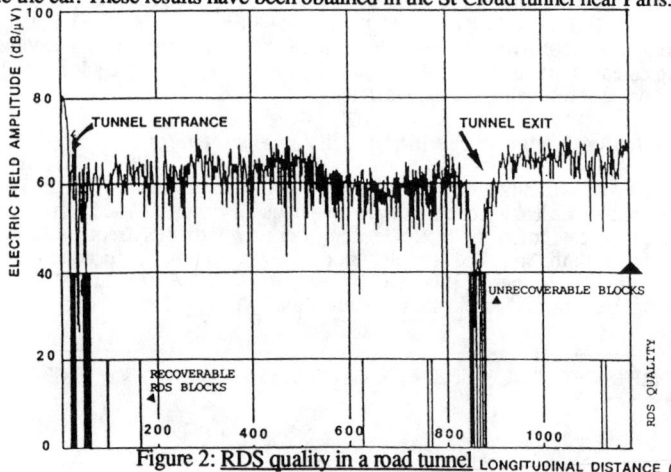

Figure 2: RDS quality in a road tunnel

As we can see on this figure 2, most of the errors occur at both ends of the tunnel and are directly related to the amplitude of the electric field fluctuations. In fact there is a beating phenomena between the wave radiated by the leaky feeder and the wave coming from the distant broadcasting VHF transmitter. This will be a crucial problem for the development of the RDS beacon to be now introduced.

The problem of the data rate

RDS-Traffic Message Channel is using the 8A group which is one of the thirty two possible RDS data groups. If we consider the total number of groups transmitted, the EBU specifies that 40% of them must be 0A and 0B groups carrying the program identification, the program service name, the program type ..., 10 % of them must be 1A or 1B groups, 15% 2A or 2B groups, 10% of the groups are 14A or 14B and 25% of the groups are devoted to all the other applications including TMC. Thus the effective data rate for the TMC is rather limited and a great deal of effort has been done throughout DRIVE to efficiently use this resource. Nevertheless, if we want to provide interfaces linking TMC to on board electronic equipments like road guidance systems or digital mapping it would be worthwhile to dispose locally from a much higher data rate.

RDS beacon principle

Let us consider the following figure 3:

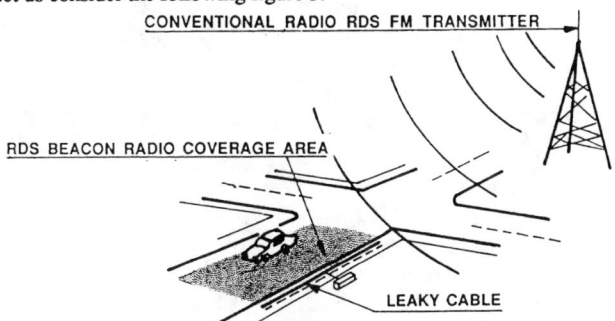

Figure 3: <u>RDS beacon principle</u>

A short length of leaky feeder or a monofilar mode cable (antenna) is laid along the road and fed by a transmitter. The transmitter feeds the cable with a signal at the **same frequency** that the VHF-FM radio program listened by the car driver moving along the road. Its output power is correctly chosen so that in the vicinity of the leaky cable, the **radiated power by the leaky feeder is higher than the received power from the conventional broadcast radio station.** We form this way a cell where the FM signal received by the car radio set is locally the one emanating from the leaky cable and no more the one coming from the distant conventional VHF broadcast station. The car antenna receives that new signal at the same frequency when going into the cell formed by the leaky feeder. Let us now modulate the leaky cable signal by **the original stereophonic audio program**. Due to the capture effect of the FM modulation the radio set now delivers the "leaky feeder" modulation to the car driver and no-more the "conventional radio" modulation. As both are identical, there will be no change for the driver from the audio point of view. But it is now possible to superimpose on the **leaky cable signal a specific RDS frame using locally the full data rate of RDS for traffic information.** Thus it is possible:
. to transmit locally a specific RDS-TMC message,
. without altering the audio program received by the driver,
. without the need of a new frequency,
. with a conventional RDS radio set in the car.

Several applications could be developped:
- to transmit specific digital informations to mobiles,
- to update informations for dead reckoning navigation systems,
- to deliver specific motorway informations,
- to send the car a localization information,
- to update locally digital maps,
- to manage parking and tunnel managment.

We have now to carefully choose the following parameters : output power, coupling losses, technology of cable, positioning... to obtain a ratio: RF level from the leaky cable to RF level from the local broadcast transmitter compatible with good performances of the beacon.

Static measurements

In laboratory, we can accurately change the ratio between the beacon power and the broadcast VHF-FM transmitter injected in a receiver using a step variable attenuator. The following figure 4 shows the results obtained.

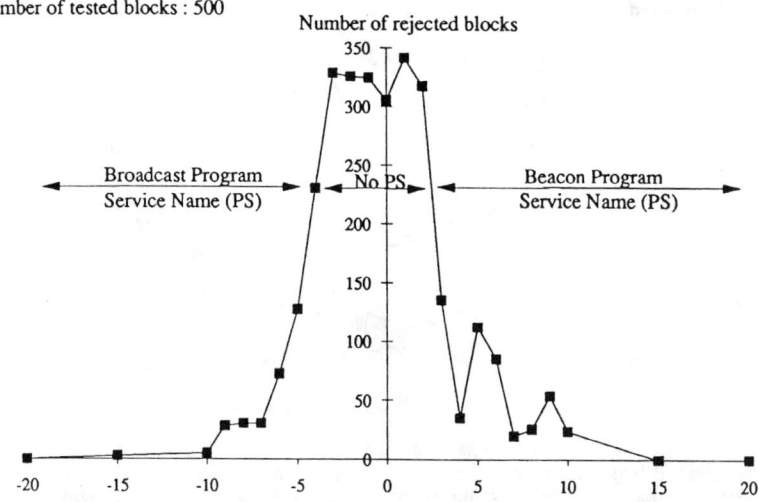

Figure 4: <u>Static RDS beacon test - Number of corrupted RDS blocks as a function of the power ratio between the beacon to broadcast input power on the receiver</u>

As we can deduce from this curve, if the ratio between the beacon power and the broadcast VHF FM transmitter equals ± 10 dB, we receive either the broadcast RDS signal either the RDS beacon signal with few block errors and a good audio quality. This measurement can be repeated for different reference input levels with similar results. It is now necessary to determine a system which enables us to radiate locally an RF level sufficient to obtain this ratio of 10 dB into the whole beacon surface with the lowest residual level outside the cell.

Electromagnetic measurements

Field trials have been performed to deduce the best suitable radiating array using either a progressive wave antenna (long wire) either a low coupling loss leaky cable. Some results are presented on the figure 5.

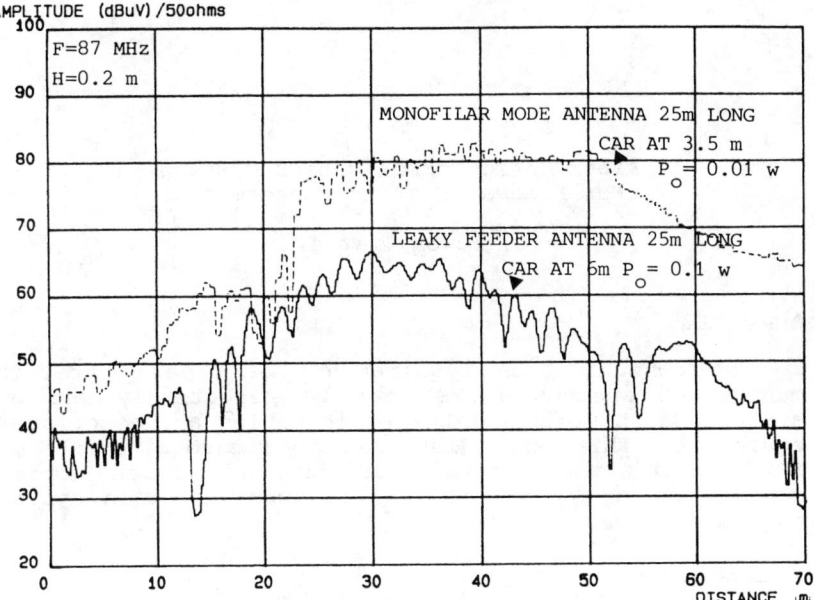

Figure 5 : <u>Electromagnetic field strength measurent radiated by a 25 meters long wire or leaky cable and received onboard a car moving along the beacon</u>

Comparing the monofilar line radiation and the leaky feeder radiation, we can deduce that both of these structures can maintain locally a higher electromagnetic field than the one radiated by the distant broadcast transmitter. Measurements have been performed for distances running from 1m to 12 m from the structure while the car moves alongside the beacon. The leaky feeder structure is preferred as it allows a faster attenuation of the beacon signal when the car leaves the extremities of the radio beacon cell.

Conclusion

The technical feasability of this RDS beacon has been shown on an experimental test site. If the necessary administrative authorizations can be obtained, this beacon could be implemented on a larger scale in a future pilot project to be developped into Europe within the framework of the DRIVE II project.

References

(1) -   European Broadcasting Union: Technical Document 3244
        "Specifications of the radio data system RDS for VHF/FM sound broadcasting"
(2) -   European Standard CENELEC EN 50067A
        "Specification of the radio data system"
        European Committee for Electrotechnical Standardization
        Brussels December 1990
(3)   Radio Data System links using leaky feeders and E.M.C. problems
        by M. Heddebaut, M. Berbineau, M. Szelag, M. Klingler,
        F. Mussino and E. Nano
        DRIVE Conference'91 - Brussels february 1991
        Proceedings pp 1358-1383

Fiber Optic Communication Systems For Freeway Traffic Management

D. Bowen Tritter [1]

Abstract

The communication requirements of freeway traffic management systems (FTMS) can be effectively met by systems which utilize fiber optic cables and associated equipment. This paper discusses why the requirements of FTMS suit the characteristics of fiber optic communication systems and various communication system implementations.

Introduction

FTMS communication systems must meet the unique requirements of FTMS in general and the specific requirements of particular freeway networks. Fiber optic communication systems offer specific advantages for FTMS. However, the application is quite different from typical communication system implementations used in the telecommunications and cable television industries. A goal in designing communication systems for FTMS is to meet the performance, environmental and other unique requirements while taking advantage of the availability of products developed for larger markets.

FTMS Communication Requirements

The communications requirements of freeway traffic management systems in most cases include video, data and voice for maintenance. The communication system should be designed with appropriate levels of reliability and performance. For example, the level of reliability and availability need not approach the availability of the telephone networks and high reliability computer communication networks. In our experience the required

---

[1] Head, Communication Section, National Engineering Technology & Delcan Corporation, 133 Wynford Drive, North York, Ontario, Canada, M4J 1T1.

reliability and the impact of failures should be evaluated and the requirements determined jointly by the operating authority and the communication system designer.

Video

The video communication requirements of FTMS are ideally suited to a fiber optic implementation since the sources (cameras in the field) are distributed throughout large geographical areas usually along a limited number of trunks. For this reason the capacity requirement of the communication network near the control center is large, while in the outlying areas it is quite low. This is significant since the number of fibers in the cable can then be tapered which reduces the overall cable cost. High video quality, although not broadcast quality, is required and some jurisdictions require color transmissions. An important user requirement is to determine if all images are required at the central control center without blocking (e.g. any camera can be viewed on any monitor). This is determined by the philosophy of the operating agency, the confidence which they have in automatic incident detection and the control center location. Video represents, by several orders of magnitude, the leading requirement in terms of communication capacity of the communication network and is the driving factor behind the selection of fiber optic communications over other media such as coaxial cable.

Data

Data represents a relatively low capacity requirement when compared to video. Bi-directional full or half duplex asynchronous communication channels are required for changeable message signs, lane control signs, incident detection, camera positioning and control and other system specific requirements. These circuits operate at a low bit rate and are usually shared between a number of field controllers (up to 20 is typical). In spite of the fact that the number of field units is high the overall capacity required is relatively low. The reliability of the data network is important, but very high reliability is not essential. It is important that the geographical effect of single failures be limited so that the entire network does not fail.

Voice

Voice circuits are used to connect highway advisor radio (HAR) transmitters to the control center, for remote citizen band (CB) radio station and many jurisdictions require voice for maintenance purposes.

The point to point voice circuits required for HAR and CB can be effectively provided though voice channel cards in the data multiplexer. Although it is feasible to provide maintenance voice circuits for service personnel through the fiber optic network we have found radio based service is more cost effective and reliable.

Characteristics Of Fiber Optic Communication Systems

Fiber optic systems have a number of highly desirable characteristics including, very high capacity, immunity from interference, large distance between repeater or amplifier locations, light weight cable and cost proportional to capacity based on the number of fibers.

For these reasons and others fiber optic systems have been the technology of choice in the telephone industry for many years and are rapidly becoming common in local area networks (LAN) and cable television systems (CATV). However, fiber optic systems for traffic management are different from industry norms due to unique requirements, as well as environmental and budget constraints. In spite of these differences equipment and techniques developed to utilize fiber optic cables in other industries can be effectively utilized for FTMS.

Typical Configurations and Components

The data network and video network are typically configured as separate systems using a common cable plant except systems which utilize digital video which use some common equipment. Typical video configurations suitable for freeway traffic management include:

a) Point to point video with a single fiber per video signal. This configuration is used for portions of the Highway 401 FTMS (COMPASS) in Toronto and in Minnesota and will be used for the entire automatic traffic control (ATC) system in Kowloon, Hong Kong.

b) Multiplexed video using frequency modulation (fm) of the video signal and frequency division multiplexing (fdm) of multiple (4 to 16) video signals on a singlemode fiber between a field node and the control center is the most common configuration. Collection of the video images at the node is through point to point links as in a) above. This configuration is used by COMPASS and the Corridor Traffic Management System (CTMS) in Toronto, for the Traffic Management System in Minnesota, and will likely be used in California and other locations.

c) Multiplexed video using fm and fdm as in b) above but

using a video drop and insert technique at each camera site similar could be used to eliminate the requirement for nodes.

d) Amplitude modulation (am) and fdm as used by the cable television (CATV) industry is an option for high capacity trunks providing 40 or more video channels on a single fiber. This will also provide compatibility with existing coaxial cable based systems.

d) Digital encoding of the video signals can be used to allow an all digital integrated network for video, data and voice. This requires high speed (and cost) video encoders, decoders, digital multiplexer and fiber optic transmission equipment. Local connections are as in a). This configuration is currently used in Taiwan and Japan. As technology improves, reducing the bandwidth required to transmit a video signal of acceptable quality, digital video will become more attractive.

There are a variety of configurations which are suitable to meet the data communication requirements as well, these are:

a) Conventional twisted pair copper cables utilizing low speed asynchronous modems and a polled multi-drop configuration as is used in most urban traffic control systems and the existing TMS in Minnesota.

b) Multiplexed data at node locations using time division multiplexer, fiber optic transmitters/receivers and redundant fiber optic cable data. The data can be collected at the node locations by conventional twisted pair cable as in a) above or a low speed asynchronous fiber network. The former approach is being done in Minnesota and the later in Toronto.

Voice can be provided through the data multiplexer on a fiber optic network, by copper twisted pair on a copper network or through a leased or dedicated radio service. The radio option is preferred for maintenance since voice service should not be interrupted during network failures when it is most needed.

Important considerations for the evaluation of the various configuration include redundancy of key equipment and cable segments, expansion requirements and provisions for future techniques such as Intelligent Vehicle Highway Systems (IVHS), performance and environmental requirements.

Trends and Standardization

The use of "standards" is encouraged to allow easier expansion of the system and to limit the reliance on single supplier products. Often telecommunications and CATV standards are not applicable to FTMS applications. The standards to be utilized must be considered carefully in conjunction with system expansion plans.

Improvements in technology and reduction in the price of fiber optic cables and equipment over the last five years have resulted in dramatic shifts in the applicability of fiber optics for a number of applications, including traffic management. This trend is expected to continue with the greatest improvements predicted in the terminal equipment, resulting in lower cost and more robust fiber optic systems. Data and video equipment being built is geared to the smaller, less sophisticated end user. Expected improvements in digital video will increase the use of "all digital" networks.

The emphasis being placed on more sophisticated traffic management techniques including communication with individual vehicles for IVHS programs and remote incident detection using video cameras will result in changes in the communication system requirements. In order to accommodate these changing demands the communication systems must maximize the use of communication standards, provide ample spare capacity and be planned in consultation with the future users with their requirements in mind.

Summary

The experience of Delcan Corporation and National Engineering Technology (NET) in designing and implementing communication systems for traffic management applications worldwide has shown that the use of optical fiber is viable and cost effective. Communication systems for FTMS have some unique characteristics and requirements which can be accommodated by careful design and a good understanding of the application.

It is our view that fiber optic communication systems will be utilized increasingly in the future as sophisticated traffic management systems are implemented and fiber optics become more accepted in traffic and transportation industries.

TRAFFIC SIGNALS AND AT-GRADE LRT

Kevin Fehon[1]

Abstract

Light rail systems are making a comeback in American cities. This often generates local concerns about traffic congestion. This paper discusses the application of computer technology to analysis, design and operations in some Californian cities, to minimize the impact of at-grade LRT operation on street traffic.

INTRODUCTION

More cities in North America are turning to light rail transit (LRT) as an economical alternative to expanding their traffic-clogged freeways and urban arterial roads. New systems have been built in San Diego, San Jose, Buffalo, Sacramento, Portland, Detroit and Los Angeles in the United States, and Calgary and Edmonton in Canada. At the same time, major extension or refurbishing has been undertaken in San Francisco, Boston and Philadelphia, as described in Fehon, et al.(1988).

The operating environment of these systems is often varied, including: single and double track sections; exclusive right-of-way (ROW), with or without grade separation; ROW shared with heavy rail; exclusive lanes at the side or in the center of roadways; and lanes shared with road traffic.

The authors have been involved in the design of new LRT systems in Sacramento, San Jose and Los Angeles, and proposed extensions in San Francisco. This paper describes the new computer-based analysis tools used in the design stages and the new control systems integrating LRT and traffic signals.

Operations Analysis

The analysis of the operational impact of LRT on street traffic cannot be undertaken using the traditional

---

[1] R.J. Nairn and Partners, 214 Northbourne Ave., Braddon, ACT 2601 Australia.

traffic engineering analysis tools. Nor can the impact of traffic, including traffic signals, on LRT operation be easily assessed with traditional tools.

New analysis tools were developed by the authors for a number of purposes:

> To estimate LRT travel times through signalized intersections
> To assess impacts on automobiles and pedestrians of LRT operation
> To determine the need for grade separations

Basic capacity analysis at intersections was achieved by extending the HCM method to include percentage of vehicles stopped and probabilistic measures of queue length (Fehon, 1987), and assuming the average impact of LRV phases per cycle. Where queue length may have a critical impact, a time series analysis was performed of LRV passage through an intersection, to illustrate the growth and decay of queues.

The impact of coordinated traffic signals on LRV travel times is assessed by a program which tracks LRV trajectories, allowing for different, pre-set levels of LRT priority.

Comprehensive operational analysis is available through ROADTEST, a microscopic rail and road traffic simulation model. It simulates the movement of all the individual vehicles (car, bus, LRV, freight trains and pedestrians) in a road and rail network. Both traffic and train signals are modelled in detail, including vehicle and train detectors. An animated display provides confirmation of performance and simplifies calibration.

Innovations in Signal Operation

The need for LRT priority is well understood (see for example Yagar et al., 1988) and implemented in various cities such as Melbourne, Australia (Cornwell, 1986).

However, the LRT lines discussed here were planned or constructed along or across many streets with high daily volumes, many of which already experienced significant peak hour congestion. The greatest fear practitioners and the public alike was that special phases at traffic signals to accommodate LRT would cause short-term congestion each time a light rail vehicle (LRV) passed through an intersection.

Signal Priority

There are two main types of active priority for LRT: full and partial. Full priority is similar to railroad preemption, except that safety timings such as pedestrian and vehicle clearance intervals are not violated. The LRT phase is run at the earliest possible opportunity. Under partial priority, phases are lengthened, shortened or skipped within operator established limits in order to reduce the potential delay to LRV's.

Design Elements

In San Jose, the LRT generally runs in the central median of arterial roads, with the downtown operation in a pedestrian mall. Along the arterial roads, the signalized intersections are sufficiently far apart that a detector on the departure side of one intersection gives ample warning to the next signal of the approach of a LRV. New controllers were installed to operate time base coordination as well as full and partial priority as appropriate.

A significant number of the signals through which the Sacramento LRT passes are closely spaced, with the block length just long enough to accommodate a stationary four car train. It was therefore necessary for advance detection to take place one or two intersections upstream of these signals. New microprocessor based controllers were installed and integrated into an existing interconnect system.

The Long Beach to Los Angeles LRT line provided additional challenges. The line includes most of the operating environments described above, and the on-street segments pass through many heavily trafficked intersections, particularly in the vicinity of the Los Angeles Coliseum and through downtown Long Beach. In addition, it passed through five different traffic jurisdictions, some more than once.

The LB/LA Signal System

System Objectives

The system design objectives developed to accommodate the various community perspectives may be summarized (Fehon et al, 1990) as to provide a system which:

Causes the minimum delay to Light Rail Vehicles (LRVs), consistent with the prevailing traffic conditions;
Permits the maximum amount of flexibility to both traffic signal and light rail operators;
Is safe and readily understood by drivers of LRVs and road vehicles;
Uses commercially available equipment;
Is compatible with the existing traffic signal systems operating along the LRT corridor; and
Has sufficient flexibility that it can be transferred to new LRT lines.

System Components

A fully integrated approach was taken to the design of the system, ensuring that no part of the traffic signal or LRT signal system operated in isolation, compromising the effectiveness of other system components. There are five major components: LRT detectors, travel time predictors, signal and sign controllers, central master computers and the LRT System Control And Data Acquisition (SCADA) center.

To effectively provide LRT priority it is necessary to predict the arrival time of an LRV sufficiently early to allow the controller to make the necessary adjustments to signal operation. To do this a travel time predictor is connected to all signal controllers between each pair of stations. When the first detector in the section is activated, generally as an LRV leaves a station, the first controller calculates the expected time before the LRV will pass through the intersection, based on the current position within the signal cycle and the level ofpriority permitted. It then calculates the expected time of arrival at the next downstream signal and informs the predictor. The predictor passes this on to the next controller. That controller calculates the time at which the LRV will pass through its intersection in the same manner.

These travel time predictions cascade through the controllers, via the predictor, generally up to the next station. As each advance detector is activated, the process is repeated, and the predicted arrival time at each intersection updated and passed on.

Conclusions

The requirement for LRV's to have some form of priority varies with the design of the system and the prevailing traffic conditions on the adjacent streets. The experience with LRT priority in Sacramento and San Jose provided the basis for the development of a complex LRT priority system in Los Angeles. The traffic conditions, existing advanced signal coordination systems and multiplicity of jurisdictions with traffic control responsibilities created many difficulties which were overcome in an innovative fashion, using the latest technology appropriate to traffic control.

REFERENCES

Cornwell, P.R. (1986) Dynamic Signal Coordination and Public Transport Priority in Proc. Second International Conference on Road Traffic Control, Institution of Electrical Engineers, London

Fehon, K. J., (1987), Signal Capacity Analysis - Beyond the 1985 HCM in Proc. Inst. Transportation Engineers, District 6 Annual Meeting, Reno.

Fehon, K. J., Tighe, W. A. and Coffey, P. L., (1988), Operational Analysis of At-Grade Light Rail Transit, in Light Rail Transit, New System Successes at Affordable Prices, Special Report 221, TRB, Washington, D.C.

Fehon, K. J., Tighe, W. A. and Albers, A. O., (1990) Advanced Integration of LRT and Traffic Signals in Proc. Third International Conference on Road Traffic Control, Institution of Electrical Engineers, London

Taylor, P.C., Lee, L.K., and Tighe, W.A., (1988), Operational Enhancements: Making the Most of Light Rail, in Light Rail Transit, New System Successes at Affordable Prices, Special Report 221, TRB, Washington, D.C.

Yagar, S. and Heydecker, B., (1988), Potential Benefits to Transit in Setting Traffic Signals, in Light Rail Transit, New System Successes at Affordable Prices, Special Report 221, TRB, Washington, D.C.

TRAFFIC CONTROL SYSTEM ON THE HANSHIN EXPRESSWAY

Tsuyoshi Yoshino[1], Takeshi Matsuo[1]
and
Toshiharu Hasegawa[2]

Abstract

The Hanshin Expressway Public Corporation, Osaka, Japan has been developing and operating its automated traffic control system since 1969. This paper deals with the latest version of the traffic control system which has been operated since April, 1990.

1. Introduction

The Hanshin Expressway Public Corporation, which is engaged in the construction and operation of the urban expressways in Osaka and Kobe area which is called Hanshin Area, the second most populated area in Japan next to the Tokyo Metropolitan area, has been developing and extending its traffic control system since 1969. It is now recognized as one of the most advanced traffic control systems in the world. The Hanshin Expressway Public Corporation has just started the operation of its entirely renewed traffic control system from April, 1990.

As of the fiscal year of 1990, the Hanshin Expressway, with the total expressway length of 152.8 kilometers, is traveled by 800,000 vehicles per day in average. On the peak day, more than one million cars traveled. In operating its traffic control system, the Corporation is incorporating traffic control strategies to maximize the utilization of the network and to ensure the safe and comfortable driving of the expressway systems. The information systems of the Hanshin Expressway play the most important role.

---

[1] Hanshin Expressway Public Corporation, Osaka, Japan.
[2] Dept. of Appl. Math. & Phys. Kyoto Univ. Kyoto, Japan.

2. Control Objectives

The traffic control system has been developed for the expressway to function at its fullest capacity, that is, to maximize its inflow traffic, or network capacity, maintaining smooth traffic flow and to ensure safe, comfortable and efficient transport. It should be noted that the smooth and efficient flow surely induces the less air and noise pollution.

The maximization of the inflow traffic is to be realized by limiting the inflow traffic in each on-ramp as a direct method and by giving rather detailed information of traffic situation of the network to the drivers by various media leaving the decision to the drivers. The inflow control is done by solving linear programming problem in every 5 minutes to maximize the total traffic along the network under the constraints not to have severe traffic jam at any link in the network in principle. In practice, however, this is not fully achieved.

3. Traffic Control System

Hardware system may be classified into Information Collection, Information Processing and Information Conveyance System. Every five minutes, the system collect data from vehicle detectors, process them and transmit information to character, graphic and travel time information boards, roadside radios, and information terminals providing automatic telephone service and personal computer communication service.

3.1 Information Collection System

Ultrasonic vehicle detectors are installed at 500-meter intervals on the main lanes to measure traffic volume and time occupancy and to calculate average speed in every five minutes. A total of 80 television cameras are installed to oversee traffic conditions. More than 60% of the whole expressway can be seen through TV monitors in the control room.

Seven pairs of automatic vehicle identification equipment are installed on the mainline to read the part of the license plates of the vehicles and then to calculate travel time. The data on travel time, calculated by the automatic vehicle identification equipment, is used to verify the data on estimated travel time derived from the vehicle detectors shown on travel time information boards.

3.2 Information Processing System

The traffic control centers are equipped with

computers for data processing and information display and consoles for control operations.

The computer systems Information Processing consists of 9 super-mini and mini-computers in a hierarchical and functionally distributed configuration in two layers. The highest system consists of two computers, one of which is a back-up. The second layer has 7 computers, two of which are back-ups. Here, five functions are maintained and one computer is assigned to each function. These functions are, i. Data collection, ii. Digital information conveyance, iii. Audio information processing, iv. Information exchange system with other systems such as surface street traffic control systems and v. Man-machine interface system in the control center.

In order to have smooth and effective interface between persons controlling the traffic and the data processing systems, many CRT displays, workstations, TV monitors and control console desk are provided at the control centers.(Fig.1)

3.3 Information Conveyance
In order to assist the decision making of the drivers, various visual and audio information concerning the traffic situations are available before and after entering the expressway.

A total of 283 character information boards are installed near the entrances and along the expressway. Each board displays up to 14 characters. Three graphic information boards are installed along the mainline to display congested areas graphically.(Fig.2)

There are 23 roadside radio stations along the expressway broadcasting automatically traffic information repeating for 60 seconds in every five minutes at 1,620KHz. Except the case of emergency broadcasting, no attendant is required for the road side radio system. As a matter of course, each one of the 23 stations can give different information according to their locations.

The following information is shown on character information boards and broadcast on roadside radio. i. The location, length and cause of congestion. ii. On-ramp closures and/or restrictions in the number of open toll booths, and the reason. iii. Mainline and off-ramp closures, and the reason. iv. Warnings for accidents, maintenance work, breakdowns, etc.

A total of 56 travel time information boards are

installed close to the entrances to display estimated travel time to a certain point. Traffic information terminals are installed at parking areas to provide character, graphic and audio information on the traffic. (Fig.3)

The automatic telephone service gives audio information and personal computer communication service transmits digital traffic information through public telephone lines.

4. Conclusion

According to survey results, 95.4% of motorists scrutinize character information boards and 91.7% listen to roadside radios. Also, 98.1% view travel time information boards, with 86.2% answering that the displayed travel time adheres closely to accuracy.

The traffic control system is always on the phase of development due to tremendous advancements in the fields of computer, electronics and communication technology in recent years. For instance, from the fiscal year of 1989, automatic incident detection system with image processing is under development and expected to be installed in near future.

The Hanshin Expressway Public Corporation wishes to expand further its traffic control systems by incorporating the latest innovative technology, with close cooperation with other institutions coping traffic problems.

Figure 1. Control Room

Figure 2. Graphic Information Board.

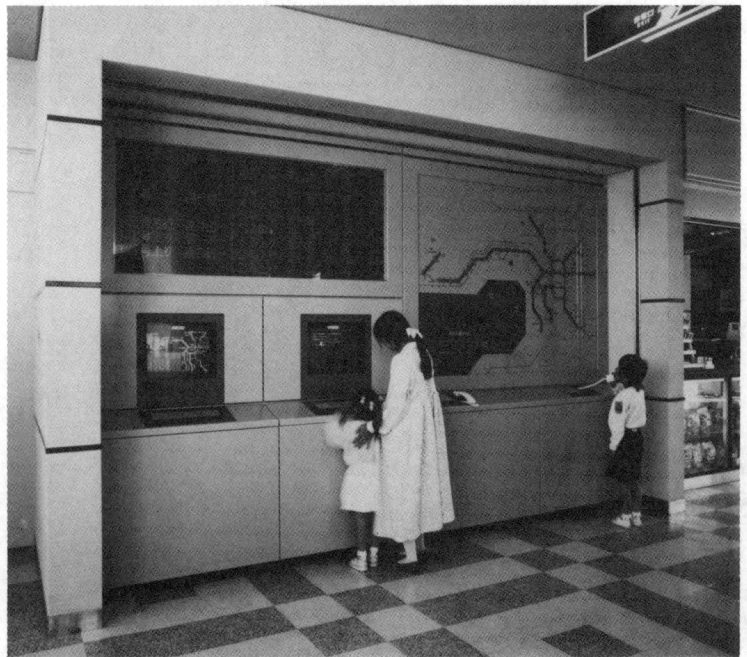

Figure 3. Traffic Information Terminal.

Computer Aided Engineering Analysis for Transportation Links & Management Options

Ashok K. Gupta[1]
PVVSS Ravi Prasad[2]

Abstract

CAD applications on Engineering Analysis have proved very effective in view of its wide applicability & the conceptualization power of interactive computer graphics (ICG). Transportation networks can be analysed for various changes in the networks & the assigned traffic on the network and each route & link. The package developed through turbo-paseal has been found to be q effeuitective & effificient, as well as, adoptable in a variety of situations.

Introduction

Transportation Systems have always been a most challenging aspect of the engineering designs, because of inherent complexity of the system and the variety of possible alternative scenerio. Computer Aided Design (CAD) has certaintly added an most important and useful dimension to the transportation system analysis and it can be visualized most vividly in the form of Computer Aided Engineering Analysis (CAEA) for transportation links and management options.

Amongst the variety of a set of widely accepted definitions of CAD, the one given by Groover (1986), applies best to the engineering analysis of the transportation links & management options and it defined CAD as the use of computer system to assist in the creation, modification, analysis or optimization of a design. Most of such designs and analysis are based on interactive computer graphics (ICG), through which the computer creates, transforms & displays the data in the form of pictures and/or symbols.

---

1. Professor & Coordinator, Centre of Transportation Engg., (COTE), Civil Engg. Deptt., University of Roorkee, ROORKEE., INDIA
2. Asstt. Executive Engineer, MOST, Govt of India, New Delhi, 110 001, INDIA

Thus, the user or the designer communicates data & commands the system and thus, performs the task of conceptualization & independent thinking.

Engineering Analysis involves a large variety of applications and the applications to analysis of transportation links & management options has been presented in this paper.

The Problem Definition & Scope of Work

The present work provides a methodology for the solution to the problem of assigning the designed O-D pairs to different routes of the network. All or nothing asignment has been adopted in the development of the package. As the work has been carried out in regional context, turn penalties have not been incorporated. In the case of an urban area, where the delays at turnings are considerable, compared to the total journey times, a modification is to be done to this work to make it more relevant & realistic for use. Within the memory capacity of a PC-AT, on which the developed package is proposed to be executed in general, network can be modified any number of times, interactively and traffic flows on the network can be observed. Highway capacity can be calculated based on the Highway Capacity Manual (1985) approach, which has been built into the package.

The economic evaluation of a set of comparative projects has been performed using the NPV method and analysing on a macro-level only. User prices have been assumed to include, all costs including travel time costs, accident costs, and discomfort cost etc. These are also linearly varied over the period upto the year of equilibrium.

In management options, a few of national objectives and incorporated on the option given by the user in the form of shadow prices. Stimulus to the reduction of unemploymeny (unskilled labour only) and development of backward regions are considered for the purpose.

Various projects with their expenditure outlays, and NPV known, can be incorporated in the budgetery analysis. It takes number of plan periods and number of projects with their NPV's and present value of expenditure in each year and maximized the output in each year in terms of net-benefit from all the projects while using up the budget allocations in various years to the maximum extent. In this computer-aided engineering analysis using linear programming, opportunity cost of budgets, priorities of projects and negative priorities of rejected projects are all evaluated. The project ranking as management options are provided.

The Methodology & Package Development

Based on the details provided in the preceding article the entire package is developed in turbo-pascal and is made fully interactive. The package has been so formed that all programs are self-explanatory and prompt the user about the missing data.

As soon as the programme is executed through a turbo-pascal-5, a display of all the programme files appear and user is directed to the logical confirmations and sequentional actions through the process of interactive computer graphics (ICG).

Conclusions & Applications of the Work

The present work presented very briefly due to limitation of space, provides following specific conclusions

(i) The package developed in turbo-pascal & combining 5-main programmes alongwith interactive computer graphics (ICG) provides a very effective & robust tool for assignment & analysis of transportation networks in a wide area.

(ii) The individual components of a given project requires their ranking before taking up the project.

(iii) This can be used as an effective tool with the input of interactive computer graphics (ICG) and all the links & nodes of a given project & the projects in general can be visualized on the computer-monitor and a number of alternatives can be conveniently tried.

Appendix : References

1. Groover, Michell, P., & Emony, W. Zimmers. Jr. (1986), "Computer Aided Design & Manufacturing", Prentice Hall (India), New Delhi.

2. Highway Capacity Manual (1985), Transportation Research Board, Special Report, TRB, USA.

3. Ravi Prasad, PVVSS, (1990), "Application of CAD in Transportation Network Planning & Economic Evaluation", (Unpublished) P.G. Thesis, University of Roorkee, Roorkee, India.

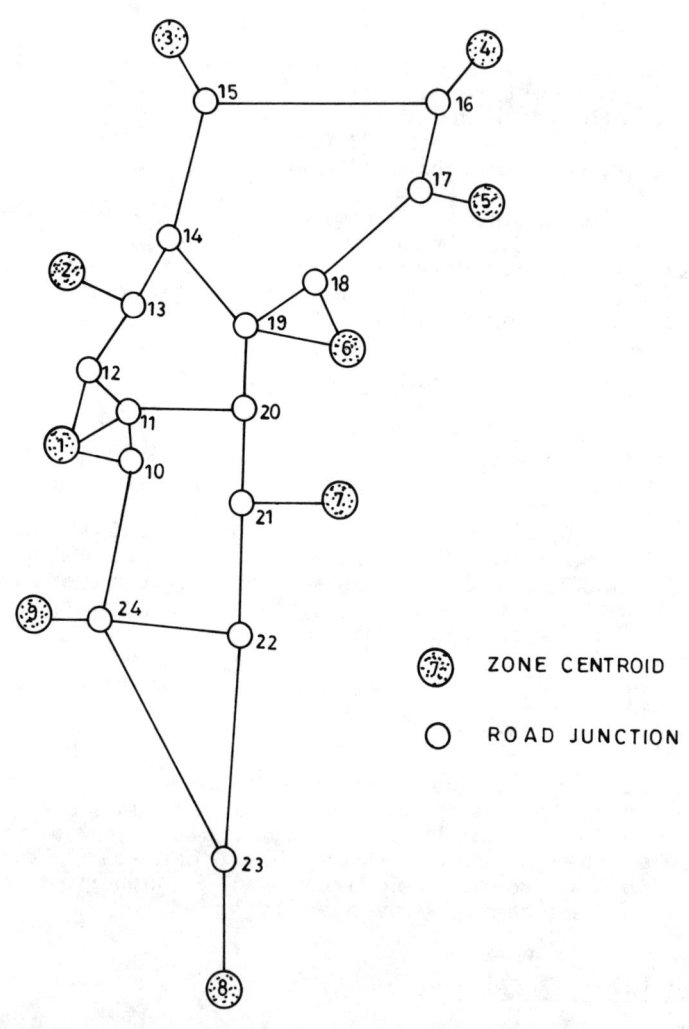

FIG. 1— ORIGINAL NETWORK TRIED WITH VARIOUS MODIFICATIONS

Safety of IVHS: Methods for Determination of Accident Levels for AVCS-1 Devices

by
Anthony Hitchcock
Program on Advanced Technology for the Highway
University of California (Berkeley)
1301 S. 46th Street, Bldg. 452
Richmond, CA 94804
(415) 231-9495

INTRODUCTION

This paper discusses ways in which the safety impact of Intelligent Vehicle/Highway System (IVHS) devices can be estimated by interpreting existing statistics. This is needed at many points during development of devices for evaluation, which improves design. A common theme will emerge - the need for specially collected data on accidents involving vehicles with special characteristics or particular classes of people.

Automatic Vehicle Control Systems-1 (AVCS-1) devices (see 1) are those driving aids which assist driving by giving warnings or taking over control in emergency. Such devices seem inherently more likely to affect safety than, for example, traveller information or commercial vehicle operation systems.

To determine a device's effect on safety, one must know what it does. The specification must say what will happen in all possible circumstances. This is not seen to be obvious in practice. "In all normal circumstances" is not sufficient - accidents often result when circumstances are not normal.

Different analyses are appropriate at different times. The choices will be affected by presently undrafted legal requirements as well as the technical situation. Clearly the situation changes radically when real-life data becomes available. (Steps should be taken before deployment to collect good data quickly.) But also, as the design progresses one needs:

    a. to test (by cost-benefit analysis or otherwise) whether a concept is worth developing;
    b. to choose between alternate realisations;
    c. to investigate the effects of manufacturing tolerances.

TYPES OF ACCIDENT

There are four kinds of change in accident rates that can usefully be distinguished in the evaluation of any AVCS device or system:

    A. reduction in accidents due to the system functioning as intended.

    B. diminution of this reduction due to changes in driver behaviour. This may be divided into:
        i. changes in driving habits;
        ii. distraction or other interference.

    C. accidents due to unexpected absence of a device.

    D. increase in accidents due to presence of IVHS devices. This may again be divided into:
        i. designed exchanges, where a device trades more serious accidents for less serious ones. These are best considered along with A above;
        ii. accidents caused by faults in the device;
        iii. accidents caused by misspecification of the device. This includes Bii above.

Accidents reduced by AVCS

At present there is great interest in this question - it affects program funding. There are two kinds of technique for this, one mainly used in the US and one in Europe; in each case currently only for initial evaluation. The latter technique is however capable of extension to the later stages of design, given appropriate data and tools (2).

One starts (See 3, 4) with a precise history of the events, physical and mental, which were the precursors to an accident. One imagines that one or more of the vehicles concerned is fitted with the device under consideration and forms judgments about whether the devices would have operated. If so one judges whether the responses would have avoided the accident. If a suitable computer modelling shell were available quantified probabilistic statements could be made, which would be used to distinguish between devices more precisely.

This process requires that:

a. There is a satisfactory set of accident data.

b. The computer model shells referred to exist.

c. There is adequate understanding of drivers' reaction times, and of the kind of response they will make to automatic warnings. This lies beyond the scope of this paper.

The European work used data of kinds not recently collected in the US. In 3 the possibility is discussed that the raw data for NHTSA's National Accident Sampling System (NASS) could be used here.

At the time of writing the author has just completed a systematic examination of NASS files to investigate this. A rigourous protocol was used in order to make the judgments as reproducible as possible. The preliminary indications are that NASS can be used to say if an IVHS device would have functioned in a particular accident situation. Precise conclusions were reached for each of the seven devices examined. If the driver's choices are limited and either the device is automatic or assumptions are made about reaction times, one can deduce whether or no the accident would have been avoided. If the censored interview data were available one may do better, for it may reveal more about the error a driver made, and his state of mind. NASS does not contain data which helps if the problem relates to the state of vehicle or driver.

This analysis will be reported fully elsewhere.

Driver Reactions to IVHS

That drivers do modify their behaviour, when offered a safety device, so as to take up some of its benefits in other ways, is well attested. There is no generally accepted method, however, of estimating the magnitude of the resulting reduction in accident savings in advance. The literature indicates reductions of zero and of 100% in different cases. Similarly there are anecdotes that a driver becomes more dangerous when he drives a car not equipped with (say) antilock braking (ABS). There is no objective evidence. The idea is persuasive, however.

It is possible to estimate the magnitude of the effect for any device which has been deployed, given a relatively small amount of data about the accidents, by using accident histories of individuals who had had relevant accidents, and a measure of their exposure to risk. It is desirable that this be done for present IVHS-like devices, for a common pattern may emerge.

Consider, for example, ABS. Suppose we had accident histories, and measures of annual mileages, of people who had had accidents involving loss of control by skidding.

Statistical tests could then confirm:
(i) Those equipped with ABS have lower accident rates than the controls, and their accidents occur in severer conditions than the others';

(ii) Those using non-ABS vehicles after experience with ABS have no more accidents in ABS vehicles than others in them, but more in non-ABS vehicles than those without ABS experience.

Such tests could also be made for other devices. But in all cases one has to identify the population to be sampled, and obtain subjects' co-operation. The latter needs tact: the former needs access to data usually regarded as confidential.

Accidents caused by IVHS

IVHS equipment may fail to give advice or take control action. A driver who has come to rely on it may have an accident in consequence. This risk is reduced if the design gives warning of impending failure - this can be of greater importance than high reliability. A more serious failure is one which gives advice to carry out a hazardous manoeuvre or takes direct control action which contributes to an accident. Another possibility is that the failure may have occurred at the design stage. A specification may be incomplete, or may fail to say what its author intended.

Failure mode effect and criticality analysis (FMECA) should enable one to say what can be the immediate consequences of IVHS equipment failure. Sometimes a fault will lead to danger to the vehicle: in other cases a warning will be given and no action result. But in other cases one cannot say whether an effect is hazardous or not. If an IVHS vehicle runs a red light, some vehicles with communication equipment, forewarned, will perform an evasive manoeuvre. Is it hazardous to falsely broadcast an intention to run a red light, and induce such manoeuvres? It depends on what the manoeuvres are, and that is not part of the design of the vehicle with the fault.

Thus hazards are defined by the whole system and there must be a search for system hazards as well as individual vehicle ones. One way of doing this is by fault tree analysis. Another is heuristic. One could construct a computer simulation of a highway and introduce an unrealistically large number of vehicles bearing randomly chosen IVHS devices. Some of these would have defined faults (discovered by FMECA as above). Then one could run until an accident occurs and determine its mechanism. The number of accidents in the simulation will increase with the number of faults, and one can extrapolate back to deduce an accident rate.

To do this one needs specifications of all devices which might be on a road. No such data-source exists.

Conclusions

Many techniques are sketched out here which address many of the problems arising in the evaluation of the safety impact of an IVHS devices. In each case it is found that there is a need for data which could be collected but is in fact not available. As indicated the author has been attempting to remedy this in one case.

Acknowledgments

The research reported here is part of the Program on Advanced Technology for the Highway (PATH), prepared under the sponsorship of the State of California, Business Transport and Housing Agency Department of Transportation, and the US Department of Transportation, National Highway Traffic Administration.

References

1 Mobility 2000. Final Report of the Working Group on Operational Benefits. Dallas 1990

2 Hitchcock, A. Intelligent Vehicle\Highway System Safety: Approaches for Driver Warning and Copilot Devices. TRB Annual Meeting. Washington DC 1991

3 Hitchcock A. Road User Safety: Possible European Research Cooperation. In Roads and Traffic Safety on Two Continents. Gothenburg 1987 pp 187-201 VTI Linkoping 1988

4 Fontaine H, G Malaterre & P van Elsande. Evaluation de l'Efficacité des Aides à la Conduite. Rapport INRETS 85 Paris 1989

DESIGN FOR IVHS SAFETY: TECHNOLOGY ASSESSMENT

Wei-Bin Zhang [1]

Introduction

The development of an Intelligent Vehicle/Highway System (IVHS) will require methodologies for predicting and controlling system safety. In principle, these methodologies should include various safety control technologies which assure that both the frequency and magnitude of inadvertent events inherent in the system (or system hazards) are acceptable.

A previous work discussed the safety problems of an IVHS (Zhang, January 1991). It revealed that an IVHS may confront unsafe conditions due to possible component failures, physical disturbances, new operating environments, design errors and human interference. It also indicates that in an IVHS, components that are directly relevant to maneuvering a vehicle (specifically the components related to an Advanced Vehicle Control Systems -- AVCS) are defined as safety critical, therefore hazard reduction design is required.

This paper assesses hazard reduction techniques including: fault monitoring techniques, reliability techniques, safety techniques, and "Saving-Life" techniques.

I. Hazard Reduction Principle

An automatic controlled vehicle can be described by a Markovian Model (Zhang, June 1991) in which four vehicle states are defined. These states include: Normal Operation (NO) state, Graceful Degradation (GD) state, Fail-Hazard (F-H) state and Pause (P) state. The vehicle operation can then be defined as a system which evolves in time while undergoing state transition. This transition may be controlled if hazard reduction strategies are incorporated into the design of the safety-critical components of the IVHS so that the likelihood of a transition to the F-H state can be minimized. The principle of hazard reduction refers to (a) reduction of the inherent failure rate of the system, (b) prevention of hazards as result of system failures or (c) reduction of the impacts of the failure incidents.

The techniques which correspond to the hazard reduction principle can be

---

[1]. Associate Research Specialist, Program on Advanced Technology for the Highway, Institute of Transportation Studies, University of California at Berkeley, 1301 S. 46th Street, Richmond, CA 94804 (415) 231-9538

grouped as follows: (1) reliability techniques that reduce the inherent failure rate of the system, (2) safety techniques which assure that the IVHS (or vehicles) will remain safe when failures occur (Fail-Safe (F-S) approaches), and (3) countermeasures that reduce the severity of hazards, thereby assuring that the consequences of inadvertent events do not exceed a predetermined acceptable level.

Prior to applying hazard reduction techniques, hazard analysis should be conducted to determine safety impact on specified actuating functions and to identify the components required for accomplishing these predefined functions. Further, Fault Tree Analysis (FTA) and Failure Mode Effect and Criticality Analysis (FMECA) need to be pursued during the design process in order to evaluate the cause-consequence relationships between system failures and hazardous events. Consequently, suitable techniques can be chosen to improve the safety level of the components and eventually that of the system.

II. Fault Monitoring Techniques

The fault monitoring techniques are essential prerequisites for both Fault-Protective and Fail-Safe techniques. They allow the supervising devices to monitor the system performance and to detect or diagnose the process of a fault. Fault monitoring can be accomplished through either online or offline monitoring. The offline monitoring of an automatic vehicle control system can be, for example, executed periodically by a roadside diagnosis device(s) which performs a series of tests while the vehicle is temporary suspended from normal operation. The online monitoring can be continuously active while the system operates. It is almost certain that online monitoring will be an essential monitoring approach for an automatic control system because the offline monitoring may not detect transient faults and also because the delay between offline monitoring opportunities would allow the consequence of a failure become widespread, thus the isolation of failures and hazard reduction would become more difficult.

The online monitoring techniques are composed of fault detection technique and failure diagnosis techniques. (a) Fault detection can be accomplished by strategically installing sensing devices to directly observe the output and/or input of the targeted component. The sensing devices are able to provide predefined signal to the control system if the performance of the observed component is outside of the specified range. (b) Fault detection can also be achieved using a redundancy technique. In a redundant system, hardware items are either duplicated or triplicated. Comparison or 2-out-of-3 voting is used to determine the operating status, thus the faulty component can be easily detected (provided that no failure of the comparitor or the voter occurs) . (3) In order to monitor the nonmeasurable variables, methodologies have been developed using microcomputers with the aid of process models, estimation and decision methods (Iserman, 1984). The detected faults can be further diagnosed to identify or locate the failure sources. However, the diagnosis only appears to be important where the system is still controllable and F-S control can be applied in different levels according to the severity of the possible consequence.

III. Fault Protection Techniques

The Fault-Protection techniques primarily refer to reliability techniques which ensure that the designed component possesses sufficient endurance against progressive weakening in order to withstand fatigue, corrosion, wear, etc., thus reducing the failure frequency of the system.

The reliability of a component or system can be enhanced through a systematic design which should be conducted in three stages: functional design, quality design and tolerance or sensitivity analysis. In designing a reliable system, the following techniques are often applied: (1) Robust design improves the ability of the system to tolerate the expected variations in the operating environment and production accuracy. Hence the adaptation of components or the overall system against parameter and process changes is provided. (2) Derating technique specifies critical loads of components in such a way that the actual operating value is lower than the designed value so that the failure rate can be significantly reduced. (3) Backup components which supplement those which perform safety critical function also improve the reliability. (4) Protection devices should be provided for possible instantaneous overloads. (5) Careful consideration of the software specification and fault tolerance design will enhance the "strength" of the software.

IV. Fail-Safe Techniques

Fail-Safe techniques can ensure that the system does not become a safety hazard should the system reach conditions outside specified tolerances. By Fail-Safe, it means that the system can perform operations necessary to avoid hazardous conditions. These operations typically involve graceful degradation or operation that leads the system to the P state according to the severity of the system failure. Fail-Safe techniques may be summarized as follows:

(1) **Technique for configuring an asymmetric failure pattern.** Failures of components can be grouped into two categories: safe failures and hazardous failures. The safe failures may result in performance degradation but create no safety impact while hazardous failures can generate an unexpected condition in which an accident may occur and no action by the control system can prevent it. When the rate of safe failure is not equal to the rate of hazardous

IVHS SAFETY DESIGN 305

failure, an asymmetric failure pattern is formed. The asymmetric failure pattern is favorable if the rate of safe failures is smaller than the rate of hazardous failures. It has been proven that a favorable asymmetric failure pattern can improve the overall system safety (Zhang, June, 1991).

An asymmetric failure patterns can be obtained by carefully choosing or designing the components in such a way that their outputs which are less influenced by component failures are assigned to correspond to an operation mode that would lead to functional escalation. By using this methodology, the hazardous failure rate of a system would be much smaller than the safe failure rate.

(2) **Fail-Safe control strategies.** Fail-Safe (F-S) control strategies are used to isolate failures and cause a system to transit to a F-S state, should a hazardous failure(s) occur. In accordance to the severity of failures, the control system with built-in F-S control strategies can apply proper control actions in order to prevent hazards. The F-S control strategies have to be designed in such a way that failures are effectively isolated and the reduction of the efficiency of the overall system is minimized.

F-S control strategies are executed by a so called "hardcore" (-- the portion of a system that must be functioning in order for the fault-protective mechanism to be functioning), i.e. computing or logic devices and sensors. By using fault monitoring techniques, detected failures are diagnosed and the severity of failures are evaluated. According to the estimated impacts, the F-S control strategies may take the following control actions: (a) maintain the original control if the failure is not hazardous; (b) degrade the performance if the failure is hazardous but gradual and can be isolated so that no further safety impact can be induced (-- a Fail-Soft control); (c) immediately lead the system into the P state if the failure is sudden and Fail-Soft control would not help the system to survive from a hazard.

(3) **F-S components.** The F-S control strategies can only be effective when the hardcore correctly perform the intended functions. In fact, failures of the hardcore represent a hazard of which the severity is usually difficult to foresee. Therefore, the hardcore itself should consist of a number of F-S components which can provide predefined outputs as a failure occurs. These outputs can then be used to lead the system to a F-S state.

(4) **Interlocking technique.** When designing a system with multi-task functions, the tasks which conflict should be interlocked. For example, the braking and throttle controls should not be applied simultaneously. Interlocking can be achieved through software and hardware. However, hardware interlocking is recommended where the conflict of functions is determined to be critical. The

interlocking technique is also commonly used for Fool Proof-techniques to enhance the system safety against human errors.

V. Techniques for Reducing Impacts of Incidents

The techniques for reducing the severity of hazards have been thoroughly applied in designing today's automobiles. These include so called "Life-Saving" technologies which largely consist of crush or impact protection designs (Stempel, R. 1989).

VI. Concluding Remarks

The technologies assessed in this paper form a basis for designing a safer IVHS system. The discussion shows that the safety of a system can be enhanced by improving system reliability or by applying special safety techniques. The method application of the assessed technologies remains to be studied. In future studies, an in-depth analysis will be conducted to understand both the failures inherent in the AVCS and their contributions to the overall IVHS performance and determine the technique(s) required for satisfying the predetermined target safety levels.

Acknowledgement

The research reported herein is part of the Program on Advanced Technology for the Highway (PATH), prepared under the sponsorship of the Business, Transportation and Housing Agency, Department of Transportation, State of California. The author wishes to thank the following people for their contribution, technical or otherwise, who helped to shape the final form of this paper: Mr. A. Arai, Dr. A. Hitchcock and Dr. S. Shladover.

Reference

Fail-Safe Technology Co.; System Design and Analysis consideration for Safety-Sensitive Digital Electronics; Los Angeles, 1989

Isermam, R.; Process Fault Detection Based on Modelling and Estimation - A Survey", Automatica, 20(4), pp. 309-326, 1984

Stemple, R.; Where auto Safety Stands Today; Media Update on Passive Restraints, Warren, Michigan, July 18, 1989

Zhang, Wei-Bin; Engineering Design Concept for IVHS Safety; Transportation Research Board 70th Annual Meeting, Paper No. 910745, Washington, D.C., January 13-17, 1991

Zhang, Wei-Bin; Analysis for Establishing Target Safety Levels for IVHS; 1991 American Control Conference, Boston, June 26-28, 1991

TAMING THE SILICON STEED - *Perceptions of Risk*

John C. Keller[1] and Dr. Paul P. Jovanis[2]

Abstract

This paper draws from contemporary research in the psychological literature to argue that transportation engineers and decision makers should consider the issue of *perceived* risk when developing Intelligent Vehicle-Highway Systems (IVHS). These perceptions will influence public acceptance of IVHS technologies and the consequent product market. Current research could be affected in at least two areas: the design standards of IVHS systems and IVHS benefit estimation.

Introduction

Many people can recall the crash of an airliner in a distant part of the United States, but they might have trouble recalling any recent fatal highway accidents in their own town. Citizens of many countries recall clearly the concern about the nuclear accident at Chernobyl in 1986, but might have trouble guessing, within an order of magnitude, last year's highway death toll in their own country.

People are concerned about highway safety, but there is something about certain calamities that makes them more memorable. These memories can influence the perceived risk of travel. Engineers generally view risk, the probability of an accident, as something to be managed (Hauer, 1982). Highway engineers use design standards to promulgate uniformity that embodies some minimum level of safety. System and product engineers use mean time between failures as a means to describe the risk of component failure.

Another dimension to risk, however, involves the perception of risk by potential users and non-users. Rather than be concerned with an analysis of the <u>actual</u> probability of an accident, risk perception studies deal with how individuals <u>perceive</u> the risk of accidents in different situations. This paper seeks to apply the concept of risk perception to intelligent vehicles and automated highways (Keller and Jovanis, 1990). This application is particularly important because:

- Intelligent vehicles and automated highways are likely to represent rather novel travel attributes to the motorist, particularly with respect to risk. It is of interest to understand the motorist's perception of these risks as a means of understanding a person's willingness-to-pay for such systems.

---

[1]Research Manager, Office of Special Projects, California Highway Patrol, 2555 First Avenue, Sacramento, CA 95814

[2]Associate Professor, Department of Civil Engineering and Transportation Research Group, University of California, Davis, CA 95616

- Public decision makers must decide about standards and operating procedures for these systems. Although their personal perception of risk is likely to be based more on objective probabilities of failures, they too must consider subjective perceptions of motorists in the allocation of public funds and the setting of design standards because they have to be concerned about usage of IVHS systems (cost/benefit tradeoffs) and legal liabilities.

Issues of Concern

Underlying the rationale for risk perception analysis were the fundamental questions of who was going to buy and use IVHS technologies, and who would be accountable if it failed. A variety of issues are related to these questions, including:

1. *Task Simplification* - One of the intended IVHS objectives is to simplify the driver's task. Risk compensation theory, as well as Mahalel and Szternfeld (1986) suggest that simplification is not necessarily beneficial; the driver may be lulled into a sense of complacency about the ability of the IVHS technologies to rescue him from the vagaries of nature and chance. This factor becomes more important if "system" failure is catastrophic because catastrophic potential is an important element in the perceived risk of a system (Slovic, 1987).
2. *Higher Performance Envelopes* - Many IVHS technologies have the potential to reduce risk because the technology is capable of performing at a higher performance level than most drivers (eg. anti-lock brakes, headway control). These higher performance envelopes may be difficult for older and other drivers to accept.
3. *Loss of driver control/technological limitations* - Although technologically advanced, many IVHS systems will be limited in their flexibility to deal with all operating conditions present in the United States (eg. some anti-lock systems have a disabling switch to allow the driver to adequately respond in some driving circumstances). At the same time, many IVHS technologies will require the driver to yield some control to on-board or external computers, or other types of assistance systems. Technological limitations and yielding control also imply some transference of responsibility.
4. *Vehicular maintenance and repair* - Liability is also an issue when an accident-causing private vehicle was allowed by a public entity to operate on an IVHS facility. If these standards are set strictly enough to minimize the catastrophic perception of crashes, the pressures of equity and denied access are heightened.

Risk Perception Analytical Techniques

Lowrance (1976) did pioneering work in this field that was developed by Fischhoff et al. (1978). Lowrance outlined ten considerations that seemed related to safety judgements by lay people. These ranged from whether the risk was assumed voluntarily, to the timing, to whether the consequences were reversible.

Fischhoff et al. refined this list slightly to nine characteristics, and added a seven point rating scale for each. The characteristics were used to judge the risk from relatively discrete, but broadly stated hazards. These included Food Irradiation, Pesticides, Nuclear Power, Motor Vehicles, and various others. Some of the authors of the Fischhoff et al. study developed an expanded scale of 18 risk characteristics that captured three main dimensions of perceived risk (Slovic et al., 1980).

- **Dread** - This was the most influential factor and included 12 of the 18 characteristics. These included uncontrollability, catastrophic, hard to

prevent, fatal, inequitable, not easily reduced, involuntary, and threatening to the subject personally.
• **Familiarity** - Five characteristics contributed to this factor, including: observability, knowledge, immediacy of consequences, and familiarity.
• **Exposure** - This factor was judged to be independent of the other two and consisted only of the characteristic relating to the number of people exposed to the hazard.

Several of the questions and rating scales from Slovic et al. (1980) are included in Figure 1 to illustrate the types of questions asked of respondents. Those marked with an asterisk are questions that could be used to gauge public perceptions of the risk of broad IVHS system concepts such as completely automated vehicle-highway systems, and advanced driver information systems

Perception of More Specified Risks

Using the approach developed to study the perception of general risks, MacGregor and Slovic (1989) constructed a study of how lay persons perceived the risks of specific automotive systems. They used a thirty item classification of the various systems found in automobiles (engine, steering, brakes, etc.) and asked respondents to rate these systems on 10 point scales. The scales covered 11 aspects of risk, including four items similar to those used in prior studies (anticipatory knowledge, severity, control, observability). Examples of these rating scales are contained in Figure 2.

MacGregor and Slovic concluded that two composite factors were highly related to judgements of risk and desire for risk regulation. These factors involved the controllability of consequences and the observability of system failures. *Controllability* consists of two elements, the degree to which the manufacturer is able to anticipate system defects, and the driver's ability to control a vehicle with a defective system. *Observability* refers to the perceived ability of a driver to observe a potential failure before it occurs. Brakes and steering were associated with high levels of uncontrollability and severity if they failed, but failures of these systems were thought to be moderately observable. Both of these factors could be important in assessing the risks of IVHS technology such as forward-looking obstacle detection systems, steering by wire systems, and automated longitudinal control.

Conclusions

For some early IVHS technologies (eg. anti-lock brakes, adaptive cruise control), perceptions of risk can be measured based on a description of the technology to survey respondents. For advanced IVHS systems, it may be necessary to obtain more informed perceptions after a respondent has been exposed to the system via a driving simulator. Education and experience with the technology will not necessarily diminish the difference between subjective and objective assessments of risk. Nelkin and Pollak (1980) argue that it is hard to change perceptions of risk and educate the public as to the "true" probabilities of an accident.

Several references have been cited to support the validity of risk perception analysis. A fair amount of related literature also is available, but there have been few attempts to use contemporary research from this field of psychological testing in the study of IVHS systems. A similar link may also be appropriate between risk perceptions and tort liability research.

Uncontrollability, unobservability, dread, unfamiliarity, and lack of exposure were common dimensions of perceived risk in several studies of various potential

Figure 1
Sample of Rating Scales for Perceived General Risks[a]

*1. *Voluntariness of risk* - Do people get into these risky situations voluntarily?
    risk assumed   risk assumed
    voluntarily   1 2 3 4 5 6 7   involuntarily

*2. *Control over risk* - If you are exposed to the risk of each activity or technology, to what extent can you, by personal skill or diligence, avoid death while engaging in the activity?
    personal risk   personal risk
    can't be   1 2 3 4 5 6 7   can be
    controlled   controlled

*3. *Newness* - Are these risks new, novel ones or old, familiar ones?
    new   1 2 3 4 5 6 7   old

*4. *Chronic-catastrophic* - Is this a risk that kills people one at a time (chronic risk) or a risk that kills large numbers of people at once (catastrophic risk)?
    chronic   1 2 3 4 5 6 7   catastrophic

*5. *Common-dread* - Is this a risk that people have learned to live with and can think about reasonably calmly, or is it one that people have great dread for -- on the level of a gut reaction?
    common   1 2 3 4 5 6 7   dread

(a) From Fischhoff et al. (1978) and Slovic et al. (1980)

Figure 2
Sample of Rating Scales for Perceived Specific Risks
(Automotive Systems)[b]

*1. *Anticipatory Knowledge* - To what extent should the manufacturer be able to anticipate defects in this system when the automobile is designed?
    manufacturer   manufacturer should
    can't anticipate   1 2 3 4 5 6 7 8 9 10   anticipate

*2. *Severity of Consequences* - How likely is it that a mishap resulting from a failure of this system would case severe injury or death to vehicle occupants or bystanders? Consider the entire range of possible consequences for the driver of the vehicle, passengers, occupants of other vehicles, bystanders, and so on.
    consequences   consequences
    not severe   1 2 3 4 5 6 7 8 9 10   severe

*3. *Control of Vehicle* - How likely is it that the driver can remain in control of the vehicle (steer it, stop it, control speed, etc.) in the event that this system should fail?
    driver has   driver has
    little control   1 2 3 4 5 6 7 8 9 10   much control

*4. *Observability* - How likely is it that the driver would notice a failure of this system well before it could cause an accident?
    extremely unlikely   extremely likely
    defect would be   defect would be
    noticed   1 2 3 4 5 6 7 8 9 10   noticed

*5. *Overall Riskiness* - Overall, how risky is driving an automobile when this system fails to operate properly?
    not risky   1 2 3 4 5 6 7 8 9 10   extremely risky

(b) Based on MacGregor and Slovic (1989)

hazards. Early evidence of their importance in IVHS research comes from a study on adaptive cruise control (Turrentine et al., 1991). During the focus group discussions of that study, respondents did not agree that automated acceleration and braking were conveniences because of the complexity of driving decisions, perceived limitations of technology, and concern about diminished driver vigilance. Many of these dimensions of perceived risk may also become important as public transit systems incorporate IVHS technology more broadly.

Perceptions like these have the potential to hinder public acceptance of IVHS technology, diminish potential usage, and limit the ultimate benefits from such a system. Better information on the perceived risks of IVHS systems could help in the setting of design and operating standards, as well as improve the estimation of benefits and potentially clarify aspects of the legal liability question.

Disclaimer

The opinions and views expressed in this paper are those of the individual authors and are not intended to represent the views of their employers or any other organization.

References

Fischhoff, Baruch, Slovic, Paul, Lichtenstein, Sarah, Read, Stephen, and Combs, Barbara (1978) How Safe is Safe Enough? A Psychometric Study of Attitudes Towards Technological Risks and Benefits. *Policy Sciences, Vol. 9, No. 2.*

Hauer, E. (1982) Traffic Conflicts and Exposure. *Accident Analysis and Prevention, Vol. 14, No. 5.*

Keller, John C. and Jovanis, Paul P. (1990) *Taming the Silicon Steed; Assessing Public Perceptions of Risk Associated with Intelligent Vehicles and Automated Highways. ITS Research Report #90-11.* Institute of Transportation Studies. University of California, Davis, California.

Lowrance, William W. (1976) *Of Acceptable Risk: Science and the Determination of Safety.* William Kaufmann, Inc. Los Altos, California.

MacGregor, Donald G. and Slovic, Paul (1989) Perception of Risk in Automotive Systems. *Human Factors, Vol. 31, No. 4.*

Mahalel, David and Szternfeld, Zvi (1986) Safety Improvements and Driver Perception. *Accident Analysis and Prevention, Vol. 18, No. 1.*

Nelkin, Dorothy, and Pollak, Michael (1980) Problems and Procedures in the Regulation of Technological Risk. *Societal Risk Assessment; How Safe Is Safe Enough?* Plenum Press. New York / London.

Slovic, Paul, Fischhoff, Baruch, and Lichtenstein, Sarah (1980) Facts and Fears: Understanding Perceived Risk. *Societal Risk Assessment. How Safe is Safe Enough?* Plenum Press. New York / London.

Slovic, Paul, Fischhoff, Baruch, and Lichtenstein, Sarah (1985) Characterizing Perceived Risk. *Perilous Progress. Managing the Hazards of Technology.* Westview Press / Boulder and London.

Slovic, Paul (1987) Perception of Risk. *Science, Vol. 236, No. 4799.*

Turrentine, Thomas, Sperling, Daniel, and Hungerford, David (1991) Consumer Acceptance of Adaptive Cruise Control and Collision Avoidance Systems. Pending publication in the *Transportation Research Record.* National Research Council. Washington, D.C.

Exploring Headsup Displays for Driver Workload Management in Intelligent Vehicle Highway Systems

M. Coyle†, S. Meir†, S. Shekhar†, A. Yang†, J. Caird‡, P. Hancock‡, and S. Johnson‡.

ABSTRACT

New transportation system provide more information to drivers to improve their effectiveness in tasks like routing, avoiding congestion. We are examining the headsup displays as means of information display. A vital question becomes the synthesis and management of this information so as to not overload the driver and create safety hazards. We have designed a driving simulator to design and evaluate the safety and effectiveness of headsup display.

1. Introduction

Intelligent Vehicle Highway Systems (IVHS) are being proposed to deal with the traffic safety and congestion problem. Intelligent vehicle will aid the drivers in various driving tasks such as routing, navigation, collision avoidance, monitoring vehicle and traffic status. IVHS will also manage the global traffic information for better decisions by the drivers and traffic controllers to reduce traffic congestion without expanding the capacity of the highways. The system will collect the information about traffic on different highways to detect and predict congestions. The control algorithms for traffic lights and entrance ramps will utilize the information to manage traffic. The drivers will receive the information via electronic signboards on highways as well as via the onboard computer.

One approach to IVHS is to make available all relevant information to the driver. For example, the traffic density on various highways could be broadcast to every vehicle. Information presentation interfaces are important to convey the relevant information to the drivers in a safe manner. The extra information presented to the drivers to help them select least congested routes increase the volume of the information to be processed by the driver tremendously. It can lead to the overloading of the drivers resulting in increased time to gather information and increased time to respond to traffic events. The volume of information also increases the time to detect driving events due to the increased searching time to access relevant information. Thus the information based approach to IVHS raises safety issue related to the delayed response of the drivers to the driving events.

Headsup display are being examined as the means to deliver IVHS information to the drivers. Headsup displays are being used in aircraft to present information to the pilot superimposed on his forward view of the external world[1]. They are used to represent two types of information in aircraft: (a) contact information about impending collisions, and (b) housekeeping information. Headsup displays save a pilot's time in acquiring the necessary information by reducing head movements. In automobile driving collision information may not be available in time to be useful. However, headsup displays can help in simple navigational tasks that indicate the next routing decision

---

† Computer Science Department, University of Minnesota, Minneapolis, MN 55455.

‡ Human Factors Laboratory, University of Minnesota, Minneapolis, MN 55455.

(turn left/right, exit freeway etc.) They can also be useful for housekeeping information (i.e. speed) as it can reduce eye movements thereby reducing the time it takes to assimilate information. The speed of car is lower, but the distances involved are smaller also, yielding similar for decision time in flying and driving. HUD technology makes it possible to display information in a wide range starting with speedometer reading (simple) to the computer generated road-map (complex). Relevant questions in the design of HUD are the placement of displays, the priority scheme, and the method of presentation (i.e. alphanumeric vs. iconic headsup displays).

The use of headsup display in aircrafts has revealed safety hazards: occlusion of objects by the displays, cognitive capture, and information overload. These safety hazards become more critical in the driving world due to higher density of objects on roads and requirements to follow road signs and signals. A potential headsup display design must be validated under realistic driving conditions to ensure safety. The driving simulator can be used as design and validation tools for headsup displays[2]. Driver simulators use graphic computers and display devices to produce pictures viewed by the driver. Driver sits in a car to operate and feel the response of the actual steering mechanism, the accelerator pedal, the brake pedal and other interfaces featured in the vehicle[3].

We have extended the driving simulator at Human Factors Lab in Minnesota to create arbitrary headsup displays and validate them. The validation is of two types: effectiveness and safety. The effectiveness relates to the improvement of driver's performance in driving tasks like navigation. Safety relates to reduction of mistakes made by drivers. In this paper we will describe the tool to create headsup displays and to design experiments to validate them.

2. Driving Simulator: The environment

The Simulation Hardware at the Human Factors Lab consists of a 1990 Accord which faces a large projection screen as shown in Figure 1. The picture consisisting of road objects is displayed on the screen by a high resolution projector. The accelerator, brake pedal, and steering wheel are connected to the A/D converter. Signals from the converter are used for real-time simulation in moving the driver through the environment. The computing system hardware uses Falcon Graphics to provide a real-time graphic system. The Falcon hardware contains three boards to be installed in the host PC. These boards are used for the real-time calculation and projection of the simulation environment.

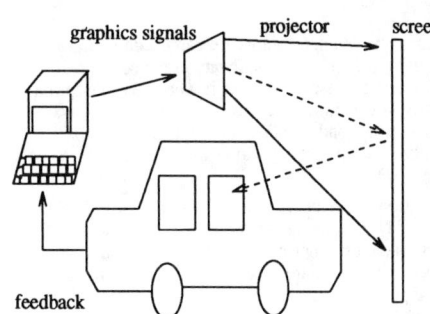

1. Get attributes for new trial
2. Get input (acceleration, brake, steering)
3. Find next position of the driver in the world
4. Find next positions for all moving objects
5. Calculate viewport (i.e. set of visible objects)
6. Update headsup displays
7. Position the driver
8. Draw the new frame
9. loop back to step 1.

Figure 1: Driving Simulator Hardware   Figure 2: Simulation Algorithm

Figure 2 describes the main loop in the driving simulator software. A frame (static picture of roadview of the driver) is created in each iteration of the loop. In each iteration, driver input (accelerator, brake, etc) are collected and analyzed. Based on the equation of motion for the car, the next position of the car is calculated. Next, the position of moving objects(cars) which move independently of the driver are calculated. The viewport of the driver is the area of the world visible to him. The next step calculates which objects are visible to him based on the new input, and a new frame is drawn. This completes one loop (frame), and this loop is executed approximately twenty times per second.

The picture of the roadview of driver are generated from the current position of the driver, and the graphic description of various objects in the world. Figure 3 shows the process to create objects for use in the simulation world. These objects are created in a graphics package such as Autocad. They are then processed by 3DVIEWIII format conversion tools to transform them to a format usable by our graphics system. These objects are stored in the database used by the driving simulator.

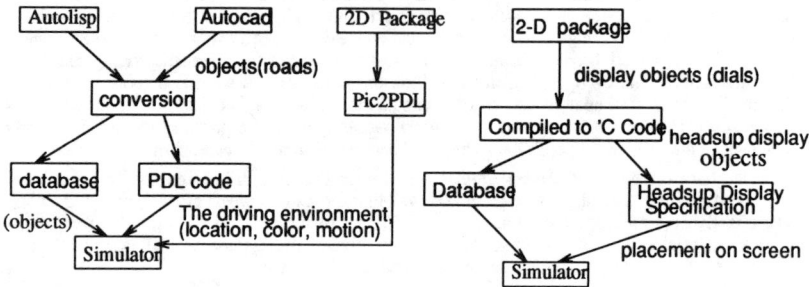

Figure 3.: Process to create driving world

Figure 4.: Process to create headsup display

The design of Headsup displays (i.e. two-dimensional displays drawn directly to the screen) required a departure from the method of designing real-world objects (i.e. three dimensional objects in drivers world). Objects created in Autocad are placed into the database world according to three dimensional coordinates, whereas the headsup displays in the flight simulator used pixel locations of the screen to directly draw displays superimposed onto the screen. Figure 4 describes the additional processes to design headsup displays.

3. Design and Implementation

We have successfully extended the simulator at the Human Factors Research Lab to allow for validation and testing of Headsup displays. The flight demonstration software provided the method to simulate headsup displays, and to simplify the design process a translator has been written to convert a drawing from Xfig to the 'C' code necessary in the simulator. The simulation world consists of additional objects modeled from Minneapolis highways and buildings providing more realism. A new method of object design, Level of Detail (LOD) programming allows complex designs to be used without the subsequent processor slowdown.

The driving simulator module receives data from three sources as shown in Figure 5. First, The driving world module specifies the location of objects in the world. Second, the database of basic objects contains object description described in the driving world specification module. Third, Headsup display description, such as dials for speed, fuel, etc., are used by the driving simulator to create and display information based on driver inputs. The driving simulator calculates the next frame to be displayed and passes this to the PG2000 module, which draws the new frame.

The headsup displays represent housekeeping information (speed, fuel) and navigational information (compass and map). Two representations for each display have been designed, alphanumeric displays present the information as text and iconic displays have been designed that capture the information graphically and present this image superimposed over the windshield. The headsup display icons are specified by creating a drawing on computer screen with the help of a drawing package. The drawing is then translated to a C program and linked to the simulation software. Figure 6 shows the view of driver with headsup displays for navigation and speed superimposed on the windshield. Navigation display specify the position and direction of travel and locations of traffic blockages. Speed is represented by three icons representing high, medium and low range of speeds. This scheme allows the driver to obtain a quick estimate of speed. Actual speed can be read from the alphanumerics below the icons.

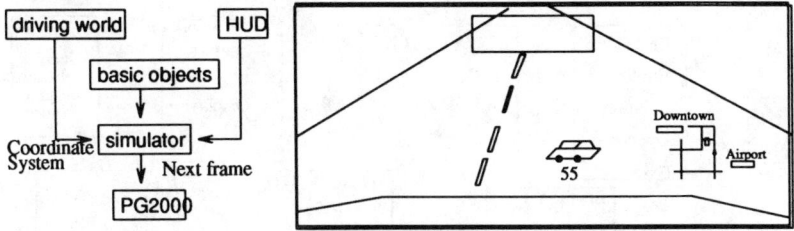

Fig. 5. Block Diagram of Simulator   Fig. 6.: Driver's view

A driving environment is designed by creating a library of 3-dimensional drawings and a 2-dimensional map showing the location of the objects. A library of 3-dimensional primitives (e.g. road-segment, cubes, pyramids, prisms) are created using AutoLisp. Buildings, roads and other interesting 3-dimensional objects can be created by composing the primitives. The 2-dimensional map is specified on computer screen using picture drawing tool. We created the skyline of Minneapolis with our software to provide realism in driving world. The driving world consists of, roads and buildings representing the the highway system from the Minneapolis International Airport to downtown Minneapolis.

Level of Detail (LOD) programming was used to achieve real-time performance in a complex environment. A LOD object is actually a database of several objects with varying degree of complexity. The parameters of the object dictate which object will be visible from the driver's viewport. For example, if a house were designed as a LOD object, it might have three different drawings associated with it. The first would be the house with full detail, the second would be the house as a cubic block with windows only, and the final drawing would be a cubic shape alone. The first drawing might appear for distances less than 1000 feet, the second for distances 1000-2000 feet, etc. We decomposed the roads into a collection of 1/8the mile segments. Each segment was designed as a level of detail object. Road segments 1/4 mile away is not drawn to lowering the amount of drawing required by the graphics processor.

4. Using Simulator for Headsup Display Design

The simulator is being used to create headsup displays and evaluate their safety and effectiveness. For example, measurement of assimilation and response time to new traffic blockage information for radio broadcast and headsup display was compared using the simulator. The parameters of experiment include distance of traffic blockage, speed of driving, and distance of routing decision point to reroute to avoid traffic blockage.

Experiment was setup in 4 steps. First, the driving environment was created. We created a section of highways from the Minneapolis Airport to downtown Minneapolis. The Minneapolis skyline was created to add familiar terrain to the environment. Second, the headsup displays were created. We created headsup displays showing speed, navigational map including current position of driver, location of blockage and suggested rerouting. Third, the experimenter specified the multiple trials to be used for for each subject by specifying the values of experiment parameters for each trial. This was accomplished by remembering the experiment parameter values in a structure for each trial This structure has the following information: control-group, goal_speed, traffic blocks with locations, distance from rerouting decision point to present traffic blockage information. Sixteen combination were stored to represent sixteen trials. This can be done by choosing the attributes from the trial design menu(figure 7).

Lastly, the experimenter specify the data to be collected from experiments via a menu. The menu represents the entire state of simulation with the help of an entity attribute model, listing each object with attributes. For our experiment the data to be sampled include trial name with the correctness of routing decisions by subjects. To record the correctness of rerouting decisions, the path followed by subject is sampled and analyzed. Automatic replay of a trial is supported to observe the trial for verification purposes. The data collection menu (fig 8) is used to provide a flexible user interface for the experimenter. For example, to store the position of the subjects position the experimenter could select the "entities-cars-self-xpos-sample" path.

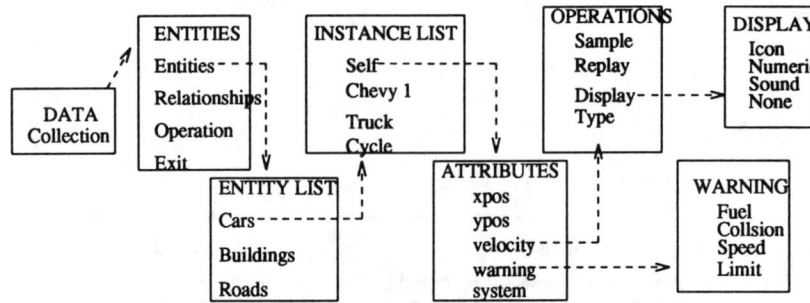

Figure 7: Menu to create experiment trials

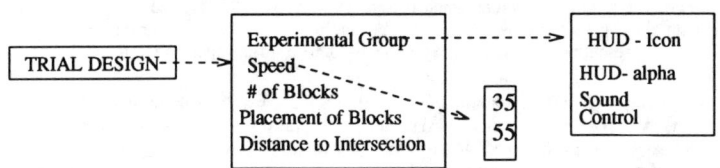

Figure 8: Menu to sample data

To run our experiment the experimenter first chooses the group type that will be tested. The trials have been created to reflect a random arrangement of the variables. He is able to simply press a start and quit button the begin and end each experimental trial. The data collection is done by recording correct, incorrect decisions for each trial. The entire experiment is saved to allow an automatic replay of the trials for each subject. This can be used to verify decisions or to analyze driver behavior.

5. Results and Impact

We have developed a system that can be used to study alternative methods of information presentation including headsup displays. We have designed icon and alphanumeric based HUD for navigation and speed. The simulator is being used for several experiments in human factor laboratory.

6. Acknowledgements

This work was supported by UROP, CTS and MinDot. The students from courses Csci 5504 and Csci 5106 in 1990-91 developed many pieces of the software. The hardware and a small part of software was acquired from American Honda, ESI, Ohio and XTAR, California.

7. References

1. D. W. Swift and F. H. Freeman, The Application of Headsup Displays to Cars, *Vision in Vehicles (ed. A. G. Gale)*, (1986). Elsevier Science Publishers (North-Holland)
2. K. Kurokawa and W. W. Wierwille, Validation of a Driving Simulation Facility for Instrument Panel Task Performance, *Proc. Human Factors Society 34th Annual Meeting*, (1990).
3. P. A. Hancock, J. K. Caird, and H. G. White, The Use of Driving Simulator For the Assessment, Training and Testing of Older Drivers, *HFRL Report NIA 90-01*, Human Factor Research Laboratory, Univ. of Minnesota, (1990).

INFRARED DETECTOR DEVELOPMENTS

M.A.G. CLARK[1] & A. HODGE[2]

INTRODUCTION

Modern road transport systems rely on information from sensors to operate safely and efficiently. The sensors are typically used to provide vehicle counting, occupancy, speed measurement and classification information.

Where pedestrians and vehicles share road space, e.g. at signaled crossings, the delays that result can be frustrating to both parties. It is thought that considerable improvements in efficiency and safety can be achieved by the inclusion of pedestrian detection facilities in these control systems [1].

The most commonly used vehicle sensor is the inductive loop. Its main limitation, which also broadly applies to all "below ground" and "surface mounted" sensors, resultsfrom the requirement to bury the sensor in the carriageway. The vulnerability of such devices to pavement deformation and damage by utility contractors is high. Annual fault rates can be in excess of 50%, especially in urban areas [2].

The optical sensors discussed in this paper offer several advantages, including well defined detection areas, fast response time and true presence capability. Two technologies have been developed based on passive and active infrared. Applications include both vehicle and pedestrian detection.

Passive Sensors: Passive infrared detectors contain an optical matrix system which focuses infrared energy onto a transducer. The transitions between adjacent zones are used to determine the existence of a target.

---
[1] Managing Director [2] Detector Systems Group Leader
Microsense Systems Ltd, Meon House, 10 Barnes Wallis Road, Segensworth, Fareham, Hants, PO15 5TT, U.K.

This is illustrated in Figure 1. These sensors are best suited to detecting moving targets. They have well defined detection zones and can detect targets up to 100m range. Being passive there is no health hazard or licensing requirement.

<u>Active Sensors</u>: Sensors using this technique comprise one or more transmitter/receiver pairs within a single housing. The transmitter section emits infrared light into well defined regions along the road. The receiver collects the radiation and signal processing is performed such that the presence of a vehicle in the zone can be determined. This is illustrated in Figure 2.

Active sensors can provide true presence detecting capability combined with a fast, well defined response. This makes them suitable for most applications where conventionally a loop detector would be employed.

PEDESTRIAN SENSORS

<u>System Overview</u>: The system comprises two types of detector. A kerbside unit confirms a demand for a pedestrian phase from the push-button while a crossing detector extends the pedestrian phase from a minimum up to a maximum preset time. Both sensors are located on the top bracket supporting the primary signal head. This is illustrated in Figure 3. Several trial sites using these detectors have been set up under the European DRIVE program.

VEHICLE SENSORS

<u>Signal Detector</u>: Utilising passive infrared technology, the main application for this sensor is at intersections and pedestrian crossings to detect approaching vehicles. A typical arrangement is shown in Figure 4.

<u>Gantry Detector</u>: This detector is intended for mounting directly over a carriageway, as shown in Figure 5. Utilising active technology, two detection zones are provided that enable vehicle speed, count, length and occupancy to be determined on a per lane basis.

One example of the application of this unit is on the M25 London Orbital Motorway Traffic Information System "Traffic Master". Several hundred battery powered sensor units are employed in this scheme, which is fully operational. It is intended to expand the system over the entire UK motorway network.

INFRARED DETECTOR DEVELOPMENTS

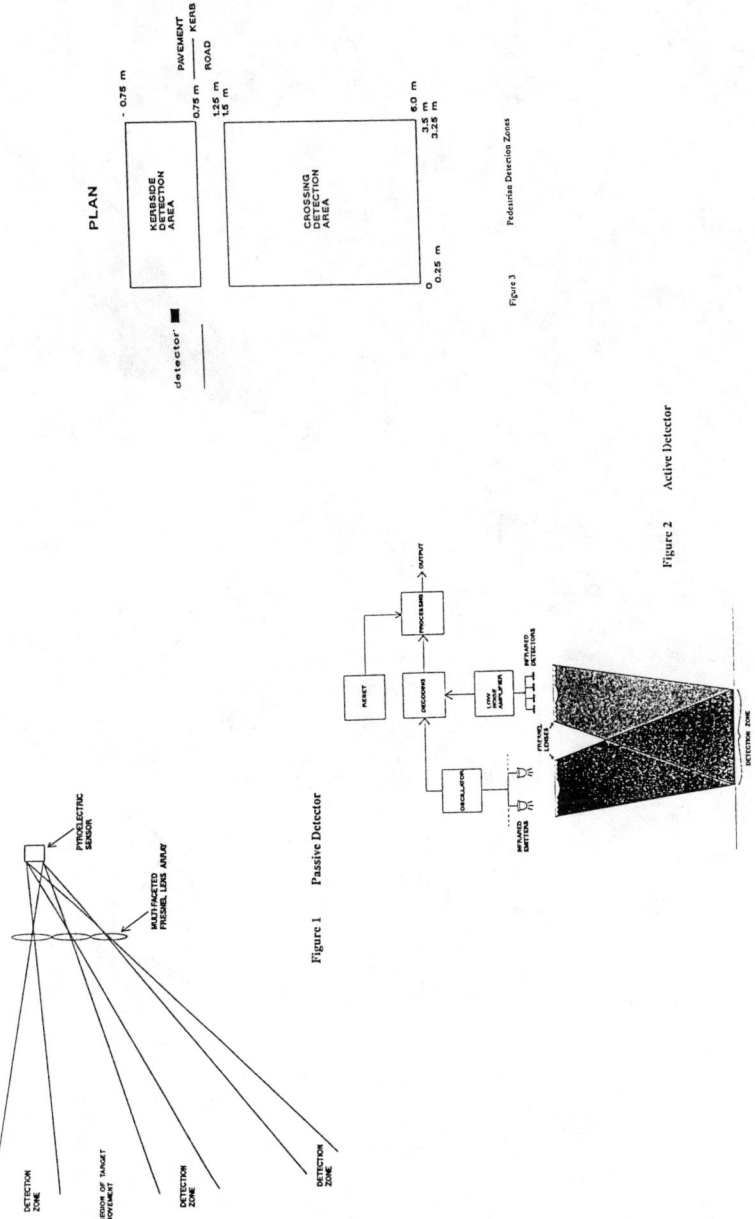

Figure 1 Passive Detector

Figure 2 Active Detector

Figure 3 Pedestrian Detection Zones

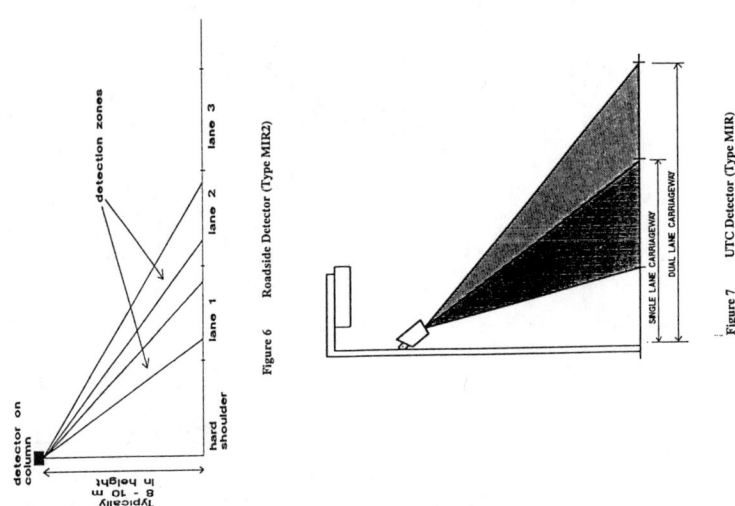

Figure 6 Roadside Detector (Type MIR2)

Figure 7 UTC Detector (Type MIR)

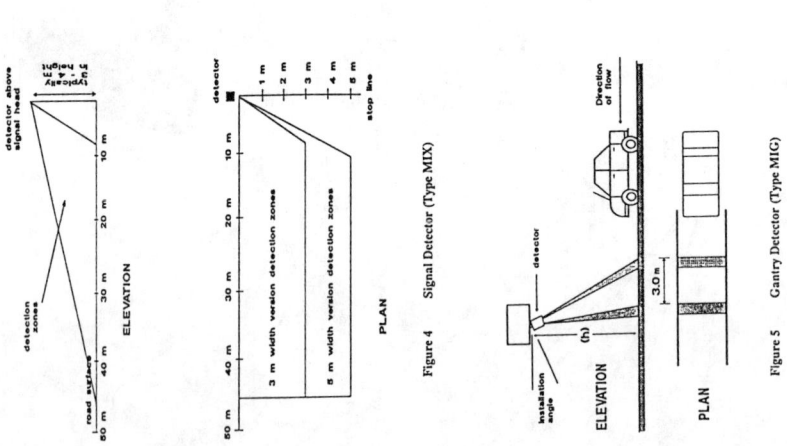

Figure 4 Signal Detector (Type MIX)

Figure 5 Gantry Detector (Type MIG)

Roadside Detector: This two zone, column mounted, unit utilises active infrared technology and gives individual outputs for the nearside and second lanes as illustrated in Figure 6.

Motorway trials have been undertaken by the UK DTp in which sensors were linked to a HIOCC [3] automatic incident detection computer and the output compared with the existing loop based system. The results demonstrate that the sensors are suitable for this application.

UTC Detector: This active infrared sensor is primarily intended for use in urban traffic control systems, SCOOT [4] in the UK. The unit is mounted at the roadside on a lighting column or similar structure as illustrated in Figure 7. Monitoring has shown that these units can replace loops without any requirement to revalidate the segment.

SUMMARY

As congestion on roads becomes more acute, the infrared detectors described are providing attractive alternative detection solutions with advantages particularly in terms of ease of installation and maintainability, resulting in a low cost of ownership. The units in service are demonstrating the viability of the techniques and that they can often provide the better and in some cases the only solution to a detection requirement.

The M25 scheme demonstrates the potential of the combination of low power detection and radio communication. It is hoped to further develop these techniques for use in urban areas.

References

[1] "Improvements of Pedestrian Safety and Comfort at Traffic Lights". Annual Review Report 1990, DRIVE Project V1061: Pussycats.

[2] "Data Acquisition and Communication techniques and their Assessment for Road Transport". Sensor Survey Report 1989, DRIVE Project V1039 DACAR.

[3] Collins J.F., Hopkins C.M., Martin J.A., "Automatic Incident Detection - TRRL algorithms HIOCC and PATREG" TRRL Report 526, 1979.

[4] Hunt P.B., Robertson D.I., Bretherton R.D., and Winton. "SCOOT; A traffic responsive method of co-ordinating signals" TRRL Report LR1014, 1981.

The California Inductive Loop Radio Demonstration Project

Sam Taff[1], Walt Winter[2],
Clint Staley[3] and Ronald Nodder[4]

Abstract

There is growing interest at both the national and international level in the application of new technologies to the highway system. While many of these new technologies are revolutionary in nature, other strategies represent a small scale, incremental approach to improving the operation of the existing road network. The Inductive Loop Radio (**INRAD**) demonstration project examines the application of one set of small scale technologies to highway transportation.

---

1 Lecturer and Research Engineer, Department of Civil and Environmental Engineering, California Polytechnic State University, San Luis Obispo, CA 93407.

2 Senior Materials and Research Engineer, Division of New Technology, Transportation Materials Research Laboratory, California Department of Transportation, 5900 Folsom Blvd., Sacramento, CA 95819.

3 Assistant Professor, Department of Computer Science, California Polytechnic State University, San Luis Obispo, CA 93407.

4 Research Engineer, Department of Civil and Environmental Engineering, California Polytechnic State University, San Luis Obispo, CA 93407.

INRAD was designed to test the functionality and utility of inductive radio communication in highway operations. Loop detectors installed in the pavement were equipped to both perform standard vehicle detection tasks and to exchange information between specially equipped vehicles and traffic operations centers. Information is gathered and messages updated in real time, with in-vehicle displays relaying information to the drivers.

INRAD technology can provide benefits to a variety of user groups, including highway maintenance personnel, traffic enforcement and emergency service staff, transit and taxi services, and parcel pick-up and delivery firms. **INRAD** may also serve as a "bridge" technology, providing a smooth transition path toward larger Intelligent Vehicle Highway Systems.

Introduction

Highway travel is likely to remain the dominant mode of personal transportation in the U.S. throughout the rest of the 20th century. Given today's levels of congestion and anticipated growth in travel demand, additional capacity will be needed throughout the highway system. While the construction of new facilities will satisfy some of the anticipated needs, it is also necessary to improve the operation of existing facilities.

Intelligent Vehicle Highway System (IVHS) technologies have the potential to significantly improve the operation of both existing and future highway facilities. The emphasis in much of the research associated with IVHS has centered on major revisions in existing highway system design and operations. While the "long range" technologies are important, there are a number of off-the-shelf component technologies that may be integrated with existing roadway and vehicle systems. The **INRAD** project investigates the feasibility of using existing sensor, computer and communication technologies with minor infrastructure changes and minimal disruption to the present system.

Project Objectives

The inductive radio demonstration project is designed to evaluate the usefulness of short range radio communications between vehicles and the roadway, using inductive loops already installed for traffic detection. The project objective is to demonstrate the potential of inductive radio technology in the IVHS context. The **INRAD** technology will be tested in a difficult environment where transmitters and receivers are below optimal design standards.

The **INRAD** project emphasizes the use of low-cost, straightforward technology that is suitable for rapid implementation, without elaborate development or research stages. The technology developed by the project will rely heavily on commercially available components. Those components designed especially for the project will be as simple and cost-effective as possible.

Methodology

The **INRAD** project is divided into four major phases:

1. Design and development of infrastructure requirements
2. Sub-project experimental design
3. Data collection
4. Data analysis and final report

1. Infrastructure Requirements
This phase involves designing and developing electronic hardware and software to allow simple two-way radio communications from some central facility to **INRAD**-equipped vehicles in the field, and to provide easy-to-use computer monitoring and control of the radio communications. The **INRAD** software and hardware are divided into four components, which work closely together, communicating either via radio or telephone lines. The four components are:

In-vehicle hardware and software -
The hardware and software for installation in test vehicles includes an on-board computer with an elementary liquid crystal display for posting messages and information sent by the **INRAD** system (see Figure 1).

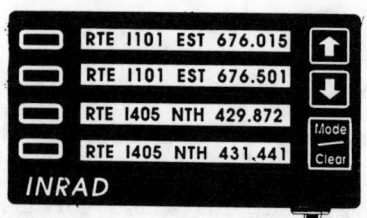

Figure 1: On-Board Computer Display

INDUCTIVE LOOP RADIO PROJECT   325

This computer is connected to a radio transceiver attached to the bottom of the vehicle. The radio transceiver has very limited range; it is able to communicate simple pre-programmed messages to and from an inductive loop as the vehicle crosses the loop. The message prompts take the form of six-byte digital codes. One code may be transmitted in each direction at each loop crossing.

In-cabinet hardware and software -
The roadside controllers for the loops in the **INRAD** demonstration will be supplemented by a computer using an STD bus, an industry standard, which will allow the sending and receiving of radio messages between vehicles and the loop. The STD system will use a 1200 baud phone line to communicate with a central location. It will use this line to send traffic data, plus **INRAD** radio messages received from vehicles, and to receive new messages to relay to the vehicles.

Communications processor -
The communications processor (CP) is a VME bus computer, running the OS-9 operating system. This computer will monitor all of the STD loop controllers and manage communications between the controllers and the central control computers described below. The CP records all communications between the central control and STD computers on disk. These archives can be copied to another medium for later analysis.

Central control computers
The central control (CC) computers provide the main control of the **INRAD** system. Up to two CC computers can communicate with the CP at a time. They may be tied directly to the CP via cables, or may communicate at long distances via 2400 baud telephone lines.

Two different version of the CC computers will be developed. The first will be an IBM PC system running the MS-DOS operating system. It will provide a simple screen-based user interface, without special graphics. It will allow the user to track **INRAD** loop and vehicle activity, receive messages from vehicles, and send messages to vehicles. This version of the CC will provide an inexpensive, simple CC computer for distribution to a number of Caltrans locations.

The second CC system will be a Sun workstation running the Unix operating system under the X window environment. This will provide a more sophisticated graphical user interface, displaying a pictorial map of the freeway, with **INRAD** loops and vehicles marked on the map. The user will control the system through the keyboard and via a mouse, and a number of the more tedious tasks of **INRAD** control will be automated. This second CC system will be built using software already constructed at Cal Poly as a part of other Caltrans projects.

2. Sub-project experimental design
In order to demonstrate the potential of this technology , several different "usage scenarios", have been developed. These have been designed to maximize the hardware and software capabilities in order to send and receive pertinent real-time information between the **INRAD**-equipped vehicles and the central control facility. The scenarios are currently under development include highway advisory radio alert, congestion ahead warnings, AVI/AVL demonstrations, and highway inventory tasks.

3. Data collection
This involves demonstration of the technology by means of application of the usage scenarios discussed earlier, using test vehicles along a selected test route. Initial testing will be conducted on the Cal Poly Test Track, with field testing scheduled in the Los Angeles area on selected freeway segments.

4. Data analysis and final report
The final phase of the project is the analysis of data collected and the reporting of the project results. Particular emphasis will be on the potential short term applications and "market niche" use of INRAD related technologies.

Schedule
The **INRAD** project began in August of 1990. Prototype testing of software and hardware began in early 1991 at the Caltrans Lab and the Cal Poly Applied Research Facility. Field installation and testing are scheduled for the end of 1991.

The Contribution of Dynamic Route Guidance and Information Systems to More Efficient Traffic Management

Jürg M. Sparmann, Ph. D.[1]

The environmental and traffic-related problems expected as a result of the continued increase in motorization require new solutions if our towns and cities are to survive and their attractiveness as places for people to visit and live is not to deteriorate still further. As the improved quality of life brings with it an increased need for mobility, conflicting aims arise which can only be adequately reconciled by considering the entire displacement chain as an integrated whole. Due to the complexity of the decision-making process with regard to local changes and the ever-increasing intervention resulting from traffic control measures, there arises a need for additional information on the part of those participating in the traffic process. But new logistical concepts also have to be developed for freight transport so that in the future it will be possible to handle the transportation of goods at lower cost and in line with urban requirements.

The following paper is intended to show the possibilities and limitations of incorporating individual traffic control systems in traffic management and in addition to show the expected effects of these systems.

New Aims for Traffic Management

In the past, the aims of traffic control were mainly geared to improving traffic flow and increasing road safety, although they also included the reduction of environmental pollution. As a result of dramatic developments in recent years and gloomy forecasts for the years to come, in future greater weight will have to be given to objectives which cater to the demands of ur-

---

[1]General Manager, SNV Studiengesellschaft Nahverkehr mbH, Auguste-Viktoria-Str. 62, D-1000 Berlin 33, Germany

ban politics and ecological considerations. As a consequence of this, the main objectives of traffic management will be less the maximization of vehicle flow than the minimization of environmental impacts. This in turn means that evaluation criteria such as, for example, the permissible traffic capacity of road networks from the point of view of urban and environmental compatibility, will play a much more important role.

The technical facilities installed for the purpose of controlling traffic are at present very limited in their effects because in many cases they are not operated in response to real-time traffic situations and because the information displayed is of a very general nature. Furthermore, very little consideration is given to the personal needs of the individual road user. Individual traffic guidance systems are trying to fill this gap by offering navigational aids to drivers. These are geared to the needs of the individual and help to influence the distribution of traffic throughout the road network. In this way it is possible to make more effective use of the available traffic infrastructure.

Individual Traffic Control

When routes are determined by a central control point and take real-time traffic data into account, the system fulfills the prerequisites for integration into a traffic management system. The vehicles must have communication faciliites at their disposal in order to be able to receive these route recommendations. In Berlin, such a guidance system has been in operation on an experimental basis for the past 2 years under the name of LISB and its acceptance by users, as well as its effects, have been evaluated.

The vehicle equipment consists of a transmitter and a receiver, a navigation computer, a device for determining the position of the vehicle continuously and a display and input facility. Before starting a trip, the driver enters a destination, either from a memory or as coordinates. For reasons of data protection the destination is only stored in the in-vehicle unit. The digital road map necessary for navigation and the traffic-responsive route recommendations are transmitted to the vehicle via so-called beacons installed at intersections controlled by traffic lights. These beacons are stationary transmitting and receiving units which ensure data transfer to and from vehicles. All vehicles fitted with the necessary equipment receive the same data when passing a beacon. The relevant information is selected in the vehicle itself, as only the in-vehicle computer knows the destination of the vehicle. The filtered information, converted into navigational aids, is shown to the driver on a display with acoustic back-up.

Traffic-responsive route recommendations can only be as good as the actual traffic situations recorded. Due to the inadequate equipment in the road network for the purpose of collecting the real-time traffic data, such as, for example, loop detectors, in the LISB system the vehicles themselves are used as detectors. They measure the travel and congestion times when driving through the individual route sections. When they pass a beacon, this data is anonymously transmitted and forwarded to a central traffic control computer. There the travel time data received is compared with the expectation value determined from the reference travel-time pattern known from historical data. If the values received are significantly higher than these, allowing for statistical adjustment procedures, there is a disruption in traffic. Based on these reports, new route recommendations are determined, these then being displayed in the following vehicles.

Route recommendations are provided only in the main road network in order to avoid driving through residential areas and traffic-calming zones. The density of the beacon network varies according to the network structure and the possibilities of directing vehicles to alternative routes. Protection of the residential sector is ensured by selecting the network for route recommendations with this need in mind. Central determination of the route has the advantage that both traffic-political and urban-political objectives can be taken into consideration when controlling traffic. In the central control point the distribution of traffic volumes within the network can be influenced by the selection of suitable criteria for determining routes, provided that there is sufficiently high acceptance of the alternative routes and that recommendations are followed. Here a compromise between the optimization of individual benefit and the needs of society at large should be aimed at.

Approximately 650 persons with their cars have taken part in the Berlin field trial. The following shows the results of surveys on the acceptance and benefits of the system. Investigations were also carried out into whether acceptable alternatives exist at all and whether the route recommendations were followed. On the benefit side, travel time gains and increased safety predominate.

Asked about the frequency with which alternative routes were displayed, 13 % of those asked said that these were given frequently and a further 56 % said at least occasionally. The willingness to accept the system is supported by the statement that a good 9 % of participants in the experiment said they had got to know several alternative routes which they intend to include in their choice of routes in future, even without the guidance system. A further 50 % stated that at least they got to know some alternative routes. One can therefore say that individual guidance systems contribute towards optimising behaviour with regard to route selection.

The frequency with which route recommendations are followed depends on the driver's familiarity with the locality. 95 % of the persons asked stated that they always or at least mostly followed route recommendations on seldom used routes. This assessment was much lower in the case of more familiar routes between frequently travalled origin-destination relationships. The evaluation of an investigation into the objective behaviour of the drivers shows that drivers of privately used vehicles always followed the route recommendations in the case of over half the routes examined. This value is a little lower for commercially used vehicles, a fact which is due to the still inadequate dynamics of the system and to the still rather complicated destination entry, both of which have a negative effect, especially in the case of trip chains.

It can be assumed that there is a direct relation between following route recommendations and the perceived travel time savings, whereby the experience from each previous trip considerably influences readiness to follow the route recommendations. The results of the surveys show that one in three car drivers is of the opinion that he/she is faster as a result of LISB on frequently used routes; on less frequently used routes as many as 60 % stated that they felt they had saved time. Parallel studies on actual trip times comparing vehicles with and without LISB travelling the same origin-destination pairs reveal that gains in travel time are overestimated. Evidently the real time-gains are less important than the subjective feeling of having saved time with regard to the acceptance of route recommendations. This is expressed in the high degree of adherence to route recommendations, an essential factor if dynamic individual traffic guidance systems are to contribute towards improved traffic management.

Furthermore, the Berlin system has generally been awarded ratings ranging from good to very good regarding the assessment of the benefit-potential anticipated for LISB when it is ready for mass production. Here it becomes evident that the greatest benefit lies in facilitating orientation and destination-finding. However, the possibility of shortening travel time and so avoiding congestion is also rated very high, although the number of very good evaluations is low. On the other hand, the system is regarded as having less promising potential for increasing road safety and energy-saving.

Accompanying studies on road safety, however, have shown that drivers using the LISB system on frequently travelled routes make slightly fewer driving mistakes in comparison to those trips undertaken without route recommendations. This effect has not proved to be significant, however. One can, though, draw the opposite conclusion, namely that there are certainly no negative or harmful effects on road safety resulting from use of the guidance system.

The less familiar a driver is with the locality he is driving in, the greater the road safety potential accruing from the use of a guidance system. With trips to unknown destinations, participants in the LISB experiment, taken altogether, have made no more mistakes than when driving on familiar routes. However, if one looks closely at individual driving errors, differences become evident in various driving maneuvers. Compared with the so-called outsider risk according to a study undertaken by the University of Cologne, the expectation value for error frequency among drivers unfamiliar with the locality shows a significant increase in safety.

When examining the benefit it was not possible to take into consideration how recognition of congestion and allowing for route recommendations affects it. Precisely the inclusion of unpredictable disruptions with considerable effects on travel time can substantially increase the benefits of the system. For this reason a procedure for congestion management has been incorporated in the control strategy for determining routes. In addition to this, the real-time database created via this guidance system can be made use of for dynamic updating of the display of collective control systems. This thus also provides benefits for those vehicles not equipped with guidance systems.

Individual Traffic Information and Route Guidance from RDS-TMC until Beacon Communication

Wolf Zechnall[1]

Abstract

A complete system of traffic management consisting of both collective and individual components is described. As a result of German and European initiatives and programs to solve the steadily increasing traffic problems, a system introduction in three steps is proposed, i.e. the digital traffic message channel (RDS-TMC) (Braegas et al., 1986) in the first step, information and access-control beacons in a second step and finally a beacon assisted infrastructure which handles all type of vehicle-roadside communication needed for automatic debiting, pre-booking of parking space and individual driver guidance including situation-dependent recommendations to change over to public transport.

Introduction

The steadily increasing traffic density both in cities and highways requires an optimal usage of the existing infrastructure such as roads and highways as well as public transportation systems. Precondition to this is the development and installation of complete traffic management systems, which inform and guide the drivers effectively. The Bosch proposal for introduction of such complete traffic management systems is a bottom up approach in three steps starting with the area covering introduction of the Radio Data System-Traffic Message Channel (RDS-TMC) and subsequent introduction of beacon infrastructures in two further steps. Description of this three step approach will be the topic of this paper.

---
[1]Manager, Research Institute Communications, Robert Bosch GmbH, Robert Bosch-Str. 200, D-3200 Hildesheim, Germany

National and International Programmes for Traffic Management and Control

In Germany the Department of Transport initiated the programme for an "Integrated Dynamic System for both Collective and Individual Traffic Information and Vehicle Guidance", which is based on results of earlier field trials of traffic management systems, e.g. the ALI-System (Braegas, 1980). Its main objectives are:

1. Improvement of up-to-dateness of traffic information
2. Regional traffic information
3. Individual traffic information and guidance

Objectives 1 and 2 can be achieved by RDS-TMC, Objective 3 needs installation of a beacon infrastructure, enabling two way communication between vehicles and roadside. In the Berlin LISB (Leit-und Informations-System Berlin) field trial such a beacon infrastructure (ALI-Scout) was tested (Hoffmann et al., 1988).

In Europe the Commisssion of European Communities (CEC) runs the 3-years (1989-1991) DRIVE programme (DRIVE = Dedicated Road Infrastructure for Vehicle Safety in Europe), whose objectives are:

- Improvement of traffic safety
- Improvement of traffic efficiency
- Reduction of air pollution caused by traffic

The subsequent ATT-(Advanced Transport Telematics) programme is scheduled to start 1992. It is intended to fund projects, which use DRIVE-results for trials and pilot installations in testsites located both in urban and inter-urban areas.

Besides DRIVE the EUREKA-Project "PROMETHEUS" of the European Car Industry is to be mentioned, which runs since 1987. The ATT-programme will be strongly influenced by PROMETHEUS results as well.

Besides DRIVE, PROMETHEUS and the scheduled ATT-programme the POLIS and CORRIDOR initiatives have to be considered:

POLIS was initiated by 15 maior European cities, which decided to cooperate in testing and development of traffic control systems being able to manage their urban traffic problems. The systems should be able to cover the following tasks:

- Automatic acces control and debiting
- Route guidance
- Parking management and public transport information
- Traffic control and driver information
- Integrated traffic management

The CORRIDOR - initiative of the CEC covers the interurban areas. The CORRIDOR objectives are complementary to the POLIS-objectives and cover the following topics:

- automatic toll collection
- individual route guidance
- Traffic and travel information (RDS-TMC)
- Freight and fleet management (surveillance of hazardeous goods transports)

Requirements to Traffic Guidance Systems

Traffic management and individual driver guidance systems will be accepted by the public only, if they meet the following 7 basic requirements:

1. Automatic debiting, access control and pre-reservation of parking space
2. Installation of infrastructures and in-vehicle equipment in discrete steps
3. Individual route guidance considering public transport according to traffic situation
4. Integration of RDS-TMC and individual route guidance
5. Agreement on a standardized communication interface vehicle/roadside
6. User-friendly operation of in car equipment, e.g text- input of destination-addresses.
7. Modular product line of in car equipment.

The Complete System of Traffic Management

The Bosch system approach, which is in accordance with the 7 requirements listed above, consists of three steps.

Step 1: Collective traffic information via variable message signs and the Radio Data System-Traffic Message Channel.

Traffic and driver guidance information is distributed collectively. The traffic information is generated in traffic control centers by processing traffic data collected by means of induction loops installed in the road surface. Data inputs from parking management control computers and public transport control centers are added to the road traffic

information. The distribution of the resulting traffic and guidance information is done collectively via variable message signs or via the digital broadcast traffic information service Radio Data System-Traffic Message Channel (RDS-TMC). As the area coverage of RDS-TMC is determined by the area coverage of FM- (or AM-) broadcast service, RDS-TMC service will be available everywhere from the very beginning.
The European standard for traffic message distribution via RDS-TMC is finished and a first pilot installation of RDS-TMC has been started this year in the Cologne-Dusseldorf area. More pilot installations in Europe are scheduled for 1992 and the regular service for 1994.

Step 2: Information beacon for toll collection and access control

In the second step more detailed individual traffic information, automatic toll payment and automatic handling of access control will be enabled via information beacons. A short range 2-way communication link using microwave or infrared is used for exchange of information between the beacon- and the in vehicle transceiver (OBU=On Board Unit). Information beacons will be installed at entrances or exits of parking lots or limits of downtown areas with resticted access. Usually they will be controlled by local computers for parking fee collection or checking of access permits. A wired link to traffic control centers will be provided, if economically feasible, otherwise traffic information will be kept up to date by continuous reception of RDS-TMC-information.
Fully operating information beacons can be realized quickly, because no area coverage of beacon installation is required for their full function. It is planned to realize pilot installations of information beacons 1992 at the entrances of parking areas located at the Hanover fairground.

Step 3: Guidance beacon for individual traffic dependent driver guidance information, automatic debiting and pre-reservation of parking space.

The hardware of the guidance beacon and information beacon is the same. The guidance beacon differs from the information beacon by exchange of more sophisticated information, e.g. individual traffic dependent guidance information including public transport connections, improved toll payment procedures, pre- reservation of parking space and feedback of individually detected traffic data. Precondition for this enhanced data exchange is the link to a network of different traffic computers which enable a real-time exchange of road traffic-, public transport,- parking- and other data. First pilot installations will start 1993/94, more widespread installations in metropolitan areas are expected to be realized in the second half of the nineties.

In vehicle equipment

The acceptance of traffic management systems is strongly influenced by the in car equipment. Therefore 5 classes of in car units have been designed according to the described three step approach.

1. Car radio with RDS-TMC-decoder (Step1)
2. On Board Unit (OBU) for access control and toll collection (Step 2)
3. Car radio with RDS-TMC-decoder and OBU (Step 2)
4. Car radio with RDS-TMC-decoder, OBU and additional graphic display (Step 3)
5. Car radio with RDS-TMC-decoder, OBU and autonomous navigation system (Travelpilot) (Step 3)

Conclusion

The described concept of traffic information and route guidance is the result of almost 20 years of international research in the field of communication techniques, informatics and traffic engineering. It is expected, that the proposed three step introduction will have a positive effect on ecology, safety and economy very soon.

References

Braegas, P. Function, equipment and field testing of a route guidance and information system for drivers (ALI) IEEE Transactions, Vol. VT-29 (1980), No.2

Braegas,P. et al. "Die Übertragung von codierten Verkehrshinweisen über UKW-Rundfunksender mittels RDS" Bosch Technische Berichte (1986), Vol. 8

Hoffmann, G. et al. LISB research project "Guidance and Information System Berlin" Proceedings "Roads and Traffic 2000", Berlin (1988), Vol. 4A-1

DYNAMIC ROUTE GUIDANCE - THE "DRIVE" PROJECT CAR-GOES
John D. Turner, MCIT, MBCS.

ABSTRACT

The DRIVE project CAR-GOES has studied strategies for integrating road traffic management and central systems for dynamic route guidance and traffic signal control. This paper summarises the recommended strategies for centralised and de-centralised DRG working in co-ordination with fixed-time and traffic-responsive signal control systems.

1. INTRODUCTION

In recent years, the traffic situation in urban areas throughout the world has been getting more and more problematical. The high traffic volumes on the restricted road network, environmental sensibility and the lack of effective driver warning and information systems have made the development of dynamic route guidance (DRG) systems a welcome new aspect. From their operation, new perspectives are opening up for traffic authorities, traffic management bodies and individual car drivers themselves.

The DRG system can make every equipped vehicle a data source for the benefit of all drivers in the network.

In parallel, urban traffic control (UTC) systems have developed from (initially) safety devices into, more and more, tools of traffic management. Their disadvantage is often the lack of timely information. Their main source of information, loop detectors, is available only at discrete locations and provide a limited set of data types.

Technical Manager, Advanced Traffic Systems Group, Siemens Plessey Controls Ltd., Sopers Lane, Poole, Dorset. BH17 7ER. ENGLAND.

Investment in UTC systems has been high, whereas investment in DRG systems has yet to take off. In order to maximise the benefits of both systems, the optimisation of traffic control and vehicle routeing should be undertaken as a single, interactive and integrated process.

The CAR-GOES project has researched several discrete aspects of traffic control and management and two main strands of integrated traffic management:-

i) The optimum design of a DRG system, and
ii) The possibilities of integrating DRG and UTC systems, to benefit and improve both management systems and thus road traffic.

The project concludes in 1991 and this paper will present its preliminary recommendations for both strands as above. It has provided the essential analysis of users' requirements and focussed on integration of the systems at the following levels:-

1. The use of data from   1'. The use of data from
   DRG systems for   UTC systems for
   improved UTC.   improved DRG.

    2. The combination of 1 and 1', with both systems running independently.

    3. The potential for combining 1 and 1' in a full-scale on-line integration.

The work has been undertaken within the European DRIVE Research Programme by a consortium of 12 organisations (5 Industrial; 7 Research Institutions) from 5 European Countries.

2. CAR-GOES WORK PACKAGES.

Significant work packages have studied:-

    i) Improving UTC operations by information from DRG systems, considering strategies as well as traffic control operation.

    ii) Advanced route optimisation strategies, involving criteria other than individual need. The optimisation of system criteria for all guided vehicles was particularly examined. The desires for individual route choice are supplemented by routeing strategies to minimise accidents and to control access to environmentally unsuitable roads.

iii) Robust and stable route recommendations by improving journey time prediction methods.

iv) Behavioural responses of drivers.

v) Detailed analyses of user requirements of a DRG system (targetted at vehicle types, route choice and added-value features, such as car park/public transport information, fleet management and multilingual operations).

vi) Data encryption for commercially-operated DRG schemes.

vii) Data transmission methods and protocols for beacon to vehicle communications, focussing on the ISO 7-layer model and encompassing functions over and above route guidance.

viii) Integration strategies, which forms the core of this paper and presentation.

3. UTC-DRG INTEGRATION STRATEGIES.

Research work has encompassed three forms of integration:-

* Centralisd DRG with Fixed-Time UTC.

* Centralised DRG with Traffic-Responsive UTC.

* De-centralised DRG with Traffic-Responsive UTC.

In the assessments and modelling, a set of pentration rates of DRG-equipped vehicles has been used:-

```
Very Low     :    up to 1%.
Low          :    up to 10%.
Medium       :    up to 20%.
 High        :     up to 50%.
Very high    :    up to 100%.
```

The form of integration considered most likely is at level 2:-

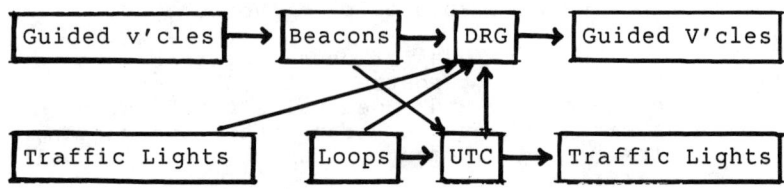

with the ultimate development being the integration of DRG and UTC into a single control system with a fully-integrated set of algorithms.

3.1. Centralised DRG with Fixed-Time UTC.

Practical data was received from (West) Berlin, where a 230-beacon trial DRG scheme was implemented alongside a fixed-time UTC system. It was found that reliable and stable information on queueing times as a control parameter would only be achieved at the high penetration level. One key item of data which may be used would be the time within a traffic signal cycle when a DRG-equipped vehicle joins a queue, thus enabling queue- or delay-balancing algorithms to be improved.

3.2. Centralised DRG with Traffic-Responsive UTC.

At low DRG penetration levels, the predominant flow of data will be from UTC into DRG, with a shift in balance as penetration levels increase. Both types of system are capable of measuring link journey times and will benefit from exchanging data in order to stabilise the timings and ensure their accuracy.

Data items investigated and found potentially useful for exchange are:-

(UTC to DRG).
- Journey time on links (low DRG penetration).
- Number of vehicles per link (all penetration levels).
- Degree of saturation (all penetration levels).
- Incident detection and cessation (all penetration levels).
- Traffic flows (high DRG penetration).

(DRG to UTC).
- Number of DRG-equipped vehicles per link (all penetration levels).
- Journey time on links (all penetration levels).
- Entry link data, at UTC perimeter (all penetration levels).
- Neighbouring link data (all penetration levels).
- Routeing advice (high DRG penetration).

The key issues are: to identify points of congestion and provide the opportunity for routeing around those points, using knowledge of spare capacity on other links, in order to achieve congestion and network stability. This concept may require the DRG routeing criteria to extend beyond journey time to items such as congestion, degree of saturation and environmental monitoring.

High-level integration may be considered at three levels:- the inter-zonal level, where whole route trees and O-D information support strategic planning data and decision-making; the zonal level, where signal co-ordination is determined on the basis of assigned path flows; and the junction level, at high penetration levels, where DRG information supplements (or even replaces) loop information.

Tests in France, with the PRODYN UTC system, at the junction level have shown improvements in travel time (3%) using DRG data to estimate queues (with high DRG penetration levels) and have shown greater robustness of saturation flow variations at penetration levels as low as 5%.

3.3. De-Centralised DRG with Traffic-Responsive UTC.

At the zonal level, a common "reference trajectory" (dynamic equilibrium) is defined for both UTC and DRG. This trajectory determines desired traffic volumes and turning flows, for a whole network. At the local level, each DRG node contains a "controller" which splits the traffic proportions assigned to each exit for each destination, according to impedance functions (updated every 1 minute, since they are linked to signal cycle times). This strategy is designed to optimise network equilibrium by localised, dynamic adjustments to perturbations.

For DRG systems, this de-centralising approach is expected to provide a method of "proportional routeing" when high penetration levels are reached, and it is inadvisable to direct all guided vehicles onto one set of links.

4. RECOMMENDED DRG SYSTEM DESIGN (RSD).

The objective of work packages 2 (iv) - 2 (vii), above, is to determine the optimum structure of a DRG system, which enables both a better response to users' requirements and a framework for more stable, wider-ranging and integrated route guidance strategies.

The outcome of the CAR-GOES project is a detailed RSD document which it is hoped to distibute to interested parties at the earliest opportunity.

5. CONCLUSION.

The CAR-Goes project, under the DRIVE-1 programme, has achieved significant progress in identifying the opportunities and ways forward for integrating two fundamental systems of general traffic management. Practical applications have been necessarily limited by the lack of operational DRG systems - however, the knowledge and experience of UTC systems has enabled a clear picture to emerge.

The next steps will be to introduce DRG systems on a variety of scales and road networks and, after an initial period of assessment at individual system level, introduce integrations at low DRG penetration levels. Several European cities have expressed strong wishes/plans to implement such schemes over the coming two to three years.

ACKNOWLEDGEMENTS
The partners in project CAR-GOES:-
1. Siemens AG, Munich, Germany.
2. CGA-HBS ALCATEL, Paris, France.
3. ITALTEL, Milan, Italy.
4. ETRA-LISITT, Valencia, Spain.
5. Siemens Plessey Controls Ltd., Poole, England.
6. Technische Universität Berlin, Germany.
7. Institut National de Recherche sur les Transports et leur Sécurité, Arcueil Cedex, France.
8. Centre d'Etudes et de Recherches de Toulouse, France.
9. Mizar Automazione S.P.A., Turin, Italy.
10. Transport and Road Research Laboratory, Crowthorne, England.
11. Transportation Research Group, University of Southampton, England.
12. Institute for Transport Studies, University of Leeds, England.

Special thanks to the Directors of Siemens Plessey Controls Ltd. for permission to publish this paper.

WEGWIJS-AMSTERDAM
A dual mode routeguidance plan

Job J.Klijnhout

Abstract.

Wegwijs-Amsterdam stands for a plan to test the feasibility of a dual-mode routeguidance system based on a network of beacons which give the driver routeguidance to reach a pre-selected destination either by car or using a P+R facility by public transport.

Foreword.

Wegwijs is a plan "under way". Some studies have been finished, some are ongoing. Gradually the form of the project is being shaped. This presentation therefore gives some information on plans which may change somewhat with time. Work on the plan is done by consultants, colleagues from the city of Amsterdam and my own colleagues M.Coëmet and G.van Leusden from Rijkswaterstaat. It's their work that is reported here, and I'm grateful for their contribution to this abstract.

Improvement of dual mode travel is one of the approved Dutch policy objectives. Dynamic travel information for drivers - including parking guidance - is expected to be a suitable tool in helping to achieve this. Dynamic travel information for drivers should contribute to keeping traffic demand within acceptable limits by stimulating the use of public transport. It can help traffic management by giving route guidance and parking guidance. The possibilities for a pilot scheme with a dynamic travel information system in a city and its surroundings, i.e. in greater Amsterdam have been investigated. The result of the work so far, the results of studies into the effects to be expected and into system layout will be presented. Main issue of the pilot will be to find out in practice what effect Wegwijs-Amsterdam will have on dual mode trips. Wegwijs-Amsterdam will be a pilot scheme. After evaluating the pilot and if results are positive, "Wegwijs" plans are to extend this system on the Dutch freeway network with as objectives benefiting from its advantages on a larger scale and increase of market penetration.

Job J.Klijnhout. Netherlands DOT, Transp. & Traffic Eng.Div. P.O.Box 1031 3000BA Rotterdam Netherlands.

1. INTRODUCTION.

The main objectives of the Dutch Traffic and Transport policy are:
-a decrease in growth of car mobility,
-less congestion and predictable travel times,
-improved traffic safety and
-a reduction of the environmental impact of traffic.

Political awareness of the environmental aspects makes that a simple extension of the road infrastructure is out of question. A combination of demand management to stimulate a more responsible user's-choice of the mode of communication and traffic management to control the traffic as effectively as possible is used to achieve these objectives. A complementary solution by the two approaches is essential.

Information is playing an increasingly important role. Information can be used not only before but also during a trip to influence the choice of mode of transport. Information is also essential in controlling traffic as effectively and efficiently as possible. Telematics in particular allow real-time traffic monitoring and provision of real-time information to road users. It also enables real-time information on public transport.

2. BACKGROUND.

A survey carried out in the United Kingdom [2] shows that inefficiency in route choice by road users on average is about 7% in mileage and travel time. Only half of the drivers seeking either the quickest or shortest routes succeed in finding the route desired [3]. The Transport and Road Research Laboratory (T.R.R.L.) concludes that Autoguide - a dynamic route guidance system for London - could achieve travel time savings averaging about 10% [4]. Their simulation study also suggests that traffic will be diverted away from non-primary roads. From the point of view of Dutch policy, the latter is an important advantage.

From the T.R.R.L. results, one may conclude that dynamic route guidance as such is a suitable instrument for the Dutch policy objectives. In particular, it cuts down mileage and travelling time, increases road capacity through greater efficiency (less need for extension of the infrastructure), improves road safety (directly and indirectly) and reduces the adverse environmental effects of traffic (indirectly).

After these studies gradually more and more field trials with beacon based dynamic route guidance systems are planned or carried out. Two, London and Berlin, have already provided very valuable information about which more is given in other contributions to this conference. Several pilot schemes based on the application of dynamic route guidance are now being outlined in the scope of the European program DRIVE II. The Dutch project Wegwijs is to be a part of this program. Cooperation with other related projects is seen as an important factor for success. Networking with other cities in Europe is therefore on the way.

3. WEGWIJS.

Environmental considerations make a reduction of car mobility imperative. Information can play an important part to achieve this objective. The Wegwijs system aims at going one step further than giving purely dynamic route guidance for drivers. Selective dynamic route guidance to parking sites is integrated in the system. The route guidance system knows what the destination is and will select a P+R facility which allows a simple public transport trip to reach the destination from the P+R site. User-friendly information on public transport should be available at the parking sites included in the Wegwijs system. Users are given a description of the route by public transport to their destination (and back). In a later phase this information should be provided in the vehicle. This makes Wegwijs a dynamic travel information system for drivers.

Wegwijs offers:
 -Dynamic route guidance for a car journey to the user's destination.
 -Dynamic route guidance to parking sites with P+R facilities.
 -User-friendly information at these parking sites for (dynamic) guidance to the user's destination with public transport.
A crucial point in this approach is the reliability of the information in all three cases.
To the authorities Wegwijs offers travel time information and a tool to reroute traffic.

Other functions, like route guidance for buses, the transportation of dangerous goods, fleet management, additional tourist information, integration with road-pricing and payment and reservation (e.g. of parking) are also part of the concept and could be integrated later [8]. They form no part of the pilot Wegwijs-Amsterdam.

Key aspects of Wegwijs are:
 -Communication and information makes it possible to improve the choice of mode of transport and route before as well as during the trip.
 -In-car information systems can perform two complementary roles:
  -Encouragement and promotion of the use of public transport.
  -Effective control of the remaining road traffic.

Wegwijs can therefore contribute to all of the objectives of the Dutch national long-term transport-policy plan.

Although Wegwijs can provide benefits both for inter-urban traffic as well as for urban traffic, a pilot in a urban area was chosen as a first step. A separate study for stepwise implementation country-wide has been performed to get an idea of the feasibility of a freeway-wide implementation.

4. THE PILOT SCHEME WEGWIJS IN AN URBAN AREA.

As part of feasibility and definition studies several aspects concerning the implementation of Wegwijs and concerning the urban pilot scheme especially have been investigated. Some of the results will be discussed in the paragraphs below.
Greater Amsterdam is targeted to become the location for the pilot scheme. The urban area involved in the pilot is depicted in figure 1. Five freeway's end in a freeway ring.
A radial network of fast and frequent trams links the hart of the city with the outskirts, crossing the freeway ring. Schiphol, Amsterdam Airport in the south-west corner of the map has a railway link with the city. The inner-city of Amsterdam is a maze of streets, in some cases the famous canals are wider than the road. (-fig.2)

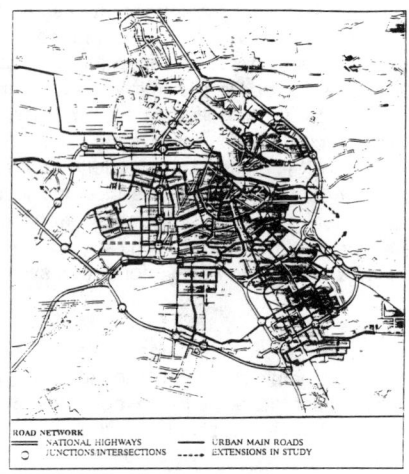

Fig.1 Greater Amsterdam's Road network

4.1. DEMAND ANALYSIS.

Two categories of actors/users have been recognised:
-Category 1: National and local authorities, operators, industries and public transport companies.
-Category 2: Specific users groups, including commuters, business drivers and leisure drivers.
Two studies were devoted to the attitude of these users.

Attitude of actors/users included in category 1:
A study describing an implementation scheme for routeguidance in the city of Amsterdam [9] made clear that just dynamic route guidance for drivers (like LISB in Berlin) would be unacceptable for the city of

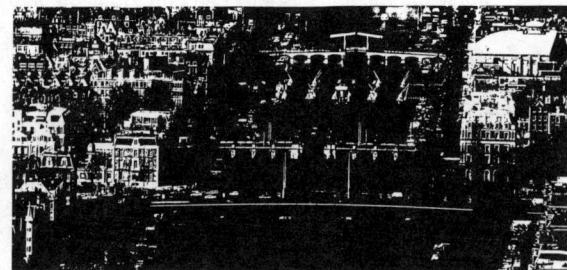

Fig.2 The very hart of Amsterdam.

Amsterdam. It was regarded as conflicting with the traffic and transport policy of the city. The feeling was that any improvement in traffic flow would immediately be used by the underlying demand and create more traffic. The aspect of demand management facilities, P+R guidance, therefore is essential for the city council of

Amsterdam. Promoting the use of public transport is perceived as vital in meeting the overall transport objectives of the city. The city of Amsterdam has responded positively towards the pilot scheme in the existing form described in this paper, where emphasis is put on parking guidance and public transport information. Vehicle leasing companies have a positive attitude to the pilot. But taxi-operators regard the proposal as a potential danger to their privacy and decline initial co-operation while expressing an interest in further development.

Attitude of users included in category 2:

As part of a study of opinions of drivers on traffic information systems [10], a survey is carried out among some 300 drivers. The study shows that 80% of the people interviewed think pre-trip planning information to be a desirable feature. 59% say that *dynamic route guidance integrated with public transport information is desirable. Note how high the percentage is for business-drivers (56%).* It is expected that Wegwijs will be welcomed by drivers particularly because in the coming times parking policy in cities will become more restrictive. Figure 3 shows the percentage of users finding a specific information system desirable or very desirable.

Desirable/ very desirable (%)	Commuters	Business drivers	Leisure drivers	Total
pre-trip info.	77	81	84	80
RDS-TMC	85	94	90	93
Dynamic route guidance	68	64	53	63
Wegwijs concept	63	56	54	59

Fig. 3: Users attitude towards dynamic travel information.

Being informed about a 30 minutes delay, 90% of the people interviewed would follow another route, 44% would be prepared to change to public transport if they were informed about a 30 minutes delay or definite problems finding a parking place in the area of their destination. I believe these results show unexpected and encouraging possibilities for the Wegwijs-Amsterdam concept.

4.2. BENEFITS OF THE DYNAMIC ROUTE GUIDANCE OPTION.

In a study, carried out for the urban area of Amsterdam [11], the benefits of fully dynamic route guidance are estimated by separating:
- A semi-dynamic component: based on static and historical data. Semi-dynamic because there is a time-dependency.
- A dynamic guidance component: based on actual traffic data.

4.2.1. Semi-Dynamic Route Guidance.
The benefits of the semi-dynamic component are calculated using results available from other studies. 6.4% potential distance saving and 9.6% time saving are used as input. Four different scenarios are calculated concerning the route choice of the users. The results presented in figure 4 concern a scenario where 63 % of

the users are supposed to use the least time route and 2% the least distance route (figures are related to a study from the U.S. project ERGS) and a scenario where 85% of the users are supposed to use the least time route and 35% the least distance route with overlap of both (based on recent studies by Wootton Jeffreys [3]). The other two scenarios are also based on this study but lower values for the overlap of least time route and least distance route are used. The benefits for the non-users are estimated to be about 6% of the users benefits.
Data provided by the city of Amsterdam and by Rijkswaterstaat is used to calculate the specific benefits for the built-up area of Amsterdam. The estimates are for the year 2000 and are specified for different percentages of equipped vehicles.

% vehicles equipped		Value of time saving		Value of distance saving		Total users benefits	Total benefits all vehicles
		million hours per year	million guilders per year	million kilometres per year	million guilders per year	million guilders per year	million guilders per year
7	L	1.14	21.2	0.91	0.2	21.4	22.3
	H	1.54	28.6	15.91	2.8	31.4	32.9
14	L	2.28	42.4	1.82	0.3	42.8	45.6
	H	3.08	57.3	31.81	5.6	62.9	66.8
25	L	4.08	75.8	3.25	0.6	76.4	80.1
	H	5.50	102.3	56.81	10.0	112.3	118.8

L: low benefit scenario,   H: high benefit scenario

Fig. 4: Benefits from route guidance in Amsterdam (estimate for the year 2000).

4.2.2. Dynamic Route Guidance - Impact of incidents.
Fully dynamic route guidance systems offer additional benefits over semi-dynamic route guidance as they can respond to the random fluctuations of actual traffic conditions. These fluctuations may be caused by incidents or variations over the days in the number of drivers taking different routes. These fluctuations cannot be anticipated from historical data alone. Fully dynamic route guidance can provide benefits directly to system users and indirectly to non-users as well. A traffic assignment computer model is used to simulate the benefits of the dynamic guidance component. Eight main corridors were selected. Each corridor has a main route and an alternative route. Benefits of the dynamic component are shown in figure 5. The left hand or lower bound figures
correspond with an information lag of the monitoring system of 10 minutes. The upper bound corresponding with an information lag of 5 minutes.

These benefits can be substantial, especially in high-density traffic corridors and in situations with many incidents. The benefits are higher during peak hours. In any case, adequate spare capacity on alternative routes is required. An interesting note of discussion is the effect of time delay between an increase in travel time, caused by an accident for instance, and the update of the information to the driver. Especially as other systems are being developed which by means of cellular radio allow a more or less continuous link with the vehicles. The main conclusion of such studies should be that a proper distribution of beacons is essential. Then vehicles can be informed of a changed route there were such a change can be effectuated.

Percentage equipped vehicles (%)	Benefits of the dynamic guidance component all drivers million guilders per year
7	5.3 - 8.1
14	12.9 - 17.5
25	21.1 - 28.9

Fig.6: Effect of dynamic guidance (estimate for the year 2000)

It is clear that substantial benefits can be obtained with dynamic route guidance. This concerns environmental benefits (less kilometres, less travel time) as well as economic benefits. Economic benefits, by the way, come mainly from the time saving effect.

4.3 PARKING GUIDANCE AND PUBLIC TRANSPORT INFORMATION.

Improvement of public transport facilities in the near future is an important objective of the Dutch Traffic and Transport policy, considering the negative effects of traffic on the environment. It is also anticipated that in the near future parking policies of cities will become much more restrictive. Especially commuter traffic will be restricted strongly while accessability for business traffic should be improved. Wegwijs plans to provide drivers with dynamic guidance to parking sites. Three schemes can be distinguished:
    - Dynamic guidance to (park and ride) parking sites at the city border near stops of trains, underground or trams. At these P+R type parking sites public transport information will be available to help planning the forward journey by public transport.
    - Dynamic guidance to parking sites in the urban area at stops of underground, trams or busses. At these parking sites public transport information will be available as well. It is questionable wether all in all in this scheme travel time will be less than in the first scheme. A choice between the two options will mainly depend on the possibilities to reach a destination easily by using public transport.
    - Dynamic guidance to parking sites near their destination for those users which insist on parking close by their destination. Cities' parking policies will make this parking definitely more expensive than in the first two schemes.

As part of the general transport policy in other projects work is done on a traffic and transport policy where integrated travel information supports travellers with information about traffic conditions, parking possibilities and public transport information as well before the trip (pre-trip planning) as during the trip.

4.3.1. Simulation of Wegwijs including Parking Guidance.

To get a better idea of the effectiveness of the intended route guidance system a simulation study was carried out, for the city of Amsterdam and its surrounding freeways [12]. Several studies concerning the effects of in-vehicle route guidance systems have been published, most of them based on modified versions of existing models, each with its own advantages and disadvantages. An example of a study which was already mentioned is the one for London, in which CONTRAM was used [4]. For our study it was decided to develop a new simulation module, to be incorporated in the existing transportation planning package MINUTP. The new software combines all features necessary to simulate the scenario's discussed below and is able to handle large networks.

The network of Amsterdam consists of 450 zones and 6000 links. The trip matrix used represents an estimate of an average evening peak period (4 to 6 pm) for the year 1995 and is split up in one-minute intervals. A departure profile to represent the building up and subsequent decline of the peak. The following scenario's are considered in the simulation study:
- guidance under normal traffic conditions.
- guidance restricted to the primary roads.
- guidance in case of incidents.
- guidance to park-and-ride stations.

The first scenario is meant to show the benefits that may result from providing the equipped vehicles their optimal route. It is known that besides incomplete knowledge of the network, day-to-day variations in demand lead to many drivers not using their optimal route [3]. In a T.R.R.L. study of the effects of route guidance in London [4] it is concluded that both travel time and kilometres driven decrease and that these benefits increase with the level of take-up (percentage of equipped vehicles). Another conclusion is that equipped vehicles achieve about the same benefit for any level of take-up, they don't lose significantly as a result of the presence of larger number of guided vehicles. In our study the possible benefits in case 0-20% of the vehicles are equipped is studied for the Amsterdam network.

The second scenario is chosen to get an idea of the effects of a restriction to guide only via the primary roads (freeways + urban main roads). In the T.R.R.L. study, using the whole network, it was concluded that guidance results in a decrease of kilometres driven on the secondary roads [4]. The second scenario with guidance on the primary network only gave an overall increase of kilometres driven. These results may be network-dependent and it is therefore of interest to further investigate these results for the Amsterdam network.

The third scenario is used to find the advantages of traffic information and guidance in the extreme (but not uncommon) situation of lane blockages as a result of an incident. It is expected that a high benefit may be achieved when part of the traffic is rerouted in these situations. Three incident situations are studied varying in duration (15-30 min), severity (50 to 70 % of reduction in road capacity) and location (freeway/city network).

For the fourth scenario the city of Amsterdam was divided in 5 districts: Central, North, South, East and West. The idea is that traffic with a destination in one of these area's can be guided to a P+R facility in that area. At the moment 9 such P+R sites are available. Any such P+R trip would mean a reduction in the mileage in the urban area. In case of commuting traffic the simulated effect in the evening peak hour would of course have a mirrored effect in the morning peak. In the simulation 5% of car trips to from the city would change into dual mode trips this way.

The results of the simulation of these scenario's are expressed in average travel time and total vehicle kilometres travelled, both per road class (city/main roads) and separate for guided and unguided vehicles. The effect on queue length and location are also recorded. The results of the simulations will be soon available.

4.4. TIME-PLAN FOR THE PILOT.

- Phase 1: Feasibility and definition of the pilot scheme (1989-1991)
- Phase 2: Design of the pilot scheme (1991)
- Phase 3: Implementation and evaluation of the pilot scheme (1992-1993)

5. WEGWIJS ON THE DUTCH FREEWAY NETWORK.

Introduction of Wegwijs on the freeway network is considered desirable. In a first phase, this would give static route guidance and guidance to park and ride sites on all freeway sections. At the major freeway intersections information would be dynamic. Each site or beacon would give information for the trips which continue to use the freeways as well as information about the surrounding area of some 10 km in diameter. The level of detail would there be a village and a district in urban areas. Also near some often congested freeway-sections, some degree of dynamics should be introduced. Dynamic information could be down-loaded via telephone lines or existing traffic systems communication lines.

At a later stage, route guidance on the freeway network should become completely dynamic. When the necessary market penetration has been achieved, the system can be used to monitor freeway traffic and full dynamic guidance will then become possible.

This incremental implementation strategy aims at kicking off the process of market penetration. Already in the first stage, users will benefit from travel information and there will be a knock-on effect for traffic in general.

The introduction of travel information on the freeway network is also expected to have beneficial effects on the pilot scheme in a city. Users being attracted by the fact that the same on-board unit can be used for both purposes. The results of the survey in Amsterdam [10] show that drivers will probably welcome both options.

6. REFERENCES.

[1] Wegwijs - een informatiesysteem dat de automobilist(e) de weg wijst (1989). J.J. Klijnhout. Rijkswaterstaat. DVK-CX. Rotterdam.

[2] Jeffery, D.J., (1981). The potential benefits of route guidance. Department of Transport, Transport and Road Research Laboratory, LR 997. Crowthorne.

[3] Wootton, J. and Ness, M., (1989). The experience of developing and providing driver route information systems. Wootton Jeffreys Consultants Limited. NVIS Conference.

[4] Smith, J.C. and Russam, K., (1989). Some possible effects of Autoguide on traffic in London. NVIS Conference.

[5] Catling, I. and Belcher, P., (1989). Autoguide - Route Guidance in the United Kingdom. NVIS Conference.

[6] Sparmann, J.M., (1989). LISB, Route Guidance and Information System; first results of the field trial. NVIS Conference.

[7] Autofreie Stadt? Parken mit Vernunft! 1990. Volkswagen Dokumentation.

[8] Notitie Wegwijs: Aandachts- en discussiepunten. December 1989. J.J. Klijnhout, G.C. van Leusden en M.J. Coëmet. DVK-CX.

[9] Implementatiemogelijkheden Wegwijs Amsterdam. December 1989. Plangroep BV/IBN-Groep - DVK-CX.

[10] Dynamische Verkeersinformatiesystemen. January 1990. Bureau Goudappel Coffeng - DVK-CX.

[11] Dynamic Route Guidance Systems. January 1990. Castle Rock Consultants - DVK-CX.

[12] Simulation of Wegwijs. Februari 1991. Hague Consulting Group - DVK-CX.

VIDEO IMAGE PROCESSING FOR TOLL OPERATION EVALUATION

Bruno DAVIET[1] - Jean Marc MORIN[2] - Jean Marc BLOSSEVILLE[3] - Vincent MOTYKA[4]

Abstract

Toll motorway companies are very much concerned with toll plaza operation and its appreciation by customers. The problem is two-fold :
- define a desired level of service (queue lengths, delays,...)
- operate tool-booths so as to achieve it.

Both items necessitate the use of good automatic means for queue lengths and delays measurements. For this purpose, and given the inadequacy of classical means such as magnetic loops on such large areas as toll plazas, video image processing has been chosen.

The Autoroutes ESTEREL-COTE-D'AZUR (ESCOTA) company have asked ISIS to implement this system on their plaza at ANTIBES. Video cameras have been settled, each of them being processed by a micro-PC, and special algorithms have been designed by INRETS for the evaluation of queue lengths and queuing times. This paper gives first results about the use of this system which is intended to help assessment of the impact of automatic toll collection introduction at ANTIBES and eventually to give an aid to the toll plaza chief operator for achieving his level of service objectives, through both the surveillance of past operations and the planning of ressources (manual toll booths).

Introduction

French toll motorway company ESCOTA was recently faced to several problems concerning toll plaza operation, due to the evergrowing traffic demand and spatial and environmental constraints.

----

1 Assistant General Manager, ESCOTA
2 Head of Research and Development Department, ISIS, 2 Rue Stephenson - 78181 St QUENTIN EN YVELINES - FRANCE
3 Researcher, INRETS, 2 Avenue du Général Malleret Joinville - 94114 ARCUEIL - FRANCE
4 Researcher, INRETS, 2 Avenue du Général Malleret Joinville - 94114 ARCUEIL - FRANCE

It was decided to choose the ANTIBES plaza as a test site and to act in two directions :
- introduce the new technology of automatic toll collection (AVI system) : the AMTECH tag was chosen and a special lane equipement developed (borne SOPHIA), so as to boost the toll traffic volume.
- enhance the management of toll booths in order to optimize the ressource allocation for achieving satisfactory levels of service (to be defined in accordance with customers' reactions).

Queue lengths and queuing times automatic measurements are key issues for both assessing the impact of automatic toll collection and implementing enhanced "classical" toll operation. Taking into account the current research of the French Institut de Recherche sur les Transports et leur Sécurité (INRETS), it was decided to use video image processing for this purpose, with preference to magnetic loops which would have entailed large investment and maintenance costs and probably great inaccuracies because of the erratic car trajectories on toll plazas.

Scheme description

ANTIBES-EST toll system is made of 2 toll plazas (entrance and exit), each of them being made of 5 lanes (2 manual and 3 automatic). The implementation of automatic toll collection (ATC) is made by means of equipping automatic lanes with additional beacons (mixed lanes).

The video system is based on 4 image processing cameras (2 per toll plaza) and 2 surveillance cameras (1 per toll plaza).

On each toll plaza, the surveillance area is divided into 2 areas (the approach area and the area close to the toll line),each of whose being surveyed by one image processing camera. Each pair of cameras is directed towards approaching traffic and is normally settled on a 3m50 pole on the toll shield (see figure 1).

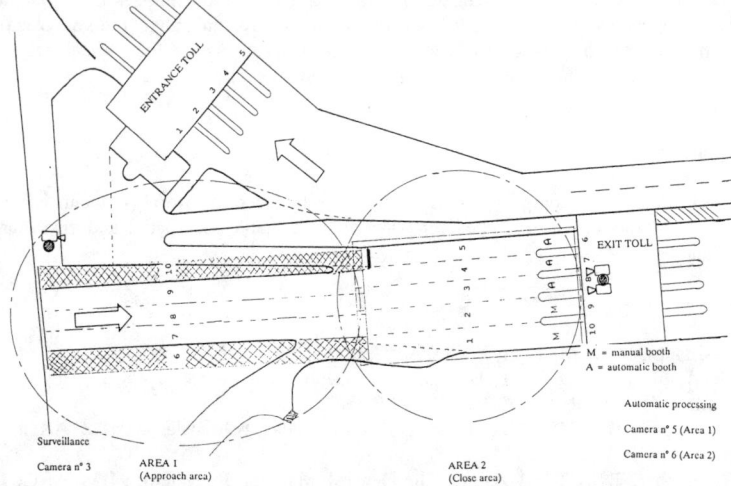

Fig 1 Scheme description (exit toll details)

Image processing cameras are standard type, fixed, monochrom, CCD-type, with 8mm lenses and automatic gain correction. Each of them is connected to a conventional 386-33Mbytes micro-PC called an Image Processing Unit (IPU), hosting a digital imaging card and running a special algorithm. A master-PC, connected to the 4 IPUs, allows initialization and parameters changes, together with on-line direct visualization of processing being made on video images displayed on a color monitor and selected on whichever camera, thanks to a video switching board. Each IPU delivers, every minute, the lengths and travel times of queues which are present in its surveillance area.

The traffic data collection system is based on magnetic loops placed at the entrance of entrance and exit slips ; these loops measure flows and occupancy rates (1mm basis) after correction of double counts (over-lapping vehicles) through a special algorithm.

Toll data are collected from the already exiting toll collection system through a dedicated computer. They consist of volumes per lane, payment types and vehicle types (1mn basis), together with transaction times per lane and per payment types (5mn basis).

All the components of the system have been installed in a temporary moveable air-conditioned unit, connected to the dedicated toll computer via a multiplexed line. Software developments for the special ATC evaluation have been made on this computer ; a special terminal (VDU and keepad) and a printer have been provided for operator's needs.

On-line data (for short-term monitoring) as well as time evolution curves (for longer-term monitoring) are available on the VDU and printer : volumes and occupancy rates of traffic demand, toll volumes, transaction times, queue lengths, queuing times, congestion indicator.

Video image processing principle

The video image is digitalized so as to get a 256 x 256 pixels definition. Grey levels are coded on 6 bits from o (absolute dark) to 63 (absolute white). Knowing the height, inclination and focus of the camera, a relationship is established between each real point and its image on the screen ; then, knowledge of the height of any point above the ground enables to know its distance from the foot of the camera.

The algorithm detects areas of the image which are changing ; as a result, each pixel can be allocated with a binary presence/absence information (a vehicle stopping more than 40 s is lost). An axis is drawn for each lane to be measured, in accordance with its geometric characteristics and the usual lateral position of cars within it. Then, from the bottom of the image, the algorithm proceeds upwards and stops its search as soon as both an interval greater than "INTERVALMAX" is found and at least one stop has been detected during the minute (see picture 1).

Then, the average queue length in the minute is calculated as the arithmetic mean value of the 120 corresponding algorithm values (2 samplings per second are made). Concerning queuing time measurements, the algorithm uses the marking of vehicles (several markers per vehicle can occur), and the building of space-time vehicle trajectories. Vehicles are followed from their entrance in the queue to their exit at the bottom of the image ; only vehicles which have stopped are kept ; the average queuing time is the mean value of queuing times of entering vehicles which have been sampled in the minute (see picture 2).

This approach has proved to be effective during daylight as well as at night.

Picture 1 : queue length measurement
(exit toll, camera n° 6)

Picture 2 : queuing time measurement
(exit toll, camera n° 6)

Results of evaluation are given in the following table.

EXIT TOLL - CLOSE AREA (CAMERA N° 6)

QUEUE LENGTH

Lane number		1	2	3	4	5
Number of minutes observed		469	468	379	306	281
Automatic measurements	m	18,8	15,4	10,6	7,1	3,8
	s	6,9	7,8	8,2	7,0	5,2
Manual measurements	m	18,0	15,7	11,0	7,6	4,2
	s	7,8	8,4	9,2	7,8	6,0
Error (meters)		+ 0,8	- 0,3	- 0,4	- 0,5	- 0,4

m = average value (in meters)
s = standard deviation (in meters)

QUEUING TRAVEL TIME

Lane number		1	2	3	4	5
Number of vehicles observed		20	23	44	40	20
Automatic measurements	m	28,5	22,0	29,3	25,0	17,8
	s	2,2	3,6	8,5	5,6	2,5
Manual measurements	m	29,0	21,4	30,6	27,9	19,5
	s	4,1	3,2	7,4	6,1	3,0
Error (seconds)		- 0,5	+ 0,6	- 1,3	- 2,9	- 1,7

m = average value (in seconds)
s = standard deviation (in seconds)

Automatic toll collection introduction

ATC was put into operation in November 1st, 1990. Data were collected during a "before period" of two weeks and were analysed (figure 2 gives an example of queuing pattern on a specific "before" day).

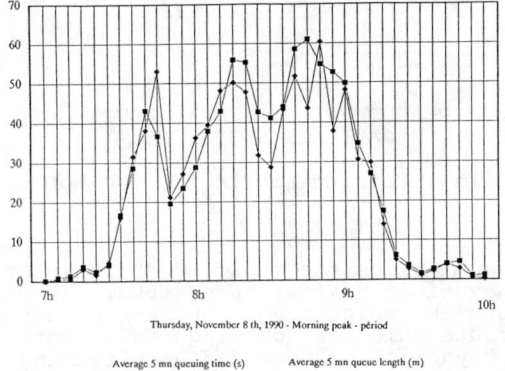

Figure 2 : Queue length and queuing time pattern example - Exit toll (same scale)

The "after period" started in March 25th, 1991. In the meantime, a regular increase in the number of purchased tags was recorded and an additional advertising campaign was made. At the beginning of the "after period", statistics were the following :

- number of tags distributed = 2803
- number of customers = 1928
- total daily number of tags at entrance toll = 1209
- total daily number of tags at exit toll = 1080

First assessment, based on toll statistics, already exhibited increased peak-hour toll capacities ; comparative assessment, based on video measurements, will confirm the enhancement of toll operation in terms of queue lengths and queuing times, transaction times, individual and overall delays, frequency of congestion stepping back onto the motorway...

The future and conclusion

Apart from the specific problem of ATC assessment, measurements of queuing characteristics using video image processing proved to be successful. ESCOTA, therefore, decided to have this prototype system made industrialized and put into day-to-day operation on ANTIBES - PLEINE VOIE toll plaza (11 toll booths in each direction). Furthermore, INRETS have made the algorithm able to monitor at the same time 3 to 4 different queues up to 120 meters long with one single camera.

A specific software is being developed to produce daily statistics about queuing on the toll plaza as well as a queue prediction model using predicted upstream demand and predetermined toll facilities. These tools will help the chief toll operator to monitor and keep to high service level the operational conditions of the toll plaza. Results of these developments will be given in next papers.

A REVIEW OF EUROPEAN DEVELOPMENTS IN ROUTE GUIDANCE AND NAVIGATION SYSTEMS

Ian Catling, Richard Harris and Bob McQueen *

Abstract

This paper summarises the rapid progress in Europe in developing in-vehicle navigation systems. It discusses the various studies which have been carried out of the benefits of route guidance, distinguishing between "static" and "dynamic" systems and highlighting the significant levels of benefit which have consistently been predicted for dynamic route guidance systems. It describes the advantages specifically for the driver who has access to in-vehicle navigation.

The autonomous systems CARIN and TRAVELPILOT are described, as is the ALI-SCOUT dynamic system which is the basis of the LISB trial in Berlin and its successor, EURO-SCOUT. The background is given to the Autoguide system proposed for London which could be the first fully dynamic route guidance system available on a commercial basis, including the legislative and implementation aspects and the current status of the pilot and full commercial systems.

Developments using the Radio Data System Traffic Message Channel (RDS-TMC) for dynamic updating of CARIN (CARMINAT) and TRAVELPILOT are described. The UK dynamic system for motorways TRAFFICMASTER is put into the context of other developments.

The DRIVE project SOCRATES is outlined, with a discussion of the role of cellular radio in providing the basis of an Integrated Road Transport Environment (IRTE). Other developments within the DRIVE programme are summarised.

The role of navigation systems within the PROMETHEUS programme is reviewed, including a description of the "Common European Demonstrator" project "Dual-Mode Route Guidance."

In conclusion the merits of infrastructure-based dynamic route guidance and navigation systems are reviewed; parallels and contrasts are drawn between the European approach of Road Transport Informatics (RTI) and the US plans for Intelligent Vehicle-Highway Systems (IVHS)."

----
* all of Ian Catling Consultancy Ltd, UK (+44 81 6434451)

Introduction

During the last five years there has been significant progress in Europe towards the widespread implementation of Road Transport Informatics (RTI), or Intelligent Vehicle Highway Systems (IVHS) as they are known in the USA. A key part of this progress had been based on developments in route guidance and navigation systems.

Static route guidance is based on mapping information, typically stored on board the vehicle, and takes no account of traffic conditions. Dynamic route guidance requires some form of communication link so that the driver can be guided on the best route taking into account actual and predicted traffic conditions.

The benefits of route guidance have been estimated to be substantial - in the UK for example a widely available route guidance system could provide travel time savings worth more than 1$bn per year.

Tests using existing demonstration systems, together with a growing number of market research studies, have shown that most drivers will welcome in-vehicle route guidance, as long as the Man-Machine Interface is well designed and the quality of the guidance recommendations (and the supporting database) is of a high enough standard to ensure system credibility.

Autonomous navigation systems

An autonomous system is self-contained within the vehicle and functions without external communication.

TRAVELPILOT is marketed by Bosch and is based on the American ETAK Navigator system, with improved functionality for European use. The vehicle's position and the selected destination are indicated on a CRT display. The driver uses this information to make subjective decisions on how best to reach the destination.

CARIN (CAR Information and Navigation system) is an in-car navigation and information system which plans the optimum route and guides the driver with spoken directions backed up with a visual pictogram display.

Both systems rely on dead reckoning combined with map-matching to maintain the vehicle's location, and both use CD-ROM for digital map storage.

Beacon-based systems

EURO-SCOUT is the latest generation, developed by Siemens, of the ALI-SCOUT system used in the LISB full scale field trial in Berlin. It uses infra-red beacons to transmit dynamic data to vehicles at specific points in the road network. The data contains both digital map information and up-to-date routeing recommendations.

The recommendations are given to the driver both aurally and visually, based on the routeing calculations carried out in the control centre within the last few minutes.

A key element of the system design is its ability to collect data - anonymously - from the participating vehicles themselves on the current traffic conditions throughout the network. This will have profound effects on many aspects of traffic control.

Negotiations are currently underway for further implementations of EURO-SCOUT in various parts of Europe, in particular in a number of German cities and regions.

Autoguide is the name given to beacon-based route guidance in the UK. A demonstration Autoguide system has been operational in London since early 1988, and the Government enacted legislation in 1989 to enable the commercial operation of Autoguide.

After a competitive bidding process, GEC were selected to negotiate with the Government the terms of a licence to promote Autoguide in London. During 1991 work is expected to begin on a large scale pilot system covering a major part of London. It is expected that the pilot system could be expanded to a fully commercially available system by 1993.

Radio Data System (RDS)

RDS is the digital sideband information channel agreed by European broadcasters primarily for programme and frequency information, but also including the Traffic Message Channel (TMC) for driver information. Within its capacity limitations, RDS-TMC will provide a common European facility for transmitting traffic information.

CARMINAT is a collaborative project, involving Philips and Renault, which investigated the use of RDS-TMC to provide dynamic traffic information to an on-board CARIN unit.

Links to dynamic route guidance systems such as Autoguide and SOCRATES will potentially provide the high quality real time data needed to take full advantage of the expected wide availability of RDS-TMC units during the 1990s.

TrafficMaster

TrafficMaster is currently the only commercially operating dynamic traffic information system. Speed sensors on the motorway network around London are connected to a control centre and drivers with the necessary on-board equipment are warned whenever congestion is detected.

The system is licensed under the UK Driver Information Act which was passed to enable Autoguide to be implemented commercially.

SOCRATES

An alternative approach to roadside beacons and RDS is being developed in the SOCRATES (System of Cellular RAdio for Traffic Efficiency and Safety) project in which major partners are Philips and Bosch.

The use of cellular radio for widespread two-way communication as a genuine dialogue will always face capacity restraints. However, the SOCRATES concept is to use specific channels from the pan-European GSM system (which will replace current cellular radio systems during the 1990s) to broadcast, in a similar way to Autoguide/EURO-SCOUT, the same data set to all equipped vehicles in a particular cell. A multiple-access protocol will provide for transmission back to the control centre from equipped vehicles.

SOCRATES will support the autonomous CARIN and TRAVELPILOT systems by providing the traffic information to enable them to provide dynamic route guidance. It may also provide the data necessary to support otherwise wholly beacon-based systems such as Autoguide and ALI-SCOUT, so that these might function in areas not equipped with beacons.

The SOCRATES communication link will also support other RTI applications such as fleet management, parking and public transport information and, in particular, emergency call based on the continuous communication link to the control centre.

Currently a demonstration scheme is operational in Gothenburg, Sweden, and a number of substantial pilot projects are planned during the period 1992 to 1994. These will be followed by full-scale commercial implementation.

Other DRIVE projects

SOCRATES is one of about 70 projects in the European Communities' $120m DRIVE (Dedicated Road Infrastructure for Vehicle safety in Europe) programme of research into RTI. Other major projects include TARDIS, developing common RTI functional specifications; SECFO which promotes standardisation and consensus formation; CAR-GOES, investigating links between dynamic route guidance and traffic control; and several projects developing models and control strategies suitable for real-time driver information systems.

The second phase of DRIVE is expected to start in 1992, concentrating on a small number of major Pilot Projects, as the next step towards the Integrated Road Transport Environment in Europe. Pilot Projects are being developed within the POLIS (urban) and CORRIDOR (inter-urban) initiatives.

PROMETHEUS

PROMETHEUS, led by the major European automotive manufacturers, is an 8-year programme of research and development of RTI systems with the objectives of increasing safety, maximising efficiency and reducing the environmental effects of traffic.

It is envisaged that eventually each vehicle will be equipped with an on-board computer to monitor vehicle operation, provide driver information and assist with driving.

The programme is currently focused on 10 "Common European Demonstrator" (CED) sub-projects, which are intended to illustrate the technical feasibility of individual applications, and their evaluation within a framework of integrated applications; the CEDs are expected to be shown in Turin in the autumn of 1991.

One CED is developing "dual-mode route guidance" in which the features of static, autonomous navigation systems are combined with the advantages of dynamic route guidance using a communication link.

Conclusion

The role of international research and development programmes is likely to be increasingly important in promoting the progress towards integrated RTI systems. In particular the European Community programme DRIVE and the Eureka programme PROMETHEUS are especially concerned with using "Road Transport Informatics" (RTI) to develop an "Integrated Road Transport Environment" (IRTE).

Dynamic route guidance will be important in the development of the IRTE, particularly when extended to include other aspects of traveller information. It is parallelled in the USA by the Advanced Traveler Information Systems (ATIS) element of IVHS.

The opportunity is available for those developing IVHS in the USA to take advantage of the lessons learnt during recent RTI field trials and demonstration schemes in Europe. One example is the need for high-quality navigation databases in order to provide credible and marketable information to the driver.

The next stage, a focus of the DRIVE Pilot Projects, is to develop the commercial and financial framework for widespread implementation of RTI systems.

Assessment of IVHS Research Efforts in Japan and Future Directions

Masahiko Katakura[1] and Mitsuru Saito[2]

Abstract

In the past twenty years, research on intelligent vehicle/highway systems (IVHS) in Japan has seen a tremendous progress. The first advanced IVHS research was conducted between 1973 and 1979 under the name of CACS (Comprehensive Automobile Control System). Since then, technologies on vehicle/road communication and information processing technologies have been advanced by both the electronic and automobile industries. From these efforts two different systems, RACS (Road Automobile Communication System) and AMTICS (Advanced Mobile Traffic Information and Communication System), have emerged as Japan's major IVHS programs. Both RACS and AMTICS have gone through initial demonstration stages. AMTICS's demonstration project was completed in June 1988 and RACS's final comprehensive field test was completed in November 1989. They are now in the practical application stage and are expected to be merged into a single system under a tentative name of VICS (Vehicle Information Communication System).

Introduction

The national R&D effort in IVHS research in Japan started with CACS. CACS was initiated under the sponsorship of the Agency of Industrial Science and Technology of the Ministry of International Trade and Industry (MITI). The project included five subsystems as the technical objectives: (1) Route guidance subsystem, (2) driving information, (3) route display board, (4) traffic

---

[1]Prof., Dept. of Civil Engrg., Faculty of Technology, Tokyo Metropolitan University, Tokyo, JAPAN

[2]Asst. Prof., Dept. of Civil Engrg. and Inst. for Transp. Systems, City Univ. of New York, New York, NY, 10031.

incident information, and (5) public vehicle priority treatment.

After the CACS project was successfully completed, part of the CACS experimental facilities was transferred to the Ministry of Construction (MOC) and the remainder to the National Police Agency (NPA). In the meantime, technology development activities to build advanced information systems have been continued by private industries and the Association of Electronic Technology for Automobile Traffic and Driving (AETATD). The latter organization was established in 1979 to expand the work done by the CACS research.

Under AETATD's guidance, a study to improve the specification of a spot communication system between the road and the vehicle was carried out from 1980 to 1985. The road-vehicle spot communication system was experimented during the Tsukuba Science Expo '85. The result was a favorable one and this communication technology based on inductive coupling is now used to monitor the location of transit buses in many large cities in Japan (Bus Location System).

Since then, communication and information processing technologies have significantly advanced and many methods have been developed which would facilitate communication links between the road and the vehicle. At the same time, the application of micro-computers to automobiles has advanced. With this background, when the communication industry was deregulated in 1985, NPA and MOC separately quickened research studies to advance traffic information systems as an expansion of the CACS project. NPA started AMTICS, while MOC started RACS.

AMTICS

AMTICS was proposed in early 1987 by the Japan Traffic Management and Technology Association (JTMTA) in cooperation with the Ministry of Post and Telecommunications (MPT) and several private corporations under the initiative of NPA. In this system, traffic information is collected by the computerized Automatic Traffic Control (ATC) center located in the police department and transmitted to the motorist by way of tele-terminals. The tele-terminal system is a radio communication system which communicates digital data between the Tele-terminal Corporation Service Center and individual mobile communication units. Public and business vehicles subscribing to the service may be provided with a bi-directional communication which allows them to send their location and other information to their companies through the Center. AMTICS, however, basically offers a uni-directional communication or transmission of data to the majority of the motorists.

AMTICS's main functions are in-vehicle navigation and dynamic route guidance. Its navigation system is fundamentally a dead-reckoning system with an on-board equipment which shows the vehicle location, using digital road maps stored on CD-ROM and map matching technique. "Signposts", however, are installed at selected locations in the system to transmit electronically the vehicle's absolute position to the in-vehicle navigator in order to correct cumulated position errors.

RACS

The RACS project was started in 1984 under the initiative of the Ministry of Construction as a cooperative R&D activity between car manufacturers, electronic industry and a group of concerned individuals. The objectives of RACS is to offer an advanced information environment to the motorist by the installation of roadside facilities and in-vehicle devices. RACS's basic functions are almost identical as those of AMTICS. The system comprises three functions: (1) Navigation (static), (2) information service, and (3) individual communication. When compared to CACS, in-vehicle communication devices are much more sophisticated. RACS was planned to have a step-by-step development of the above three functions.

The main difference between RACS and AMTICS resides in the way information communication is made. RACS uses intermittent minimum zone beacons while AMTICS uses tele-terminals. The intermittent beacon system provides spot radio communication around the beacon. Three types of beacons are used: location, information, and individual communication beacons. Location beacons are stand-alone isolated beacons emitting static navigational information to correct dead-reckoning errors. The other beacons are connected to the RACS Traffic Systems Center and the Information Control Center by fiber-optic cables. Information beacons transmit dynamic traffic and parking information. Individual-communication beacons not only allow the exchange of information between the vehicle, the RACS information center and the communication center of the subscribing private companies, but also collect vehicle identification information to measure travel time and traffic flow directly from individual vehicles. Lack of this bi-directional communication is a disadvantage of AMTICS. As RACS offers a bi-directional, digital data exchange, various data communication services can be used including character, voice, facsimile and picture transmittance. Automatic toll collection can be done if the vehicles send their ID information to the roadside antennas. Automatic vehicle monitoring (AVM) will also be possible if the passage time at beacon is recorded and monitored.

Developments for Practical Applications and VICS

Having completed their experimental stages, both RACS and AMTICS have stepped into a practical application stage. Two separate organizations were established to promote their practical implementation by private companies and institutions which have taken charge of developing the two systems. Also, the Japan Digital Road Map Association was established in 1988 and it laid out standards for digital road mapping using the research results of the RACS project. The first version of the Japan digital Map Data Base was completed in 1989.

AMTICS's practical experiment was carried out in Tokyo from April to September, 1990, at the occasion of the International Exposition of Flower and Greenery in Osaka. In this experiment, traffic congestion and regulation information, parking and event guidance of the Expo, and also railway transportation information were disseminated to the vehicles used in the Expo, which included 20 buses, 9 taxi-cabs, 1 truck, 4 passenger cars and 2 patrol cars. 20 traffic information guidance devices were installed at the Expo site, parking places, main terminals of public transportation, and several gas stations. The experimental project in Osaka has been extended for one and half years after the closure of the Expo and the numbers of experimental vehicles and parking places to be guided were increased. NPA is also planning to install electronic signposts all over the country as one of the vehicle detectors of the ATC systems to provide location information.

On the other hand, the open experiment of RACS was held in May 1990 to demonstrate the entire functions of RACS including individual communication. After the research work was completed in 1989, the practical experiment to establish the final specification of beacon dissemination of RACS was carried out by the road administrators from September to December in 1990. 103 beacons were installed on the National Highways and Expressways in three metropolitan areas of Tokyo, Osaka and Nagoya. MOC is planning to install beacons nationwide. As the first step of practical implementation of RACS they are now planning to open the on-way communication service by beacons.

Thus far, both systems have advanced to their practical application stages independently. This obviously resulted in duplicated efforts for all in-vehicle equipment manufacturers and users, and required two different communication methods to provide traffic information to the motorists. In March, 1990, in order to discuss and adjust the problems of the general mobile communication systems, the VICS committee has been formed jointly by MOC, NPA and MPT. After several organizational meetings they came to an agreement in December, 1990, that efforts of the practical application of the two systems be unified by the VICS Implementation Council (tentative name) from now on. Using the unified system, traffic information can be provided to the motorist through a single communication method which is not yet finalized at the time of this writing. For the time being, tele-terminals, beacons (one-way) and FM multi-channel radio broadcast will be used as information communication media. The FM multi-channel radio had been proposed during the AMTICS development as a replacement to tele-terminals.

Future Directions

For a while the communication media of AMTICS and RACS will be used and compete although they will be managed by one organization, VICS. RACS's method will be easily set up on expressways and inter-city trunk roads. It will be useful for road administrators and long distance line-haul transportation companies when bi-directional communication becomes available. On the other hand, the police and delivery transport companies may prefer an continuous area communication system like tele-terminals to the intermittent spot communication system of RACS. It will take longer, however, for tele-terminals system to be constructed throughout wide area. And since the number of subscribers to it is limited, it will have less probability of becoming a main communication method of AMTICS. FM multi-channel radios also provide only one-way information.

At present, many mobile communication systems are being developed and their trends have not yet been set. Bi-directional communication beacons have not been accepted for practical use at present. It is anticipated that the bi-directional beacon system will be materialized in the near future because bi-directional communication between the vehicle and the road facility will be essential for future developments of more advanced application of IVHS techniques, such as dynamic route guidance for individual vehicles, advanced signal control based on actual and estimated travel time and OD data, automatic toll collection system, and automatic warning and collision avoidance system. The present development of IVHS mainly focuses on information dissemination; however, system developments for safe and comfortable driving should be promoted in the future as well.

CAR WARS -- THE DOTs STRIKE BACK

by Jerry L. Hautamaki,[1] M. IEEE, and
Katharine S. O'Hara,[2] M. ASCE

Abstract

HNTB is developing integrated computer systems utilizing standard hardware and software platforms coupled with graphical user interfaces for traffic management systems (TMS). Rapid, user-friendly monitoring and control of numerous TMS sub-systems will improve the flow of traffic and provide the framework for a system-wide flow of information. However, while the technology is available, the procurement and implementation of state-of-the-art TMSs run contrary to many standard highway construction practices.

Introduction

In our urban centers, the traveling public is faced everyday with ever-increasing traffic congestion, diminishing highway safety and accumulating air pollution. Population growth increases pressure on already overcrowded highway corridors. Minor vehicular incidents quickly become major, taxing the ability of even the most careful driver to avoid collisions. When traffic isn't flowing, the concentrated exhaust fumes contribute to the haze that hangs over our cities. State and Federal Departments of Transportation (DOTs) are fighting a high-tech battle to keep traffic flowing by quickly informing motorists of traffic problems as they occur.

Advanced traffic management systems (ATMS) and associated advanced traveler information systems (ATIS) are the hottest weapons in the DOTs' arsenals. ATMSs use computers to control full coverage camera systems, real-time reactive (and soon to be predictive) ramp meters, variable message signs, and HOV or reversible lanes to increase the capacity of existing highway corridors. ATISs provide information for radio and television broadcasts, cable TV, emergency callbox systems, and public dial-in access. While HNTB is involved in various stages of ATMS and ATIS development for a number of DOTs, this paper will describe the system that is currently under development for the greater Seattle area.

---

[1]Manager, Electronics/Computer Systems, HNTB Engineers and Planners, 600 108th Ave., N.E., Suite 405, Bellevue, WA, 98004 (206)455-3555
[2]P.E.; HNTB Engineers and Planners, 600 108th Ave., N.E., Suite 405, Bellevue, WA, 98004 (206)455-3555

TMS Overview

HNTB has redesigned and is currently programming the computer system for Washington DOT's (WSDOT's) Traffic Management System in conjunction with the completion of Interstate 90. This TMS comprises sub-systems of over 120 closed-circuit TV (CCTV) cameras, 32 variable message signs (VMS), 50 ramp meters, 100 data collectors, reversible lane controls, a highway advisory radio system (HAR), radio dispatching and two tunnel safety systems. All of these field devices are monitored and controlled from an integrated traffic system management center. The TMS, which is expected to quadruple in size over the next ten years, will also provide voice, graphical, and statistical output for commercial radio and TV broadcasts, cable TV, public dial-in access and the state highway patrol.

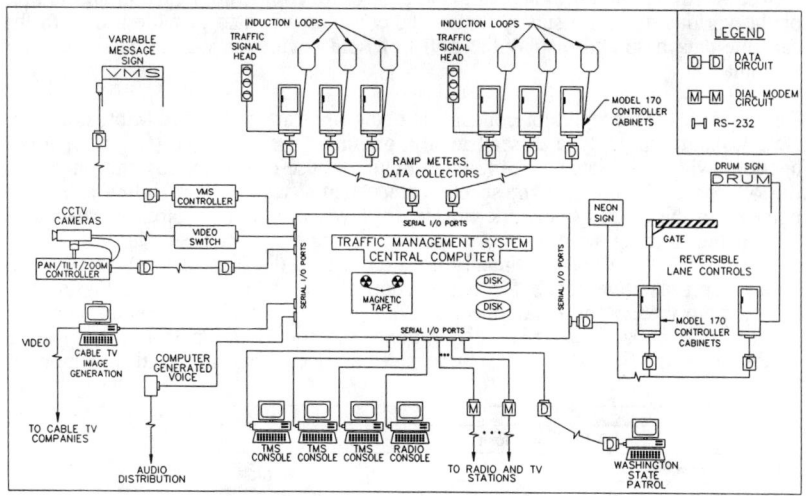

Figure 1 - Washington DOT Traffic Management System

The central computer, a 6000 series VAX[3], will direct the control commands sent from operator consoles to the various field devices. The central computer will also collect traffic data every 20 seconds from over 1,500 induction loops monitoring 100 miles of freeway in the greater Seattle area. This data will be collected, analyzed and stored in a central data base for real-time access by TMS operators and other users. The real-time data base will store one hour of 20-second data in memory. This 20-second data will be aggregated into 5-minute values that will be stored in memory for 24 hours. One year of 5-minute data will be archived on disk for on-line retrieval. This 5-minute data will also be stored off-line on magnetic tape as a permanent record for historical analysis. An operator will have immediate access to data stored in memory and rapid access to data stored on disk. The real-time and 5-minute data bases are in-memory structures that will be dynamically created each time the TMS software is started allowing the size of the data bases to change. The

---

[3]The mention of specific commercial products is not an endorsement of these products by WSDOT or HNTB. The views expressed are entirely those of the authors.

traffic analysis algorithms for detecting incidents and bottlenecks and calculating speeds are being developed as modular programs, that will allow the system to grow as technology advances. Any individual algorithm can be modified, added or deleted without impacting other algorithms.

The ramp meters, data collectors and reversible lane controls are based on Model 170 controllers. The control software in the 170 traffic controllers is being modified to interface with the new software on the VAX and to provide more efficient communication between the field devices and the central computer.

Each TMS operator will communicate with the VAX through a console using two IBM compatible 80386 microprocessor PCs with VGA monitors. Each operator console will integrate all monitoring and control functions, both current and future, for the various disparate sub-systems. The consoles will also provide access to the real-time data base and display the status of field devices as well as facilitate their manipulation.

The software for the operator consoles is being programmed under Microsoft Windows 3.0 and will provide a user friendly, graphical user interface (GUI). Operator functions will be initiated with point-and-click mouse-driven menus and on-screen control icons. The congestion, speeds or occupancy within all mainline and ramp segments in the freeway network will be displayed in multiple colors on schematic maps of the freeway system. These dynamic programs will access real-time data in the central data base and continually update the map displays. The maps will also display representative icons for every controllable field device (e.g., cameras, VMSs, ramp meters and data collectors). By pointing and clicking on an icon, the operator will activate menus and dialog boxes specific to the control or monitoring of that particular field device. Cameras can be selected and assigned to a monitor

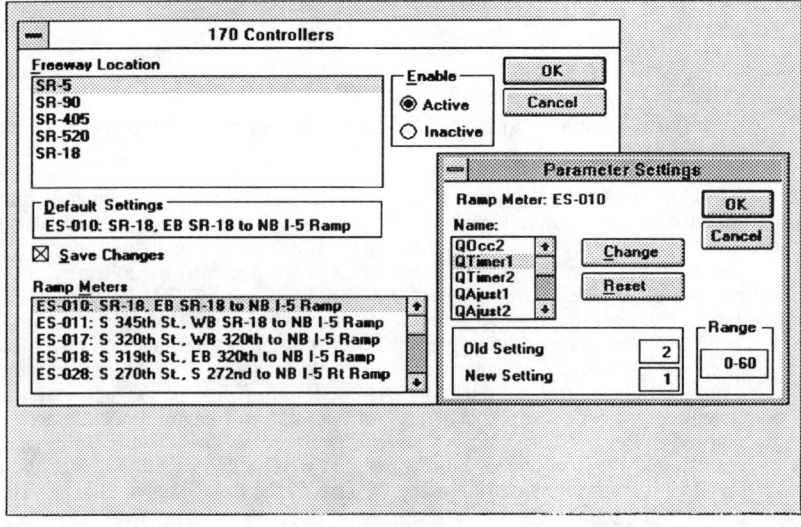

Figure 2 - Ramp Metering Control Dialog Box

or several cameras can be selected for sequencing on a specified monitor, each with a unique dwell time. After a CCTV camera is selected and assigned to a monitor; pan, tilt, zoom and focus will be controlled through a joystick. The current parameters for a ramp meter or group of ramp meters can be displayed, modified and saved through menu-driven dialog boxes. Other dialog boxes will allow manual operator override of the meter rates calculated by the traffic analysis algorithms.

Like the software for the VAX, the PC software is written in separate modules. Each module independently accesses the real-time data base for the information needed to display freeway status on the schematic maps or control a field device. This access is coordinated through communications drivers and protocol handlers developed specifically for these hardware/software platforms.

Hardware and Software Selection

The hardware selection for the central computer was predicated upon providing reliability and redundancy within a commercially available system. VAX computers are a proven technology. The VAX operating system is a well-tested, reliable and flexible platform with a wide range of system services and security features. The hardware offers a well-defined growth and upgrade path with readily available technical support. In addition, the equipment is compatible with existing WSDOT computers.

IBM compatible PCs were chosen for the user interface because they provide a readily available and relatively inexpensive hardware platform. Radio and TV stations already have access to PCs and will not have to purchase new equipment in order to tap into the sources of information made available by the TMS.

Microsoft Windows was chosen as the software platform for the PCs because of its user friendly graphical interface and its multi-tasking abilities. In 1988 when it was chosen, Microsoft Windows was the most widely supported GUI for the PC environment. Microsoft Windows also offers a rich and comprehensive set of software development tools. This selection has since been justified by the wide-spread acceptance of Microsoft Windows as a standard operating environment.

Procurement and Implementation

HNTB has learned a great deal about the hurdles and pitfalls that can be encountered in the procurement and implementation of a state-of-the-art traffic management system for a government agency. Most of the problems involve ensuring the overall compatibility of the equipment. DOTs typically spread the planning and design of highway systems over many years. Advanced computer and electronics technology progresses rapidly, and becomes constrained by planning and design that takes place more than a couple of years prior to implementation. TMS designs need to be dynamic enough to incorporate continual technological improvements during the development stage as well as after implementation.

In a typical highway project, the predominant cost element is the original construction cost. The maintenance and operation of a highway system is borne by the state DOT and the amount of maintenance is often determined by the current tax budget. In any construction involving electronics and telecommunications equipment, the long-term, yearly costs for maintenance, operations, training and support can be as high as 10% of the original equipment acquisition cost.

Another characteristic of state DOT construction contracts is that contractors must be selected for their low bid with little regard for the technology they intend to use. The prime considerations in traffic management systems are system-wide compatibility and integration of the various sub-systems. Contractors should be selected based on their understanding of the technology and their range of technical solutions by using a prequalification list, a weighted selection matrix or a multi-step evaluation process. This would permit the selection of a technically responsive contractor at the lowest overall system cost.

Lastly, but no less important, is the fact that most large highway construction contracts are broken into smaller contracts geographically. Traffic management systems must have an overall system design and integration plan to ensure equipment compatibility and optimal maintenance costs. Equipment incompatibility within WSDOT's TMS has complicated the design of camera and VMS controls. Existing cameras use various combinations of three different controllers directing two different pan/tilt devices and multiple camera bodies. Of the three different types of VMSs in the system, none can be easily controlled from the central computer. All three utilize dedicated PCs running exclusive control software packages. In order to integrate control of the VMSs at the TMS operator console, new hardware and software will have to be obtained.

By ensuring the compatibility of equipment throughout a traffic management system, the number and types of different replacement parts can be reduced. The DOTs can also minimize the training requirements for maintenance personnel and eliminate the storage requirements needed for extensive multi-vendor spare part inventories.

Advanced traffic management and traveler information systems must go far beyond identifying traffic problems and installing state-of-the-art electronics and communications. They must also consider the integration of disparate components, a user-friendly interface, and the entire system's flexibility and reliability. DOTs equipped with high-tech TMS weapons, will go a long way toward annihilating the waves of congestion that are invading our freeways.

A catastrophe theory approach to freeway incident detection

Lisa Aultman-Hall[1], Fred L. Hall[2], Yong Shi[1], and Bradley Lyall[1]

Abstract

This paper discusses an approach to incident detection on freeways that classifies traffic conditions at a single station rather than by comparing conditions at adjacent stations. The approach has been developed from a catastrophe-theory-based model of freeway traffic operations. Initial testing on two systems has provided encouraging results.

Introduction

For effective real-time control and management of traffic on freeways, it is essential that there be a means to identify traffic conditions reliably on an on-going basis. For over twenty-five years, freeway traffic management systems (FTMSs) have used a variety of incident-detection algorithms (most commonly a version of that described by Tignor and Payne, 1978). The current paper discusses an approach to incident detection based on classification of conditions at a single station using flow rates, occupancies (i.e. percent of time a detector is occupied by a vehicle), and speeds from that station.
The approach has been developed from a catastrophe-theory-based model of freeway traffic operations (Persaud, Hall, and Hall, 1990). The model states that the speed-flow-occupancy pattern can be transformed to the partially-folded surface that represents the cusp-catastrophe. Within the model, speeds experience a rapid (catastrophic) change as one moves from uncongested operation to congested operation. This accords with observation (whereas conventional traffic flow theory does not), and provides one basis for classifying traffic

---
[1]Student and [2]Professor, Departments of Civil Engineering and Geography, McMaster University, Hamilton, Ontario L8S 4L7 Canada

conditions for those systems that are able to measure speeds directly. The flow-occupancy data pattern is also well-defined, and can, on its own, identify the movement to or from congested operation for those systems that do not measure speeds. This paper discusses the logic used in the algorithm, and some results from testing. The freeway systems on which the algorithm has been developed and tested have stations spaced roughly 800 meters apart, and transmit data to the central computer every 30 seconds.

Algorithm logic

The first step in the McMaster algorithm is to classify operation at a station as congested or uncongested based on the most recent data received. The classification as congested will be made if either of two conditions are met: the volume-occupancy pair falls into a congested region; or the measured speed falls below a set threshold for the station. At stations without paired loops to measure speed, only the volume and occupancy criterion is used, because calculated speeds have been shown to be too inaccurate to rely on (Hall and Persaud, 1989).

The basic volume-occupancy criterion was initially derived from the catastrophe theory model of the three dimensional relationship between volume, occupancy and speed. Four possible states are shown in the flow-occupancy data (Figure 1): (1) normal uncongested operation; (2) operation downstream of a capacity-reducing incident; (3) operation within a queue of slow-and-go traffic; and (4) capacity operation downstream of a

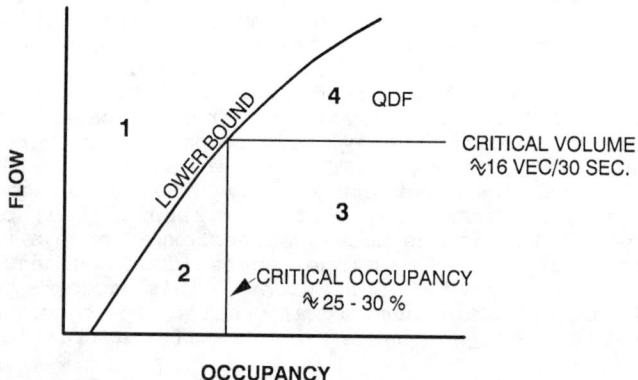

Figure 1. Four states identified for Flow-occupancy data.

recurrent bottleneck. The lower bound of uncongested flow is labeled in Figure 1, and is defined by a quadratic function. The exact location of this boundary can vary for each station depending on such characteristics as geometrics or grade.

The lower uncongested speed threshold is defined as the speed below which operations are always congested. This parameter is set for each station based on system experience and has been found to fall between approximately 50 and 60 km/h. A speed observation which falls below the set value at a given station would be classified as congested.

A persistence check is included in the logic, such that a problem is not 'declared' the first time an interval at a station is found to be congested. Only if the station is congested for a set number of consecutive intervals is the state redefined as congested. Two or three 30-second intervals are appropriate to balance the trade-off between false alarms and time to detection.

On a freeway that does not experience recurrent congestion it is not necessary to test for the cause of congestion. However, most systems will experience recurrent as well as incident-caused congestion. To distinguish between these two, the algorithm classifies data within the congested portion of the volume-occupancy template, and uses the state of operations at adjacent stations. The congested portion of the volume-occupancy template is divided into three general areas as shown in Figure 1. Area 2, the incident/lane-closure region, corresponds with low volumes under reduced speeds, resulting in occupancy values larger than uncongested. This almost always signifies an incident upstream. Area 3 represents flow within a queue, which contains high occupancy observations with relatively low flows. Operations fall into this area upstream of incidents or bottlenecks, where demand exceeds capacity. Area 4, queue discharge flow (qdf), is observed downstream of heavily used entrance ramps that have caused a queue to build up on the freeway. The high volume reflects the capacity flows experienced from the combination of entrance ramp traffic and traffic from the queue on the mainline freeway.

If congestion is detected at a station downstream of an entrance ramp known to experience recurrent congestion, the algorithm checks if the volume-occupancy observation falls within the queue discharge flow area. If it does, then the congestion is defined as recurrent and no incident is declared. If queue flow (area 3) is found at a station and no queue or queue discharge flow exists at the next station downstream, then an incident alarm is reported to the system operator. If queue flow is found at a station and a queue, incident or queue discharge flow is occurring at the downstream station, it is assumed that the queue has simply extended through the upstream station. Keeping track of the operational states at each station until congestion is ultimately declared over can also allow for incident detection within queues. In order for an alarm to be declared over, the operational state at the specific station must be found to be uncongested for a given number of consecutive intervals, currently set at three. Conditions at a station are uncongested if two conditions are met simultaneously: the volume-occupancy pair is above the lower uncongested bound (Figure 1); and the speed (if available) is above a threshold speed, usually set between 70 and 80 km/h.

Two other features warrant mention: the single-lane logic, and real-time updating of the lower uncongested boundary. The algorithm uses data from each lane separately, rather than averaging the data from all lanes. Single-lane data make the patterns clearer, and can result in faster detection. The updating of the boundary simply accounts for the fact that with inclement weather speeds tend to reduce, with the result that occupancies increase slightly for a given volume.

Results of testing

Development of this approach to incident detection has taken place in two stages. In the first, the congestion-detection qualities of the method were tested and confirmed. These tests took place on the Burlington Skyway in Ontario, Canada, where an algorithm implementing this approach has been operating on-line for over a year. The most recently analyzed results, covering the period 1990 Nov 15 to 1991 January 13, show that during medium to heavy traffic flow the algorithm is quite effective in identifying congestion (which on this system is always incident-caused), and that the false alarm rate is quite acceptable, except during snowstorms. The analysis consisted of matching the algorithm output with the operators' log of events. During the 60 days studied, the algorithm caught 6 incidents, and missed 4 others that might have been detectable, although the operator's comments indicate these 4 had no significant effect on traffic. During light traffic, several accidents occurred that the algorithm did not find, but in this it is not different than any other functioning algorithm. It did, however, detect 10 other incidents that by the time the operator logged them consisted of vehicles parked on the shoulder, which is an unexpected bonus for the algorithm. With a persistence check of 2 intervals, the false alarm rate during normal weather is one every 10 hours (over 28 stations reporting every 30 seconds, 24 hours a day); with 3 intervals it would be one false alarm every 39 hours. The exception was during a major snow storm, when there were false alarms more frequently than one every 10

minutes. Clearly the updating feature needs more work if the algorithm is to be effective under such extreme conditions.
The second stage of testing has used data from the Mississauga section of the Queen Elizabeth Way, also in Ontario. This section experiences recurrent delay, making it possible to test the ability of this algorithm to distinguish between incident-caused and recurrent congestion. Preliminary off-line results have been promising. The most recent analysis covered 39 days, during which the operators recorded 31 incidents, and the algorithm 34. The two agreed on 14. Of the 17 recorded by the operators but missed by the algorithm, 7 misses were the result of bad or no data coming from the detectors at the time. For all but one of the remaining 10, close inspection showed no visible effect of the recorded incident on the data. There were 20 algorithm alarms that have no corresponding operator-identified incident. Five of these occurred between 2200 and 0600 hours, when there is no operator on duty. Even if all of the remainder are false alarms, 15 in 39 days, over 15 stations, is certainly acceptable performance.

In conclusion, then, the McMaster single-station algorithm has been successful in tests to-date. Because it is able to classify the state of operations at a station on the basis of data from that location alone, it offers considerable promise for implementation with new technologies for highway control that rely on knowledge of the state of traffic on the freeway.

References

F.L. Hall and B.N. Persaud, 1989. "An evaluation of speed estimates made with single-detector data from freeway traffic management systems". Transportation Research Record 1232, 9-16.

B.N. Persaud, F.L. Hall, and L.M. Hall, 1990. Congestion identification aspects of the McMaster incident detection algorithm. Presented at the 1990 Annual Meeting of the Transportation Research Board, and accepted for publication in Transp. Res. Rec.

Payne, H.J., and S.C. Tignor, 1978. "Freeway incident-detection algorithms based on decision trees with states". Transportation Research Record 682, 30-36.

A LOW PASS FILTER FOR INCIDENT DETECTION

Yorgos J. Stephanedes, A.M. ASCE[1]
and
Athanasios P. Chassiakos, S.M. ASCE[2]

INTRODUCTION

Incident detection is instrumental in reducing freeway delay by acting at the highest level of freeway control and guiding traffic flow towards smooth operation. When a capacity-reducing incident is detected, a responsive ramp-metering control strategy restricts or prevents additional vehicles from entering the freeway upstream of the incident. Excess demand is diverted to alternate routes, increasing the speed with which favorable operating conditions are restored on the freeway [8].

Automatic incident detection (AID) involves two major elements, a traffic detection system that provides the information necessary for detection and an incident detection algorithm that interprets the information and ascertains the presence or absence of an incident. Presence detectors imbedded in the freeway pavement are extensively used to obtain traffic data, primarily volume and occupancy. Experimental video detectors could also be used but high cost prohibits their widespread application.

The most important AID algorithms include four major types: (i) The *comparative* "California" algorithms [7] assume that an incident significantly increases occupancy upstream while reducing occupancy downstream (see Figure 1). (ii) *Time series* forecasting models can provide short term traffic forecasts attributing any significant deviation between observed and forecast values to incidents [1,4,5]. (iii) Separating the flow-occupancy diagram into four states can detect incidents after observing specific changes in the traffic state in a short time period [6]. (iv) The HIOCC algorithm uses 1-second occupancy data to detect stationary or slow moving vehicles [3].

---

[1] Professor; [2] Graduate Student, Dept. of Civil and Mineral Engineering, University of Minnesota, Minneapolis, MN 55455

ALGORITHM DESCRIPTION

A new algorithm was developed after observing that existing algorithm performance can improve if detector data are filtered before use [2]. The new algorithm lies within the traditional framework of comparative algorithms, i.e., is based on simple comparisons extracted from loop detectors. The development focused on using filtered detector output, i.e., values averaged over short time periods, to avoid false alarms that are primarily due to short-term traffic inhomogeneities.

In particular, to filter out the high-frequency random traffic fluctuations in the data, we propose a low-pass (LP) filter on $\underline{x}_t$, the discrete-time series of traffic occupancy data, resulting in $\underline{y}_t$, a linear filtered data series. The LP filter is a moving-average smoother of the form

$$y_t = \sum_{k=0}^{M} \frac{1}{M+1} x_{t-k} . \tag{1}$$

The gain functions of the frequency response of the filter are presented in Figure 2 for two cases, M=5 and M=9.

To apply the above model for incident detection, consider the detector occupancy time series O_t^u and O_t^d at the upstream and downstream stations respectively. Consider

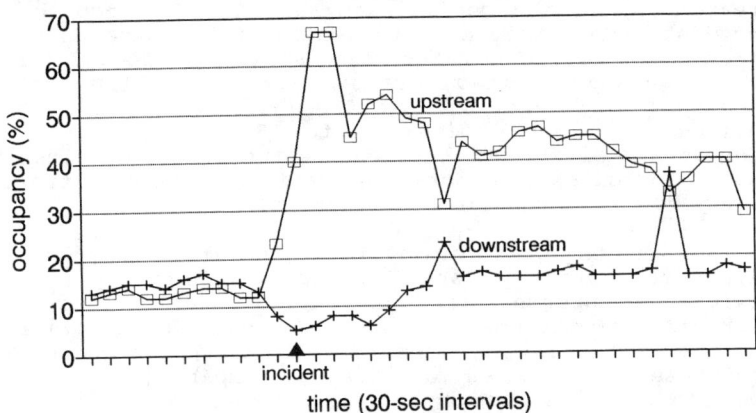

Figure 1. Traffic occupancy during incident (Minneapolis, I-35W south, 4:48pm, 20 July 89, station 026S).

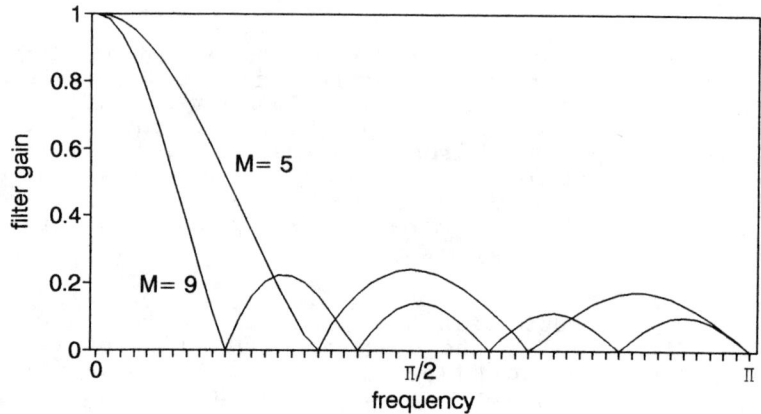

Figure 2. Gain functions of two low pass filters.

also the spatial occupancy difference time series $x_t = O_t^u - O_t^d$ and its filtered version $y_t = y_t^a$ with k=0 to M; M=5 in eqn. (1). Measurements are made every 30 seconds, so that y_t^a corresponds to a moving average of 3 minutes. Since high frequencies are filtered out, high y_t^a values cannot be random and, therefore, indicate that congestion probably exists in the segment between the two stations as a result of an incident or a bottleneck.

To capture the substantial temporal change that an incident creates in traffic conditions (and thus distinguish the incident from bottlenecks for which such change does not exist or is small and takes place slowly), a second filter is designed as the 5-minute time average of the spatial occupancy difference x_t past the first 3 minutes; it is defined as $y_t = y_t^b$ with k=6 to 15, M=9 in eqn. (1). Significant difference between the two filtered values y_t^a and y_t^b indicates a temporal change in the state of traffic signalling the probable occurrence of an incident.

To increase the transferability of the algorithm, the two filter variables, y_t^a and y_t^b, are normalized by m_t, the maximum of upstream and downstream station occupancy averaged over the most recent five-minute period prior to the incident. The algorithm logic includes two tests. The tests use the filtered variables to compare the values of two ratios against thresholds. In particular, congestion is detected if y_t^a/m_t is greater than the congestion threshold, and an incident is detected if $(y_t^a - y_t^b)/m_t$ exceeds the incident threshold.

ALGORITHM EVALUATION

The algorithm was tested against major algorithms reported in the literature. The evaluation included main features of the algorithm, i.e., detection rate, false alarm rate, and mean time-to-detect. Test results with incident and incident-free data along I-35W, a typical freeway in the Twin Cities Metropolitan Area, are presented via its operating characteristic curves (a plot of detection rate vs. false alarm rate).

The operating characteristic curves of the new algorithm and Algorithm #7 of the California series [7] are given in Figure 3. At all rates, the new algorithm is superior to Algorithm #7. It achieves a 0.15% false alarm rate at 60% detection rate, whereas Algorithm #7 results in a 0.3% false alarm rate at the same detection rate. Test results indicate that the new algorithm produces 30 - 70% fewer false alarms than the comparative algorithms, and 70 - 80% fewer than the time series algorithms. In addition, although its structure imposes a 3-minute detection delay, its mean time-to-detect remains very close (by less than 0.5 minute) to the detection times of the comparative algorithms.

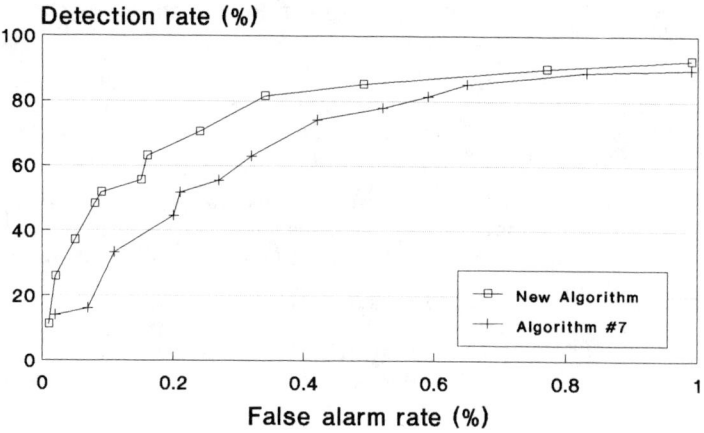

Figure 3. Detection performance comparison.

Acknowledgements - This study was supported by the National Science Foundation through project No. NSF/CES-8713277. The Minnesota Supercomputer Institute provided partial support through projects Nos mg26701 and mg29901. The Center for Transportation Studies, Department of Civil and Mineral Engineering, University of Minnesota is acknowledged for its support.

REFERENCES

1. Ahmed, S.A., and Cook, A.R. "Discrete Dynamic Models for Freeway Incident Detection Systems", Transportation Planning and Technology Vol.7, 1982, pp. 231-242.

2. Chassiakos, A.P. "Improving existing incident detection algorithms", Intern. Rep. CME-5084-89-I1, Dept. of Civil Engineering, University of Minnesota, September 1989.

3. Collins, J.F., Hopkins, C.M., and Martin, J.A. "Automatic incident detection - TRRL algorithms HIOCC and PATREG", TRRL Supplementary Report No. 526, 1979.

4. Cook, A.R., and Cleveland, D.E. "Detection of Freeway Capacity-Reducing Incidents by Traffic-Stream Measurements", Transportation Research Record No. 495, 1974, pp.1-11.

5. Dudek, C.L., and Messer, C.J. "Incident Detection on Urban Freeways", Transportation Research Record No. 495, 1974, pp.12-24.

6. Gall, A.I, and Hall, F.L. "Distinguishing Between Incident Congestion and Recurrent Congestion: A Proposed Logic", Pres. at Transportation Research Board 68th Ann. Mtg, Washington, D.C., January 1989.

7. Payne, H.J. and Tignor, S.C. "Freeway Incident Detection Algorithms Based on Decision Trees with States", Transportation Research Record No. 682, 1978, pp. 30-37.

8. Stephanedes, Y.J., and Kwon, E. "On-Line Demand-Diversion Prediction for Integrated Control of Freeway Corridors" University of Minnesota Supercomputer Institute Research Report UMSI 91/49, February 1991.

AUTOMATIC INCIDENT DETECTION USING IMAGE PROCESSING TECHNIQUES: a specific system used in INVAID

J.M. BLOSSEVILLE, V. MOTYKA, N DJEMAME, F. LENOIR

Institut National de Recherche sur les Transports
et leur Sécurité (INRETS)
BP 34 94114 ARCUEIL FRANCE

Abstract:

INRETS together with other partners, within the European DRIVE program framework is working on developing a system able to detect incidents on motorway by processing video images (Jref. JM Blosseville,91). This paper describes the methodology used and the results obtained so far. The part of the prototype which has been tested, aims at detecting incidents associated with stationary vehicles.

Introduction :

The system is attached to one video camera and is based on an image processing treatment ; it primarily aims at detecting incidents that occur within the camera field of view (usually, several hundred meters of motorway are covered by one camera). It does not produces only a binary alarm ("there is an incident") but provides a more detailed information :

"there is an incident of type T - for instance, a stationary vehicle on the shoulder lane -, and the confidence level of this alarm is X". The system was splitted in several units :

- the Image processing unit which converts images into vehicle markers.

- the measurement unit is fed with vehicle markers and produces measurements.

- the incident detection unit processes the measurements and calculates incident alarms as also alarm confidence levels.

- finally, the communication unit is specially in charge of transmitting incident alarms and confidence level to a central system.

Among these units, two are of a special interest : the measurement and the incident detection modules. The present paper describes the way we designed, implemented and tested these two modules.

Methodology Summary

The first stage of the development consisted in answering the question "What incidents to be detected and how to detect them?".

Secondly, a set of candidate traffic measurements was selected, taking into account the list of incidents to be detected and the possibilities of our image processing system. Measurement algorithms based on the TITAN sensor (cf. Blosseville and al., 89) were designed, implemented and tested.

In the third stage, we designed the incident detection module. We defined a general framework for detection algorithm.

Finally, test and evaluation have been performed. The four stages are detailed below.

Incident list :

A survey in several French TCC allowed us to draw up the following list :

- accidents, breakdowns
- stationary vehicle on the road or the shoulder
- forbidden action of a vehicle
- end, head of a congestion
- too short headways
- unusual flow
- roadworks
- objects on the road
- pedestrian, animals

The survey was also a good opportunity to get incident images including several tens of real incident views. This bank was an essential basis for developping and testing our AID system.

During the survey, it appeared that the great majority - in terms of occurence - of incidents are characterized by one or several vehicles remaining stationary (car breakdown, accident). This important remark has been taken into account when developping a practical tool of AID, as it is detailed below.

Incidents Categories

For clarification, incidents have been classified in 3 categories according to the way we can detect them :

- the category 1 includes incidents that can be directly detected. In our case, as AID is based on vehicle detection by image processing, the category 1 corresponds to specific abnormal vehicle behaviour : stationary vehicles on the shoulder lane or on the road, very fast or slow vehicle..

- Some incidents can be detected through their consequences upon traffic ; they constitute the category 2. One can expect to detect the incident by analyzing the vehicle behaviours. Such characteristic traffic situations are named "traffic patterns".

Other incidents form the category 3. This includes incidents that are only detectable at a central level, by comparing traffic measurements at two or several different sites.

In the project, we focused our attention on categories 1 and 2 for which an image processing system - as a spatial sensor - is very adapted.

Traffic measurements

A set of candidate traffic measurements was selected summarized in the table below. We distinguished integrated measurements - aggregated in time, from emergency measurements that refect the present situation almost instantaneously.

EMERGENCY MEASUREMENTS	building period	integr. period
stationarity duration	1/4s	none
extreme speeds	1/4s	none
extreme acceleration	1/4s	none
shock wave speed	1/4s	none
INTEGRATED MEASUREMENT		
Mean concentration	5s	15-30s
Concentration histogram	5s	30s
Volume	5s	15-30s
Volume histogram	5s	30s
Mean speed	5s	15-30s
Mean acceleration	5s	15-30s
Shortest headway	5s	1mn
Shoulder lane use	5s	15-30s
Lane changes	5s	15-30s
Traffic indicators	5s	3-5mn

A series of algorithms was issued, giving a precise definition for tracking vehicle detections and computing all candidate measurements.

Incident Detection

Traffic measurements feed the incident detection unit, which in return produces incident alarms and informs about the alarms degree of certainty through confidence levels.

A general format of algorithms was defined, that can be applied to detect every type of incident associated to one traffic measurement (called "main indicator"). Algorithm format is divided in three parts :

- the main indicator test is the corner stone of incident alarm triggering. It simply consists in comparing the main indicator to a threshold.

- a basic confidence level : The alarm can be triggered only if the main indicator test is satisfied. In that case, a basic confidence level is calculated. It increases as the difference between the indicator and the threshold increases.

- alarm consistency may be checked by a set of absolute constraints : each constraint is a boolean test, the rule being that an incident alarm is inactivated if at least one constraint attached to this incident type is found false.

- the last part of the algorithm deals with refining the alarm confidence level. This role is played by a set of refining conditions, composed of boolean tests and weighting factors. Conditions that are expected to happen (resp. not to) when incident occurs increase (resp. decrease) the confidence level.

Although somehow restrictive, the special role given to only one measurement for each incident type (the main indicator) intends to facilitate the human operator task : in this conditions it is easier to understand the cause of an activated alarm, or to tune the system by modification of the main indicator threshold.

A.I.D algorithms can addressed most of the known incidents : stationary vehicles, obstruction, queue end, congestion shock wave, very slow and very fast vehicles, short time headways...

Test and evaluation

This part of the research was devoted to increase the robustness, the simplicity and the generality of the method. These improvements were carried out at the three levels : image processing, measurements and incident detection.

Image processing has been improved so as to provide a wider range of application conditions : images coming from cameras placed in tunnel or at a very low locations can be processed. In this new approach, the image processing outputs are vehicle markers such as roofs, hoods, vehicle fronts... Compared to TITAN, The constraints are weaker : the device does not expect only one marker per vehicle.

Because of limited ressources ,the evaluation was focused on a few number of measurements and incident types. The measurements which have been programmed and tested are:

. the stationarity duration,
. The queue length,

Three types of incidents were included in the evaluation :

. stationary vehicle on the shoulder lane
. stationary vehicle in a flow that is not a congested flow (cf. the fig. below)
. stationary vehicle in congestion

The classical three criteria were retained during the tests : detection rate, delay, false alarm rate. The method for evaluating the detection rate and the delay consisted in running the system with, as input, incident images from our databank. On the other hand, false alarm rate were computed by feeding the system with "ordinary" traffic images. This testing bench was as large as possible and diversified (day and night, sun and rain,...)

figure 1 : example of detected incident

The results of 600 processed hours are given in the following table :

incident location	detected incidents	non detected incidents	false alarms
in the shoulder	121	8	5/24h
in the main lanes	3	1	4/24h
total	124	9	9/24h

The previous figures indicate a rate of detection greater than 93% associated with a false alarm rate of 9 false alarms per 24 hours. The detection delays range between 15 and 30 seconds with respect to the used main indicators.

Conclusion :

A prototype of a local system has been designed which is capable to detect the incidents associated with stationary vehicles.The device has given statisfactory results during extended laboratory and restricted field tests. The next step will be an installation of several devices in operational and various geometric and traffic conditions.

References:

JM Blosseville, V.Motyka, F. Lenoir
" TITAN, A traffic measurement system using image processing techniques",
IEE Conference London, fev 89

JM Blosseville , S. Guillen, V.Motyka, F. Lenoir
"Automatic Incident Detection using computer vision techniques",
Drive Conference, Brussels, feb 91

A DISTRIBUTED REAL TIME KNOWLEDGE-BASED SYSTEM USING VIDEO IMAGE PROCESSING FOR JUNCTIONS AUTOMATIC INCIDENT DETECTION

S. Sellam, A. Boulmakoul, J.C. Pierrelée
(INRETS-MAIA, Arcueil - France)

A. Introduction

A junction in an urban network is a strategic place : for road users (routes choice, pedestrians crossing...), for traffic managers (sensors data collection, traffic lights control, incidents and blocage main source...) and also for safety considerations.

The scope of the INVAID project (in the European DRIVE program) is to build an automatic incident detection system which incorporate the image processing capabilities. In the urban aspects of this project we must notice that the output of the image processing is a binary image, i.e. a black point belongs to a vehicle in the image and a white one to the background. A mock-up of this expert system is currently designed with G2™ (from GENSYM™ Corporation) a real time expert system tool *(Moore 1984)*.

We will report in this paper the main components of the system (including the Image Processing main characteristics) and the main elements of the knowledge base leading to the diagnosis process:

B. The general architecture

The three main components (see Figure 1 and 2) of the system we will present are the following:

- The Image Processing Unit (IPU) which output is a binary image of the junction giving the space occupied by vehicles,

- A Measurements Processing Unit (MPU) which processes the IPU data in order to compute indicators of the traffic behaviour,

- The Diagnosis Processing Unit (DPU) which achieves a diagnosis and, if required, launch a data seeking process to the MPU for traffic measurements.

The static knowledge used in the DPU consist of a model representing the junction (next section) and a set of rules and formulas which are controlled by a real time inference engine.

JUNCTIONS INCIDENT DETECTION

It is through the MPU that the correspondance between the view of the video camera and the representation of the modelised junction is implemented (see Figure 3).

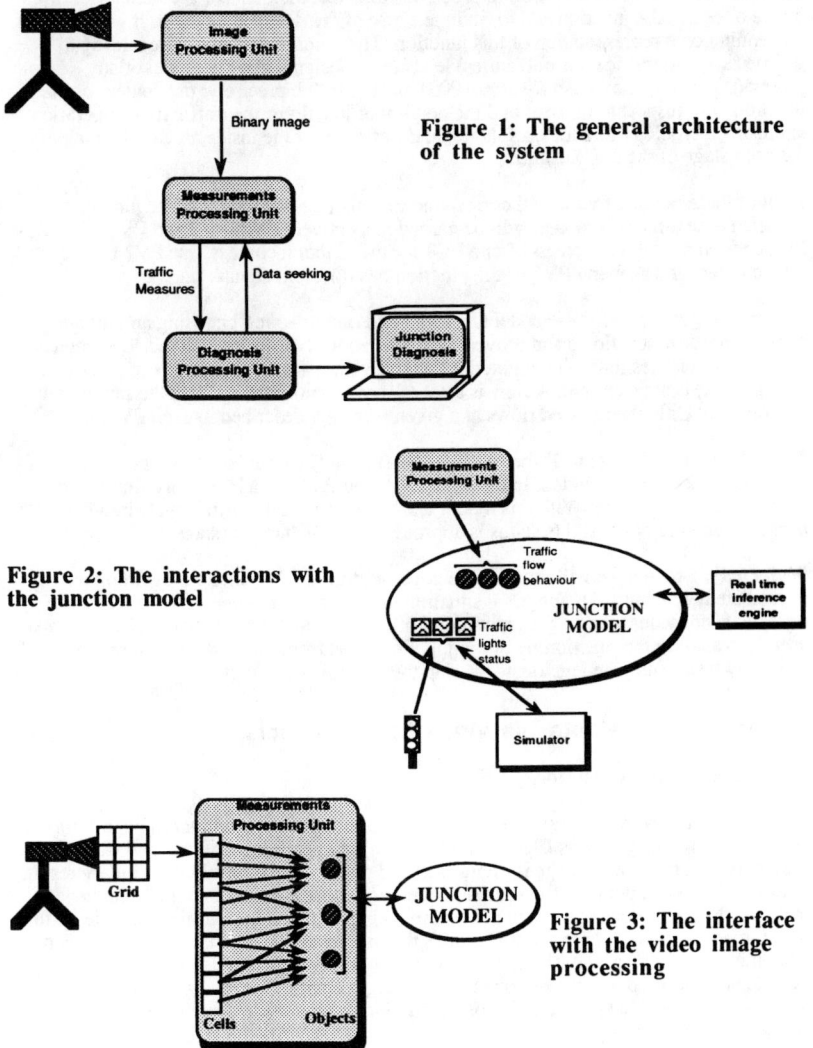

Figure 1: The general architecture of the system

Figure 2: The interactions with the junction model

Figure 3: The interface with the video image processing

C. The junction modeling

The main purpose of the monitoring process is to detect incidents or accidents occuring inside or beside the junction and to find the cause of the detected incident. It will build its reasonning on a representation of this junction. The junctions diversity (regarding to geometrical and traffic control features) leads us to design a generic model of any junction. This generic model *(Sellam.1990)* must describe precisely: the context of the junction with respect to its arms and the neighbour junctions, the traffic lights operation and using a typology of zones, a functional description of the inside area of the junction for each stage of the traffic lights.

Each of these points of view will correspond to a distinct sub-model of the junction model. These three sub-models will be named respectively PV3, PV2 and PV1 Model. A full junction model will consist of one PV3 for the junction context, one PV2 for the traffic lights, and as many PV1 as the junction has different signals stages.

The main objects of the PV1 model are: input and output zones, choosing and merging zones, conflict zones, flows and movements. The zones objects are connected to each other to form routes through the junction. A given route (from an input to an outpout of the junction) is a movement. A flow is a set of movements coming from the same input of the junction. Only the allowed flows at a given stage are described in each PV1.

The PV2 model will "control" the traffic lights status. The term "control" is exact only if the control module is included. In the other cases the PV2 model will only simulate the traffic lights functioning. With this model we will try to break down a cycle into periods during which the traffic lights status is approximately unchanged (stages).

With the PV3 model we will monitor the good working of the junction regarding its immediate neighbours. An abnormal situation occuring on only one of these may have immediate consequences on the functioning of our junction. This PV3 model will be also usefull to analyse the functioning of the junction apart from the stage type in progress. This may be usefull when incidents overlap the duration of a single stage.

D. The interface with the video based sensors

1 The image detection

The result of the type A Image Processing Unit for urban sites is a binary image *(Motyka 1990)*. The principles of this IPU are the following. Several times per second (four times is an optimum rate) each image is digitalized and processed to give finally a binary result. That is to say, each point on the source image corresponding to a "moving" vehicle is black on the result image and each point corresponding to the background is white on the result image. A parameter in the IPU allows the user to fix the time interval until which a stopping vehicle is still considered as a "moving" vehicle. After this time interval this vehicle is considered as a parked one and included as part of the background. It is important to notice that the result is not an individualized vehicle detection but only a spatial status detection.

2 The "presence-absence" image

Each point of an image can be characterized by two instantaneous status: it is **occupied** or it is **free**. But each of these two status can be characterized by a variable which measures the time interval during which this point has kept its status. Therefore, we can compute, for each binary image a "presence-absence" image. For each occupied point of this last image we will have a value reflecting the time interval during which this point has been occupied (the presence time). At the other hand, each free point will correspond to a value of its "absence time".

3 Relations between an object and its topological definition

The interface between the junction model and the junction video image will use predefined objects. Upon an image of the junction, the user has to draw the borders of the different zones. Each zone will correspond to the topological definition of objects in the junction model.

E. The diagnostic process

1 The main purposes of the diagnostic process

The two main purposes of the diagnostic process are : automatic detection of any abnormal behaviour and incident characterisation (congestion, accident) regarding its location and severity. A report of any detected event is displayed and can be sent to operators or other automatic systems.

In order to be relevant a diagnosis must be reported as soon as possible but must also save from false alarms. The indicators computed with the video based sensors (see the next section) seem to be enough reliable to reach this objective.The main result of a diagnosis is to find which traffic flow(s) (which input) or which movement(s) (which output) is (are) primarily concerned. The second one is to locate, if possible, the incident or accident. The third one is to list the possible consequences of the detected incident.

2 The measurements used in the diagnostic process

Any region (zone or area defined in the junction model) denoted R is a concatenation of a set of image points.This region will be characterized by three measures which will be directly derived from the points status indicators. Those measures are the occupancy rate $(C(R))$, the presence mean time $(P(R))$ and the absence mean time $(A(R))$: The meaning of those measures are easy to understand. A presence mean time of 5 seconds for example means that the part of the region which is now occupied has been occupied for 5 seconds ago on an average.

For diagnosis purposes we can say that the presence time is connected with a steady flow measure. At the other hand the absence time reflects the use level of a given region. A low use level can have two meanings: Low flow conditions (but this must be confirmed by the total presence time) or an incident upstream. We can achieve a diagnosis about the behaviour of the region with those three measures from which we can derive a fourth one:

$$I(R) = \frac{P(R)}{A(R)}$$

Depending on the size of the regions and the zone type (conflict zone, output zone...) we will have different tresholds on each of the previous measures. Further statistical analysis will allow us to set automatically those tresholds.

3 A hierarchical diagnosis

As we have noticed in the junction model description, we will use three different submodels in our diagnostic process. The PV3 model will allow a first level diagnosis. Inside a junction many events occur frequently like stopping vehicles in left turn movements without other consequences. An alarm is needed only if a movement seems to be blocked more than a given time interval. An abnormal status.detected on one of the areas of the PV3 model will start a second level diagnosis but the first level one will continue in order to detect any further event. Depending on the signals status (given by the PV2 model), the second level diagnosis will determine which PV1 model to activate. Depending on the behaviour of each flow of this model a third level diagnosis is started for each abnormal one.This third level will locate the zones primarily concerned by the incident. It will list the actual and potential consequences of the incident.

F. Conclusion

Two main extentions of the system are currently in progress. The first one is the integration of this "local" AID system in a centralised one. Several communication capabilities are already integrated in the system (with neighbour junctions and central systems) and it will be possible to cover partly or completly a urban network.

The second one is the interfacing of this system with a control module. The aim of the Intelligent Intersection INRETS project is to design a system able to deal with all the aspects of junction management (monitoring, control, communications) regarding all kinds of users (vehicles, pedestrians, buses...). By integrating this system in a network level the aim is to build a distributed management system which is flexible enough to support different kinds of control strategies, and to solve the conflicts between needs of different categories of users.

REFERENCES

BEUCHER S., BLOSSEVILLE J.M, LENOIR F.(1987) "Traffic spatial measurements using video image processing, Application of mathematical morphology to vehicles detection" Congress SPIE , Cambridge, Mass.,Nov 1978.

Lee D. HAN and A.D.MAY "Artificial Intelligence Approches for signalized network control" Working paper UCB-ITS-WP-88-4 Institute of transportation studies University of California Berkeley, 1988.

MOORE R.L,HAWKINSON L.B and CHURCHMAN L.M and KNICKERBOCKER C.G " A real time expert system for process control " First conf.on A.I applications, IEEE computer society, 1984.

MOTYKA V., LENOIR F., BLOSSEVILLE J.M. "Analyse d'images de trafic: un logiciel pour la détection des véhicules en milieu urbain" Journée spécialisée INRETS "Le Carrefour Intelligent", Arcueil, France, 1990

SELLAM S., BOULMAKOUL A., PIERRELEE J.C. "Intelligent Intersection: a monitoring and control integrated system" OECD Workshop on knowledge based expert systems in transportation. Espoo, Finland, June 26-28, 1990.

A Real-Time Traffic Diversion Model: A Conceptual Approach

A.G. Hobeika and Y. Zhang
Center for Transportation Research
Virginia Polytechnic Institute and State University
Blacksburg, VA

Abstract: The development of a real-time traffic diversion control system for integrated freeway and arterial networks, as part of the most currently envisioned Intelligent Vehicle/Highway System (IVHS) is the topic of this paper. The objective of the traffic control system referred to as "Smartnet" is to use real-time data to divert traffic from freeway to arterial networks so as to reduce congestion and improve drivers' travel time. The freeway network chosen for this work is the I-395 freeway corridor area in Northern Virginia, between the Capital Beltway and the District of Columbia. The arterials designated as alternative routes for diversion were primarily chosen to comply to network conditions and to conform with expected driver behavior in route selection. In addition, the route capacity, its associated ramps, and the normal flow volume were also considered in this selection process. The diversion strategies under this approach are based on the highway system performance. The aim is to optimize the system's performance as a whole.

1. Introduction

Advances in microelectronics and communications, and the use of computers in information technology have already brought about many improvements in the efficiency with which our roads are used. The Intelligent Vehicle/Highway System (IVHS) is an exciting application of information technology which has the potential for providing benefits to travellers. The IVHS program can make significant improvements in mobility, safety, and productivity on highways and urban arterials. It can help drivers keep track of their exact location, assist them in the selection of the best routes that take them directly to their destination. Through information transmitted to the vehicle from equipment at the roadside, the driver can receive route guidance based on detailed and current information on traffic conditions. The IVHS program integrates vehicles, drivers, and highways into a cohesive system so as to maximize its performance. It has been grouped into four functional subsystems [USDOT,89]:
- Advanced Traffic Management System (ATMS)
- Advanced Traveller Information System (ATIS)
- Advanced Vehicle Control (AVC) system
- Commercial Vehicle Operations (CVO) system.

A viable electronic route guidance system is currently being experimented in Japan, West Germany, Britain, and other countries. Initial reports indicate that the technology will work, and benefits could be significant.

"Smartnet" is a system being developed at Virginia Polytechnic Institute and State University to help drivers find their way to their destinations in a complex network, whenever there is congestion, blockage, or other disturbances in traffic conditions. For this purpose, it uses current information on traffic conditions. Detailed information on link capacity and other static characteristics of the network will come from the stored database in the system. Real-time information, such as current speed, volume, congestion, blockage, etc., will be collected through road loop detectors or other sensory devices. This part of the information will constitute the dynamic database of the "Smartnet" system. After it acquires all the needed information, the system can run diversion assessment models to formulate the diversion strategies in a short time. The system continuously updates the feasible routes on the basis of current traffic conditions. Drivers can therefore be given route guidance based on real-time information.

2. "Smartnet" System Model Structure

"Smartnet" is a diversion control system aimed at helping drivers find good alternate routes in case of congestion or blockage. It is considered to be the first step toward a fully electronic diversion control system which will help give drivers information at any time and under any situation. This research task concentrates on freeway corridor diversion. The effort is aimed at developing a reliable route-generating algorithm for implementing in the real-time diversion control system. In a fairly congested urban network, since alternate routes may not be able to carry all the diverted traffic, several different diversion options need to be assessed in terms of link capacities and other constraints. The data requirements for this model can be classified into four groups: static road characteristics and constraints of the freeway, dynamic traffic data of the freeway, static characteristics and other elements of arterials, and current traffic conditions of arterials. The static data are built into the model directly. The dynamic data will be collected from road sensors at each cycle of model execution.

To accomplish this, a computer program is being developed which will be able to generate diversion strategies based on up-to-date information on traffic congestions. Real-time simulation will be performed for the modification and for assessment of the model. The models will also be coupled with a graphic capability which will not only enable the display of the road network on the screen showing the affected regions, but also show the modified plans including alternate diversion routes.

The Smartnet traffic diversion control model is composed of three interrelated components, as described below:

i). Major Arterial and Freeway Monitor Component: It continuously monitors the performance of major arterials and freeways inside the corridor area and identifies traffic bottlenecks and blockages.

ii). Diversion Decision-Making Component: After congestion or blockage is detected on the freeway, this component will decide whether or not to divert freeway traffic to some major arterials. It is carried out by a knowledge-based expert system based on expertise gained in freeway and ramp-metering controls. The expert system establishes warrants for diversion based on several factors including upstream flow, downstream flow, changing occupancy levels, etc. This component is planned to be addressed in a later phase of this research.

iii). Network Traffic Control Component: This part is designed to select the best alternate routes for diversion. It has information on network characteristics and traffic routing methods and keeps track of all diverted vehicles. It also produces the measure-of-effectiveness report for the routing strategies developed.

The Smartnet system is using the current Traffic Management System (TMS) in Arlington, Virginia as a case study. The TMS has the following operating and control strategies: time restrictions, high- occupancy vehicle priority, motorist information, monitoring, and incident management. The major field equipment items installed on I-395 and I-66 are entrance ramp-metering signals, signs and detectors; main roadway automatic surveillance detectors; pole-mounted closed circuit television cameras; and variable message signs and I-395 reversible roadway barrier gates. The automatic surveillance detectors have been installed with a half-mile spacing on I-395 and I-66. These detectors can give classification measurements as well as calibration data, i.e., volume, occupancy measurements, speed and average vehicle length. The cameras on both freeways have coverage of reversible roadway access ramp areas, variable message signs, and main roadway and interchange areas. These cameras can be used to detect accident and congestion, and verify variable message signs. Variable message signs are the primary means for sending real-time information to the motorists.

The proposed research approach envisages the interchange of data transfer among the traffic control centers at Arlington County and Alexandria City, and the Traffic Management System (TMS). This will give each signal system the capability of responding to traffic and operating conditions on the TMS system and vice versa. The Arlington Co. and Alexandria traffic signal systems will be centrally controlled by computers to provide signal timing patterns that are responsive to real-time conditions. The availability of these traffic control systems and their adjacent arterials compose the first major component of the Smartnet system.

3. Diversion Points

Since there can be many points along the freeway where diversion can be implemented, the choice for diversion points for this research had to be made using some criteria. One important criterion is that the alternate routes and associate ramps and intersections have to have sufficient capacity to carry the flow. Other criteria have to conform with users' behavior. Drivers have certain criteria and habits in choosing an alternate route. Consistent with expectations, Stephanedes et al. (1989) have found that trip time is the dominant factor determining diversion at the trip origin, whereas route length and the number of intersections along the trip also play significant roles. Diversion at freeway entrance ramps depends on the perceived trip time on the freeway and arterial. Further, for commuter trips shorter than one hour, freeway drivers consider only one diversion alternative, a preferred arterial, and do not divert to downstream ramps. After careful analysis of the concerned network and the investigation of user behavior, the diversion points and the primary diversion routes were chosen. The two diversion points are at the intersection of I-395 and I-495, and the intersection of Rt 236 and I-395. The three main alternative routes are along Rt 236 and Rt 400; I-495, Rt 236 and Rt 244; and I-95 and Rt 1. Figure 1 illustrates the freeway corridor area and the chosen diversion routes.

4. Alternate Route Selection Strategies

The execution of the model has the following steps. The system first selects one route with the shortest time from among the three chosen alternate routes. Then it diverts the traffic to the chosen route for a short period (e.g., five minutes) if diversion is

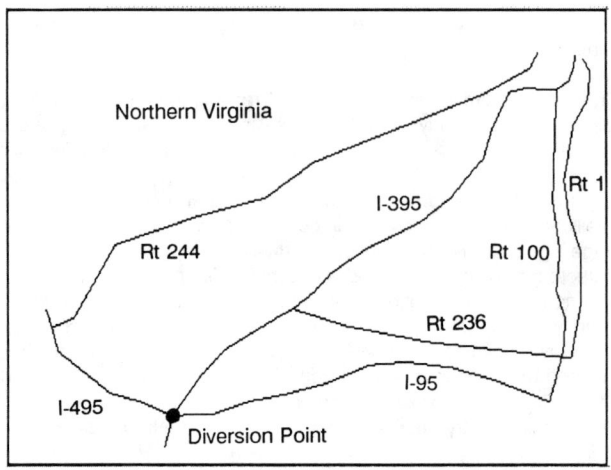

Figure 1. Illustration of Alternative Routes in I-395 Corridor Area

initially warranted. After that it calculates the travel time on all three alternate routes again. If another route's running time is less than that of the previous route, say in a 15 minutes interval, the system may divert the traffic to the new route. After a few cycles, it may start to divert to the other routes. Overall, it tries to minimize user delay by assigning traffic to different routes at different cycles.

5. Time-Varying Flow Conditions

An essential feature of Smartnet is the modeling of time-varying flow conditions. This is achieved by dividing the simulation period into a number of consecutive time intervals and estimating the average flow on each link during each of these time intervals. The intervals are chosen to be as long as possible consistent with being able to represent the broad trends in the flow pattern. An interval length of five minutes is currently being used to model the flow patterns.

In practice, the Smartnet system will be working in a real-time mode, and its working cycle can be fixed at one minute. The Smartnet system is to be built as an auxiliary system which will help the Traffic Management System (TMS) in decision making. The TMS is able to collect traffic flow data every second. Technically speaking, it is possible to make the Smartnet system work in a cycle of a few seconds. In reality, the traffic flow does not change much in a few seconds. The five-minute cycle is considered reasonable in order to be able to capture the major trends in flow pattern without loss of accuracy.

6. Network Simulation Model

A special simulation model is built for the third component. The model is designed for the analysis and evaluation of real-time traffic diversion strategies for an urban freeway corridor area. This model simulates a small highway network (freeway corridor) that might have traffic congestion problems due to heavy traffic flow or lane blockage caused by accidents. It will be used to test different traffic control and operational management strategies to improve the network performance.

The model is capable of simulating the traffic flow on a freeway and its adjacent arterial ways, identifying congestions on the freeway, and selecting an efficient arterial route to divert the freeway traffic. But this module does not support the simulation of diverting arterial traffic to freeway.

The cycle of the simulation model is five minutes. In each time interval, the number of vehicles which will be diverted depends on three factors: the traffic flow volume on the freeway, the percentage of travellers who respond to the diversion message sign, and the ramp capacity. Once the diverted volume is determined, the next important task is to choose an efficient route among all the pre-determined candidate routes. At each iteration, the dynamic flow data (like average speed, travel time, etc.) on every route are read in from the disk storage. If there is any congestion on a route, then that route is dropped from the candidate set. Also, if any route has no more available capacity, that route will also be dropped out of the choice set. Among the remaining candidates, the one with the least travel time will be chosen to be the efficient route for this interval.

The diverted vehicles are loaded onto the chosen efficient route uniformly through the time interval. The vehicles are grouped into packets of twenty vehicles. The system keeps track of all diverted vehicles through packets. At the end of each iteration, if there is any packet leaving the system, the information which the packet carries will be recorded for collecting statistical information like average vehicle travel time, total time saved, etc. Of course, at the end of the simulation, the entire system performance will be evaluated to produce a performance report.

7. Conclusion

Applying advanced technologies to control urban traffic is currently a major trend in transportation research. In this paper, a real-time traffic diversion model for an urban freeway corridor was proposed to be used as an auxiliary traffic control system for the current traffic management system. Its main objective is to relieve the congestion problems induced by accidents. It could also be extended to handle recurrent congestion problems. The "Smartnet" system is not a finished product yet. Current research is focused on decision support system, traffic forecasting, and dynamic traffic assignment.

8. References

Stephanedes, S.J., E. Kwon, and P. Michalopoulos, (1989), "Demand Diversion for Vehicle Guidance, Simulation, and Control in Freeway Corridors," Transportation Research Record 1220, Washington D.C..

US Department of Transportation, (1989) Discussion Paper on Intelligent Vehicle-Highway Systems, USDOT, Washington D.C.

A NEW APPROACH FOR REAL-TIME PREDICTION OF TRAFFIC DEMAND-DIVERSION IN FREEWAY CORRIDORS

Eil Kwon[1]

ABSTRACT: An adaptive method is developed for real-time prediction of demand and diversion of traffic flow at on-ramp areas of freeway corridors. The prediction method explicitly treats the time-variant effects of control on the traffic demand to be predicted by combining behavioral modeling with filtering. In particular, behavioral demand-diversion models and an Extended Kalman Filter are developed, with the filter continuously updating the model parameters with the most recent prediction error.

INTRODUCTION

Increasing freeway congestion, coupled with socioeconomic and environmental concerns restricting any significant increases in physical roadway capacity, has forced traffic engineers to adopt integrated corridor control strategies, which include efficient real-time management of freeway traffic diversion onto less congested arterials. Since traffic diversion is substantially affected by the traffic condition resulting from the control, such as ramp metering and intersection signal timing, the effectiveness of an integrated corridor control strategy largely depends on its ability to estimate and predict the diversion resulting from the control in real time. This paper presents a new approach for the on-line demand-diversion prediction in freeway corridors. Unlike the previous work, the new method reflects the drivers' choice behavior in a rapidly changing traffic environment and explicitly treats the time-varying effects of control on corridor demand.

BACKGROUND

Fig.1 illustrates a typical on-ramp area consisting of an intersection, a ramp and a frontage road. Two possible diversion points can be noted in this configuration, i.e., the intersection and the ramp. Depending on the traffic condition, drivers approaching the freeway can divert either at the intersection prior to joining the ramp queue, or divert to the frontage road after joining the queue. Such diversion directly affects the demand entering the freeway and concurrently, the traffic volume on the arterials in the corridor. Since diversion is influenced by the

---
[1] Post-Doctoral Research Associate, Dept. of Civil & Mineral Eng. University of Minnesota. Minneapolis, MN

control, control action should be based on the predicted behavior of traffic flow responding to the control to be applied. However, no on-line demand predictor developed to date has explicitly considered the effect of control on the demand to be predicted. To be sure, existing on-line predictors, mostly developed for intersection control, adopt a statistical trend-tracking approach, i.e., prediction is primarily based on the past trend of flow measurements. This paper develops an on-line adaptive predictor which explicitly treats the time-varying effects of control on corridor demand in real time. The resulting predictor predicts the ramp-approaching flow at the intersection and its ramp-entering proportion, i.e., the ratio of vechiles entering the ramp over the total number of vehicles approaching the ramp from the intersection. The new predictor is applied to predict traffic demand-diversion at on-ramp areas in the I-35W freeway corridor in Minneapolis, Minnesota.

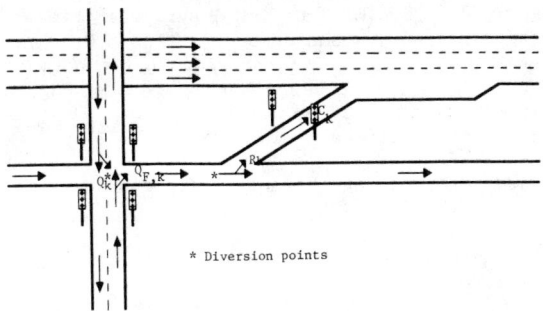

Fig.1 A typical on-ramp area

MODEL FORMULATION AND PREDICTION PROCEDURE

The predictor developed in this research combines behavioral modeling and filtering. In particular, the Extended Kalman Filter (Jazwinski, 1970) updates the parameters of models that reflect driver diversion behavior. The behavioral characteristics of commuter route choice, that represent the basis on which these models are developed, were earlier identified from an extensive questionnaire survey conducted by the research team including the author (Stephanedes, Kwon and Michalopoulos, 1989.). The earlier route-choice findings indicate that freeway demand can be treated as a decision making entity which makes diverting decisions based on maximization of trip utility determined by traffic conditions. Further, the diverting decision is reflected in the proportion of freeway users in the total traffic flow approaching the ramp area. Based on these findings, it is assumed that the total flow, Q_k, crossing the upstream intersection stopline during green time, responds to the ramp condition as reflected by the ramp-approaching proportion and the ramp-entering proportion of Q_k. Two logit models are formulated to predict the freeway ramp-approaching proportion, $P_{F,k}$, and the ramp-entering proportion, $P_{R,k}$, of Q_k for every time interval k. The unknown logit parameters, Θ, are considered as state variables following the random walk process (Young and Jakeman, 1984) and estimated in real time by applying the Extended Kalman Filter with the most recent demand-diversion information. Further, historical and current flow data are employed to predict Q_k, which is also a function of intersection green time. The resulting models and prediction procedure are as follows:

$$Q_{j,k} = \Theta_{j,k} \{E[QG_j]_k + E[QG_j]_{k-1} + QG_{j,k-1}\} G_{j,k} \quad (1)$$
$$P_{F,k} = P_{o,k}/\{1+\exp[\Theta_{4,k}+\Theta_{5,k}(x_k/C_k)]\} \quad (2)$$
$$P_{R,k} = 1/\{1+\exp[\Theta_{1,k}+\Theta_{2,k}C_k+\Theta_{3,k}(x_{k-1}+R_{k-1})]\} \quad (3)$$

where, for each time interval k,
$Q_{j,k}$ = the number of vehicles crossing stopline j of a given intersection. $G_{j,k}$ = the green time for approach j. $E[QG_j]_k$ = the past average flow crossing stopline j per green time.
$P_{o,k}$ = (original freewaybound demand)$_k$ / Q_k, $Q_k = \sum_j Q_{j,k}$.
x_k = the number of vehicles on the ramp in the beginning of the kth time interval. R_k = the number of vehicles entering the ramp during the kth time interval. C_k = metering rate. $\Theta_{i,k}$ = the unknown parameters to be identified in real time.

The procedure for updating the model parameters, $\Theta_{i,k}$, is summarized below.

1) Initialize algorithm (k=0) with any prior knowledge of model parameters,
$\hat{\Theta}_{k/k} = \Theta_0$,
$\Sigma_{k/k} = \Sigma_0$,
where $\Sigma_{k/k} = E[(\Theta_k-\Theta_{k/k})(\Theta_k-\Theta_{k/k})^T]$.
2) Set the model parameters $\Theta_{k+1/k} = \hat{\Theta}_{k/k}$.
3) Predict flow per intersection approach $Q_{j,k+1}$, freeway ramp-approaching proportion $P_{F,k+1}$, and ramp-entering proportion $P_{R,k+1}$ using prediction models (1), (2) and (3) with the parameters $\Theta_{k+1/k}$.
4) Measure actual $Q_{j,k+1}$, $P_{F,k+1}$, $P_{R,k+1}$ and obtain prediction error vector e_{k+1}, which is defined as
e_k = [measured value] - [predicted value]
5) Update model parameters $\hat{\Theta}_{k+1/k}$ using the gain and the prediction error,
$\hat{\Theta}_{k+1/k+1} = \hat{\Theta}_{k+1/k} + K_{k+1} e_k$
where
$K_{k+1} = \Sigma_{k+1/k} S_{k+1} [S_{k+1}^T \Sigma_{k+1/k} S_{k+1} + s_k]^{-1}$: gain vector,
$\Sigma_{k+1/k} = \Sigma_k + q_k$, the covariance matrix,
$S_{k+1} = [\partial Q/\partial\Theta, \partial P_F/\partial\Theta, \partial P_R/\partial\Theta]^T \big|_{\Theta=\hat{\Theta}_{k+1/k}}$, the sensitivity vector,
$\Sigma_{k+1/k+1} = (I-K_{k+1}S_{k+1}) \Sigma_{k+1/k}$, the updated covariance matrix, defined by the covariance matrices

$E[w_k w_j^T] = q_k \delta_{kj}$ $E[v_k v_j^T] = s_k \delta_{kj}$
and w,v are zero-mean Gaussian white noise sequences,

TESTING AND VALIDATION

The on-line predictor developed in the previous section is tested and validated with real traffic data collected from the I-35W freeway corridor in Minneapolis, Minnesota. The I-35W corridor, which crosses the Twin Cities in a north-south direction, often experiences severe congestion that extends through the off-peak period. Three on-ramp areas of the corridor were selected for testing the new predictor. The selected areas, illustrated in Figure 2, represent major geometric-control configurations of freeway corridors; each ramp area includes a frontage road that can accommodate traffic diverting from the ramp. For each on-ramp area, all the traffic movements including vehicles diverting to the frontage road from the ramp were measured manually for every 4 cycles (4-6 min.) during the morning rush hour over a period of 2 to 14 weekdays, depending on the location, from Oct. 1988 to Sept. 1989.

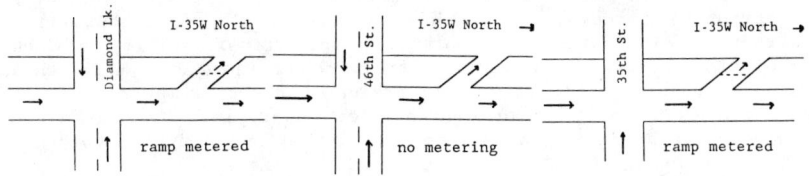

Fig.2 Configuration of selected on-ramp areas

Using the collected data, each component model of the on-line predictor is tested. Figure 3 illustrates the prediction results of one week day at the Diamond Lk. St. intersection. Two error indices are calculated to evaluate the performance of the predictor, for N predictions:

Mean one-step prediction error (MOPE, %):
$$100/N * \sum_k |[\text{Measured}]_k - [\text{Predicted}]_k|/[\text{Measured}]_k$$
Mean Absolute Error (MAE):
$$1/N * \sum_k |[\text{Measured}]_k - [\text{Predicted}]_k|$$

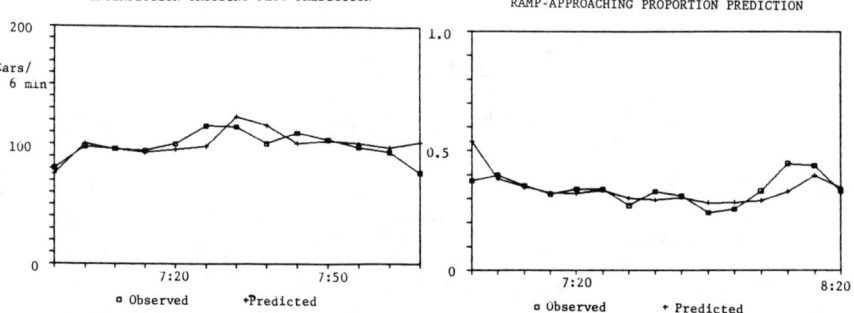

Fig.3 Prediction results at Diamond Lk. St. intersection (3/21/89)

The test results from three on-ramp areas indicate that the total flow, Q_k, prediction has the MAE ranging from 7.1 to 20.9 veh/4-6 min. and a MOPE range of 6.1 to 12.9%. Further, the prediction of $P_{F,k}$ resulted in the MAE range of 0.03 to 0.07 and the MOPE ranging from 4.1 to 19%. Finally, the prediction of $P_{R,k}$ has a MAE range of 0.04 to 0.06 and the MOPE between 5.0 and 8.8%.

CONCLUSIONS

An adaptive prediction method was developed for demand-diversion at on-ramp areas by combining behavioral modeling and filtering. The new method treats the on-ramp area as one system consisting of the freeway entrance ramp and the upstream intersection. Three dynamic predictors are developed for predicting, at 4-6 min. intervals, the flow crossing the intersection, the proportion of flow that approaches the freeway, and the proportion of freeway bound flow entering the on-ramp. Prediction models are tested with data sets different from those with which they were developed without error degradation indicating the potential for transferability of the prediction algorithm along the freeway corridor. While this research focused on the demand-diversion at the on-ramp area, it is expected that the behavioral principles underlying the models and the adaptive prediction algorithm should also be applicable to other diversion points in the corridor, such as freeway exit ramps and major arterial intersections. Such extensions of this research would include, modeling the effect of mainline freeway congestion on exit volume of the nearby off-ramp, and, further, modeling the effect of local congestion on turning movements at major arterial intersections.

Acknowledgements - This study was supported by the National Science Foundation through project No. NSF/CES-8713277. The Minnesota Supercomputer Institute provided partial support through project Nos. mg26701 and mg29901. The Center for Transportation Studies, Department of Civil and Mineral Engineering, University of Minnesota is acknowledged for its support.

REFERENCES

Jazwinski, A.H. (1970) Stochastic process and filtering theory. Academic Press, New York.

Stephanedes, Y.J., Kwon, E. and Michalopoulos, P.G. (1989) Demand diversion for vehicle guidance, simulation and control in freeway corridors. Transpn Res. Rec. No. 1220, 12-20.

Young, P.C. and Jakeman, A.J. (1984) Recursive filtering and smoothing procedures for the inversion of ill-posed causal problems. Utilitas Mathematica, Vol. 25, 351-376.

DYNAMIC STOCHASTIC EQUILIBRIUM MODEL IN MULTIPLE VEHICLE TYPE TRANSPORTATION NETWORKS

Shogo Kawakami
Civil Engineering Department, Nagoya University, Nagoya 464, Japan
Zhimin Xu
Department of Environment and Urban Planning,
Construction Technology Institute, Tokyo, Japan

Abstract
In this study, a multi-vehicle type dynamic stochastic user equilibrium model (called MDSUE model for abbreviation) is developed. By using MDSUE model, the prediction of departure rates as well as time dependent link flows of different vehicle types becomes possible.

1 Introduction

MDSUE model is considered to be an extension of Wardrop's first principle and is defined as the state at which no traveler in the same user class believes that he can increase his total utility of travel by unilaterally changing route or departure time(but different user classes can have different maximum travel utilities between the same O-D pair). The work environment of MDSUE model is shown in Fig.1.

In the formulation of the model, we also assume that the multiple vehicle types transportation networks are composed of M single vehicle type network copies with the same size(M is the number of vehicle types). From now, we will use the mth user class as a typical one in formulating the model for simplification. The interactions between the different user classes will be represented by the link performance functions with multivariables.

The network for the mth class user is represented by a directed graph that includes a set of consecutively numbered node N^m , and a set of consecutively numbered link L^m . Each O-D pair r-s is connected by a set of path, k^m, denoted by K_{rs}^m . A path $k^m \in K_{rs}^m$ will be defined as an ordered set, denoted by $L_{k,rs}^m$. An ordered set, in the context of directed network, is defined here as a set that includes links which constitute an ordered chain. Each element i of n ordered set $L_{k,rs}^m = \{i_1, i_2, ..., i_g\}$ is associated with a variable $O_k(i)$, which defines the order of i within the ordered set, thus in $L_{k,rs}^m$, $O_k(i_1) = 1$, $O_k(i_2) = 2$,...,$O_k(i_g) = g$, i_1 is connected to the origin r, and i_g is connected to the destination s.

A typical term of the O-D matrix for m user class is denoted by Q_{rs}^m which defines the total number of trips of user class m between each O-D pair $r - s$ of the network.

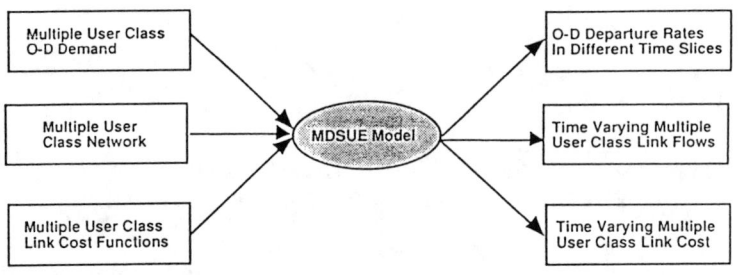

Figure 1: **The Work Environment of MDSUE Model**

Each individual user traveling between origin r and destination s has to arrive at his destination at his official work start time. But the official work start times for different user class are assumed to be different. For the user of m class, it is assumed, that he first decides on what time $t \in [T_0, T_0 + T]$ to depart (T_0 denotes the earliest and $T_0 + T$ denotes the latest possible departure time), and then which route $k^m \in K_{rs}^m$ to follows. Since the number of possible paths which connect an origin to a destination is generally very large, travelers are assumed to be able to compare only a small number of alternative routes, which they estimate as being reasonable option[1]. Next section will describe the mathematical formulation of MDSUE model. Section 3 presents the solution algorithm. Section 4 concludes the paper.

2 The Mathematical Formulation of MDSUE Model

2.1 The Link Travel Cost Model

A number of different link performance functions have been used in DSUE model. In this section, the following functions which are derived from BPR type function will be employed:

$$tt_i^m(t) = tt_{0i}^m(t)[1.0 + \alpha_m(\frac{\eta X_i(t)}{C_i^m})^{\beta_m}] \tag{1}$$

where $tt_{0i}^m(t)$ is the time needed by m class vehicle to traverse link i in free flow condition at time t, and $X_i(t) = (x_i^1(t), x_i^2(t), \cdots, x_i^M(t))$ is the average multiple vehicle type link flow vector which in fact are running the same link i. C_i^m is the capacity of link i when it is only occupied by m class vehicle. $\eta = (\eta^1, \cdots, \eta^M)$ is the coefficient vector representing the interaction between different vehicle classes. The meaning of other parameters are the same with static model.

where, μ_r^m, μ_t^m are the scale parameters associated with the departure time and the route choice decision, respectively, (μ_t, must be greater or equal to μ_r in order the above formulation to be valid),

$$V_{k,rs}^{*,m}(t) = \frac{1}{\mu_r^m} ln \sum_{k \in K_{rs}^m(t)} e^{\mu_r^m V_{k,rs}^m(t)} \tag{10}$$

and expresses the expected maximum utility from the choice of among the alternative feasible routes at time t.

3 The Equivalent Mathematical Program Formulation and Solution Algorithm

3.1 The Program Mathematical Formulation of MDSUE Model

In the special case that there are two types of vehicles in the network(for example truck and passenger car) but the interaction between the two types is symmetric, that is to say $tt_i^a(t) = tt_i^a(x_i^a, x_i^b)$, $\forall a,b \in M$ and $\partial tt_i^a(t)/\partial x_i^b(t) = \partial tt_i^b(t)/\partial x_i^a(t)$. An equivalent mathematical program as follows can also be suggested.

$$\begin{aligned} minZ(X) = & -\sum_{rs} S_{rs}[-V_{rs}(t)] + \sum_{rs}\sum_{t}\sum_{k} Q_{k,rs}(t) SD_{k,rs}(t) \\ & + \varpi \sum_{t}\sum_{i} x_a(t) tt_i(X_i(t)) \\ & - \frac{1}{2}\varpi \sum_{t}\sum_{i}[\int_0^{x_i^a(t)} tt_i^a(u, x_i^b) du + \int_0^{x_i^a(t)} tt_i^a(u, 0) du] \end{aligned} \tag{11}$$

where, $S_{rs}[-V_{rs}(t)]$ is the so-called satisfactory function. The equivalence of the above program to the MDSUE conditions and the uniqueness of the solution can be proved easily.

3.2 The Diagonalization Method for MDSUE Model

In the general cases, the link performance functions are multivariable functions, and are generally assumed to be asymmetric. Like the static MUE model, there is no equivalent mathematical program available for this kind problems. The solution algorithm adopted in this paper is a diagonalized method of the successive averages(MSA) which was proposed by Sheffi and Powell.

In the diagonalized MSA, the solution is found by using an iterative procedure, where in each iteration, a dynamic stochastic network loading is performed in order to find the current solution.

A dynamic stochastic network loading(DSNL) mechanism is defined as the process of assigning(in time and space) a set of O-D trip rates to a transportation network in which link travel times are time varying but fixed in the sense that they are not flow dependent. Thus a form of a DSNL mechanism can be defined from the modeling procedures developed in the previous section, which are used in the following sequence:

Let $LT_{j,k}^m(t)$ denote the time that a driver, who departed at time t and used route $k \in K_{rs}^m(t)$, will depart from link j, $LT_{j,k}^m(t)$ is given by the following recursive formula:

$$LT_{j,k}^m(t) = LT_{i,k}^m(t) + tt_j^m(LT_{i,k}^m(t)) \quad \forall i,j \in L_{k,rs}^m, and O_k(i) = O_k(j) - 1$$
$$= t + tt_j^m(t) \quad \forall t \in [T_0, T_0 + T], j \in L_{k,rs}^m, and O_k(j) = 1 \quad (2)$$

The arrival time at the destination, for a driver who travel from r to s, follows route k and departs at time t is denoted by $TD_{k,rs}^m(t)$, and can be defined as:

$$TD_{k,rs}^m(t) = LT_{g,k}^m(t) + tt_n^m(LT_{g,k}^m(t)) \quad (3)$$

where n is the link along route $k \in K_{rs}^m(t)$, which is connected to the destination, i.e., n is the last element of the ordered set $L_{k,rs}^m$, $g \in L_{k,rs}^m(t)$ and $O_k(g) = O_k(n) - 1$.

The O-D travel time from r to s for a driver departing at time t and following path k, denoted by $TT_{k,rs}^m(t)$, can be computed as:

$$TT_{k,rs}^m(t) = TD_{k,rs}^m(t) - t \quad (4)$$

2.2 The Mathematical Formulation of the MDSUE Conditions

Given the set of multiple vehicle O-D rates, Q_{rs}^m, the multiple vehicle dynamic stochastic user equilibrium conditions are characterized by the following Equations:

$$Q_{k,rs}^m(t) = Q_{rs}^m P_{k,rs}^m(t), \quad \forall rs, t, m, k \in K_{rs}^m(t) \quad (5)$$

where $P_{k,rs}^m(t)$ is the probability that a route k and a departure time t is selected by m class user given the set of the measured utilities $V_{k,rs}^m(t)$.

The utility function can be expressed as

$$V_{k,rs}^m(t) = -\varpi_m TT_{k,rs}^m(t) - SD_{k,rs}^m(t) \quad \forall rs, m, t, k \in K_{rs}^m(t) \quad (6)$$

where, ϖ_m is the time valuation of m class user, $SD_{k,rs}^m(t)$ is the measured disutility of m class vehicle due to schedule delay associated with a departure time t and route k, it is assumed to be continuous.

An alternative expression of $P_{k,rs}^m(t)$ is given by:

$$P_{k,rs}^m(t) = Pr(\boldsymbol{V}_{k,rs}^m(t) > \boldsymbol{V}_{l,rs}^m(t'), \quad \forall (t',l) \neq (t,k),$$
$$k \in K_{rs}^m(t), l \in K_{rs}^m(t')) \quad (7)$$

where $\boldsymbol{V}_{k,rs}^m(t)$ is a random variable representing the perceived total utility of travel associated with route k and departure time t, and $E[\boldsymbol{V}_{k,rs}^m(t)] = V_{k,rs}^m(t)$.

$$\boldsymbol{V}_{k,rs}^m(t) = V_{k,rs}^m(t) + \xi_{k,rs}^m(t) \quad \forall rs, t, m, k \in K_{rs}^m(t) \quad (8)$$

where $\xi_{k,rs}^m(t)$ is the random term of $\boldsymbol{V}_{k,rs}^m(t)$, and $E[\xi_{k,rs}^m(t)] = 0$. By assuming that $\xi_{k,rs}^m(t)$ is i.i.d Gumbel distributed, the departure rate using path k can be given by the following logit model[2]:

$$P_{k,rs}^m(t) = P_{rs}^m(t) \cdot P_k^m(t) |_{k \in K_{rs}^m(t)}$$
$$= \frac{e^{\mu_r^m V_{k,rs}^m(t)}}{\sum_{l \in K_{rs}^m(t)} e^{\mu_r V_{l,rs}^m(t)}} \cdot \frac{e^{\mu_t^m V_{k,rs}^{*,m}(t)}}{\sum_{u=T_0^m}^{T_0^m+T^m} e^{\mu_t V_{k,rs}^{*,m}(u)}} \quad (9)$$

1) Given the time varying link travel times $tt_i^m(t)$, determine the set of reasonable paths $K_{rs}^m(t)$, $\forall t, rs, m$(Dial method).
2) Calculate the O-D travel times $TT_{k,rs}^m(t)$.(Equation 4)
3) Calculate the utility function $V_{k,rs}^m(t)$(Equqtion 6), and then the time dependent departure rates $Q_{k,rs}^m(t)$(Equation (9) (5)).
4) Calculate the time dependent link flow patterns X_i.

Having defined the DSNL algorithm, the diagonalized MSA for MDSUE model can be expressed as follows:

Step A: Initialization. Perform a dynamic stochastic network loading (DSNL) based on the set of the initial free flow travel times $\{tt_i^{r_i,0}\}$. This generates a set of time dependent link flows $\{x_i^{m,1}\}$, $\forall m, i \in L, t \in [T_0, T_0 + T]$. Set $n = 1$.

Step B: Diagonalization. Fix the link flows of $M - 1$ classes but the a class.

$$tt_i(t)^{m,n} = tt_i(x_i(t)^{1,n}, \cdots, x_i(t)^{a-1,n}, x_i(t)^a, x_i(t)^{a+1,n}, \cdots, x_i(t)^{M,n})$$

solve the dynamic stochastic equilibrium with MSA method.

Step B.1: Update. Set $tt_i(t)^{m,n} = tt_i(X^{m,n}(t))$ $\forall m, i \in L, t \in [T_0, T_0 + T]$.

Step B.2: Direction finding. Perform a dynamic stochastic network loading (DSNL) procedure based on the current set of the time dependent link travel times $\{tt_i(t)^{m,n}\}$. This yields the auxiliary time dependent link flow pattern $\{y_i(t)^{m,n}\}$.

Step B.3: Move. Find the new time dependent flow pattern by setting

$$X^{m,n+1} = X^{m,n} + \lambda^{m,n} \cdot (Y^{m,n} - X^{m,n}) \quad t \in [T_0, T_0 + T]$$

Like in static MSA method, $\lambda^{m,n}$ is set to be $1/n$.

Step C: Convergence check.

4 Summary

This study proposed a dynamic multiple user class stochastic user equilibrium model by combining the existing dynamic SUE model and multiple user class UE model. A mathematical formulation of DMSUE was given at first. And then a equivalent Beckmann type mathematical program solution algorithm was suggested for the symmetric cross-class interaction cases, a diagonalization type solution algorithm was suggested for asymmetric cross-class interaction cases. The numerical simulation results and the discussion of solution properties will be presented on the conference.

References

[1] Dial, R.B.(1971). "A Probabilistic Multipath Traffic Assignment Which Obviates Paths Enumeration." Transpn. Res., Vol.5,No.2.

[2] Vythoulkas P.C.(1990) "Two Models for Predicting Dynamic Stochastic Equilibria in Urban Transportation Networks", Proceedings of the Eleventh International Symposium on Transportation and Traffic Theory.pp253-272.

Dynamic Estimation of Freeway Demand Patterns and a Stochastic Programming Approach to Freeway Ramp Metering

Gary A. Davis[1], Associate Member, ASCE

Abstract

Controlling the rate at which vehicles are allowed to enter a freeway from the freeway's on-ramps has become a standard method for reducing the impact of both recurring and nonrecurring traffic congestion. However, existing metering algorithms tend to treat the traffic flow as a deterministic process and to treat the demand for travel on the freeway as fixed and known. In practice however, the demand for travel on a freeway is more accurately modelled as the realization of a stochastic process and there will be uncertainty in the knowledge of the parameters governing this process. This paper concerns two issues: (1) How can data routinely collected by freeway surveillance and control systems be used to estimate the volume and distribution of travellers' demand for freeway travel, and particularly how can the uncertainty concerning this demand be quantified? (2) Given this uncertainty, how ought optimal ramp-metering rates be computed? The first issue is treated as a problem in online systems identification, and both recursive and nonrecursive approaches to the estimation of freeway demand are treated. The second issue is treated as a problem in stochastic programming, in which the demand parameters are treated as random variables. Particular attention is given to dual control issues, in which variable metering rates might be used to reduce uncertaintly concerning the demand parameters.

---
[1] Assistant Professor, Dept. of Civil and Mineral Engineering, University of Minnesota, 500 Pillsbury Dr. SE, Minneapolis, MN   55455

EFFECTIVENESS OF REAL-TIME INFORMATION STRATEGIES IN SITUATIONS OF NON-RECURRENT CONGESTION
Srinivas Peeta, Hani S. Mahmassani, Richard Rothery and Robert Herman[†]

Abstract

The effect of real-time information on the performance of a traffic commuting corridor in situations of non-recurrent congestion is studied using a dynamic simulation model. Non-recurrent congestion refers to situations resulting from a temporary reduction in the capacity of a section of a highway due to accidents, repairwork or construction. Simulation experiments are performed using four experimental factors- location of the incident, duration of the incident, market penetration(fraction of users with access to real-time information), and user behavioral rules- to study the overall system performance. The results illustrate the interactions between the various experimental factors and provide insights into possible control and diversion strategies.

1. Introduction

Advances in communication technology applied to the highway system provide new opportunities to alleviate both recurrent and non-recurrent congested traffic conditions. Of interest in the present study are the IVHS technologies known as Advanced Driver Information Systems (ADIS). These systems will provide vehicles in the network the ability to communicate on a real-time basis with a central controller. This provides vehicles with information, on a real-time basis, about existing traffic conditions, and/or instructions on route selection from their current location to their respective destinations.

Situations where ADIS technologies have potentially large benefits include non-recurrent congestion associated with traffic incidents and disruptions, especially during peak hours. According to a Seattle Area study (Mannering et al., 1989), an accident on the ship canal bridge (on Interstate-5) during the evening peak results in the loss of 1123 vehicle-hours in the first 10 minutes after the accident. By providing information (about the incident) to vehicles upstream of the incident, ADIS causes possible diversion of some vehicles to alternative facilities, thereby reducing motorist frustration and delay. However, the effectiveness of this information may depend critically on several factors, including the characteristics of the incident, prevailing conditions in the network, the type of information provided, the fraction of users who receive this information, and the manner in which users respond to the information. These factors are explored in this paper using computer simulation experiments.

2. Modelling Framework:

The modelling framework is based on that of Mahmassani and Jayakrishnan (1989), modified to better represent incident situations. It consists of three principal components: a traffic simulator, user decisions component and a network path processor.

2.1 Commuting Context and General Assumptions

The traffic corridor, shown in Fig. 1, consists of three major parallel highways, and is used by people residing near them to commute to work to a common destination downstream (CBD). Crossover links connect the highways and allow route switching. Traffic is generated along the first six sectors of all the highways according to a time-dependent cumulative departure function.

A total of 9600 commuters, split equally among the first six sectors, share the facilities in the corridor during the morning peak period. Sector 1 is the most distant from the CBD. Commuters in each sector split equally among the three facilities, and depart uniformly over a 20 minute period. Sector 1 starts loading first. There is a five minute lag between the start of loading of adjacent sectors. No loading is generated on the crossover links.

Department of Civil Engineering, The University of Texas at Austin. Austin, TX.

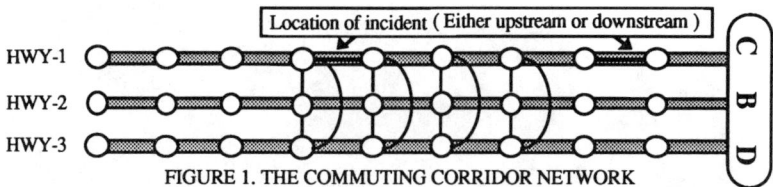

FIGURE 1. THE COMMUTING CORRIDOR NETWORK

Some users have receivers to obtain information on prevailing traffic conditions on all links of the network. This information allows each user to calculate the best (shortest) path from a current position to the destination. In the commuting context, the current path refers to staying on the present facility. An alternate path refers to switching to another highway using a crossover link, followed by the remaining portion of that highway to the destination.

2.2 User Behavior Rule

Experimental evidence by Mahmassani and Stephan (1988) suggests that commuter route choice behavior exhibits a boundedly-rational character. That is, drivers hesitate to shift unless they have some minimum time savings. If $TTC_j(k)$ represents travel time on the current path from node k to driver j's destination, and $TTB_j(k)$ the travel time along the best path, the satisfying behavioral rule can be stated mathematically as:

$$\delta_j(k) = \begin{cases} 1, & \text{if } TTC_j(k) - TTB_j(k) > \max(\eta_j \cdot TTC_j(k), \tau_j) \\ 0, & \text{otherwise} \end{cases} \quad (1)$$

where $\delta_j(k)$ is a binary variable equal to 1 if user j switches from the current path to the best alternate path, and 0 if the current path is maintained; η_j is the threshold level for user j, as a fraction of remaining trip time on the current path, with $\eta_j \geq 0$, $\forall j$; and τ_j is the absolute minimum savings in travel time below which user j will not switch routes. The relative indifference band η_j is assumed to be a random variable across the user population; the minimum threshold level τ_j is assumed to be identical for all users, for convenience.

2.3 The Simulator

The simulator is a slightly modified version of the code by Mahmassani and Jayakrishnan(1989), which is an extension of a special-purpose macro particle traffic simulator (MPSM) (Chang et al., 1985). The traffic simulator handles the movement of vehicles on the links given a time-dependent loading pattern (as described in the commuting context), as well as the movement on the crossover links. The switching from one highway to another via the crossover links requires instructions provided by the user behavior component.

The traffic simulator makes use of the conservation equation and the speed-concentration relationship to model the flow. The MPSM logic, used in plasma physics, moves vehicles in bunches, or macroparticles, consistent with the conservation equation and a speed-concentration relationship (a simple modified Greenshields model). Presently, since in-vehicle information strategies are being used, the user behavior component for route switching is applied individually to vehicles, thereby moving them with macroparticle of size one. The traffic simulation is still macroscopic as average speed-density relationship is used to model interactions in the traffic stream. Microscopic details like vehicle lengths, car following or lane changing are not considered.

The network path processor is specific to the particular structure of the commuting corridor and crossover links.The traffic simulation follows a deterministic fixed time step approach. Vehicles are moved each time step at the prevailing local speed on the link or they are switched to another link, based on the user behavior rules. The concentrations are updated and the corresponding average speeds are calculated. Queues develop when capacity is insufficient.

3. The Simulation Experiments

3.1 Corridor Network

The commuting corridor is the one shown in Fig. 1. Each of the three highways is divided into 9 equal sectors of one mile length. Crossover links from each highway to the others are provided at the third, fourth, fifth, and sixth miles from the upstream end. Equal access time to the three highways is assumed, though queuing at entry points is possible. Highway 1 is the fastest with a 55 mph free mean speed, followed by highway 2 with 45 mph and highway 3 with 35 mph. The

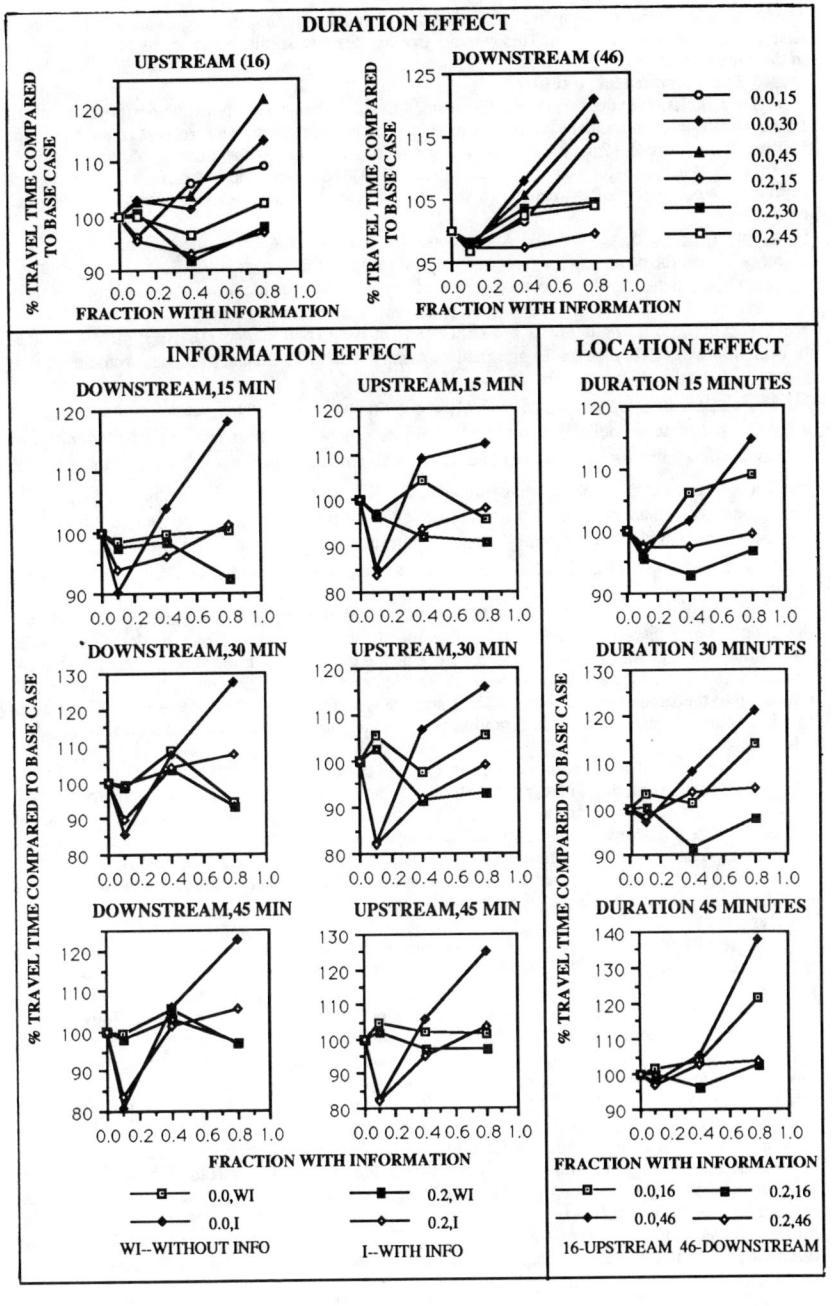

crossover links have a 45 mph free mean speed. All links including the crossovers are assumed to have two lanes in each direction. This network exhibits features similar to an actual corridor network in the Austin area.

3.2 Experimental Factors

The simulation experiments were performed to study the sensitivity of the performance of the traffic system with an incident under real-time in-vehicle information, with respect to the following factors: 1) the location of the incident on the commuting corridor, 2) the duration of the incident, 3) the fraction of drivers with access to information; and 4) the mean relative indifference band, which captures the propensity of the drivers to switch when they have information. These are described below.

Location of the incident: For this study, two locations are considered, one upstream (sector 4 of highway 1), and the other downstream (sector 8 of highway 1), as shown in Fig. 1.

Duration of the incident: To reflect the degree of severity of the incident, we consider three levels of the duration: 15 minutes (short), 30 minutes (medium), and 45 minutes (long).

Market penetration: Four levels are considered in the experiments: 0.0, 10%, 40% and 80%. Information availability status is assigned randomly and independently to each vehicle as it is generated, according to the specified fraction.

Mean Relative Indifference Band: The quantity η_j in Eq. 1 governs users' response to given information and their inclination to switch. When generated, a user is assigned randomly and independently a value for η_j. For convenience, η_j is assumed to follow a triangular distribution, with mean η and range of $\eta/2$. Here, we assume two levels of η: 0.0 and 0.2. In the no band ($\eta = 0.0$) case, all users are assumed to always switch to an alternate path offering improvement in travel time, no matter how small its magnitude. The minimum improvement τ_j (in Eq. 1) is assumed to be identical across all users, and equal to one minute, except in the zero band case where there is no such minimum.

The simulation is performed for 120 minutes. Loading starts in sector 1 sixteen minutes into the simulation. Subsequent loading of the sectors follows the loading pattern described in the commuting context. An incident is modeled by reducing the capacity of the sector where the incident occurs by half, for the duration of the incident. Incidents are assumed only on highway 1 and start 25 minutes into the simulation. Thirty-six simulation runs were conducted for cases with information. In addition, the 6 cases with no information (i.e., no switching) were simulated so as to provide a reference for comparison.

3.3 Queuing time prediction and Queue dissipation

An incident, as modeled here, represents an abrupt decrease in the capacity of the highway at the beginning (and an abrupt increase at the end) of the incident. Under such conditions, there can be abrupt increase or decrease in the queue size over very short intervals of time. In providing information to users it is important to add an estimate of the queuing time to the link trip times.

For the study, queues are predicted for the current time step by averaging the queue lengths over the past 30 time steps and using an assumed service rate.

4. Results

The results of the simulation runs are summarized in terms of the average travel times, shown in the various plots in Fig. 2. The plots present the total (or average) trip time in the system as a percent of the trip time in the no-information base case, for the various cases considered.

Switching activity increases when a greater number of people have information. However, this switching activity is tempered by the presence of an indifference band which decreases the propensity of the driver to switch. The average travel time is less when the incident occurs upstream (in our case, link 16 of highway 1) compared to downstream (in our case, link 46 of highway 1). When it occurs downstream, a greater number of segments on that highway are congested compared to the upstream situation, and the opportunities to switch to other highways are fewer as they are also filled to near capacity. On the other hand, an incident upstream reduces the flow of vehicles going downstream beyond the bottleneck point, allowing vehicles from other highways to shift to highway 1 (as this is the most efficient highway) thereby increasing the capability of the other highways to carry vehicles. From the plots, it can be observed that the system performs best when the fraction of users with information is somewhere between 0.1 and 0.4. Note that when a large fraction of users has information, the latter could be counterproductive for the system, especially under myopic driver behavior (zero indifference band).

When the incident occurs upstream, the system performs best with in-vehicle information when the incident is of medium duration. When it is short, the downstream vehicles on highway 1 enjoy the advantage of lower concentrations in those segments (and hence higher speeds) for that duration only. Only a limited number of vehicles enter highway 1 during this duration from other highways. Once the bottleneck is removed, highway 1 quickly reaches jam concentration as vehicles upstream of the bottleneck that were previously queued seize the opportunity of an increase in capacity (after the incident is over). On the other hand, when an incident of medium duration occurs, a larger number of vehicles downstream of the incident have the opportunity to make use of highway 1 (it being the most efficient). When the incident is over, jam concentrations are reached on highway 1, but at a much later time than in the case of a short duration incident. When an incident is of long duration, the advantage of lesser concentrations downstream of the incident is offset by the fact that generation of vehicles into the system stops sometime before the end of the incident. Hence, a limited number of vehicles enter highway 1 as allowed by the rate of entry from the crossover links. Also, the vehicles upstream of the incident are queued for a longer time. Hence, the system performs worse in this case than in the medium duration case. When the incident occurs downstream, the anticipated behavior is exhibited. No benefits are accrued downstream of the incident because there is no generation in the corresponding segments. The longer the incident, the greater the delay to vehicles. There are more switches from highway 1 when the incident occurs downstream, reflecting worsening conditions on highway 1.

The plots with legends labeled I and WI illustrate the incidence of benefits and costs on the various users: those with and those without information. Users with information attain the most savings when around 10% of the users have information. The comparison is made with respect to the 'base case', that is, the situation when no one has information about the traffic conditions. Also, as the fraction of users with information increases, those with information progressively perform worse after the 'optimal' savings at the 10 to 20% level, and may do worse than in the no information situation. Under the zero band situation, when users switch too readily, the situation worsens for all levels of market penetration. Users without information accrue benefits as their fraction increases compared to those users with information. The downstream plots (information effect) show that for a zero band situation and large market penetration, users with information could do far worse than in the no-information situation. These results illustrate that the relative usefulness of information to those who have access to it depends not only on market penetration, but also on user behavior.

5. Discussion and Concluding Comments

The simulation experiments provide insights into the effect of in-vehicle real time information on overall traffic system performance in the presence of non-recurrent congestion. The modelling framework allows the identification of the key underlying parameters and the exploration of their effect. The results highlight the complex interactions between the various incident parameters, real-time information strategies, and the behavioral models in determining the system performance. Some results are intuitive and some are counter-intuitive. The misconception that information will automatically lead to improvements in traffic conditions is illustrated. The importance of the location and duration of the incident in the determination of system performance is highlighted. The manner in which the users respond to information is an important factor. The results indicate that myopic switching by individual drivers can result in worse outcomes for them as well as the system.

With many parameters coming into play, control strategies and diversion strategies can help the system perform better in the presence of incidents. Such strategies involve addressing such issues as - how much information to give, whom to give, and when to give it.

References:
1. Mahmassani, H. S. and Jayakrishnan, R. (1989), "System Performance & User Response Under Real-Time Info. in a Congested Traffic Corridor", pres. at Seminar on Urban Traffic Networks, Capri, Italy.
2. Chang, G.-L., Mahmassani, H.S. & Herman, R. (1985), "A Macroparticle Traffic Simulation Model to Investigate Peak-Period Commuter Decision Dynamics", *TRR* 1005, 107-120.
3. Mahmassani, H.S. & Chang, G.-L. (1985), "Dynamic Aspects of Departure Time Choice Behavior in a Commuting System: Theoretical Framework & Experimental Analysis", *TRR* 1037, 88-101.
4. Mahmassani, H.S. & Jayakrishnan, R. (1988), "Dynamic Analysis of Lane Closure Strategies", *Jrnl. of Trans. Engn.*, Vol. 114, No. 4.
5. Mahmassani, H.S. & Stephan, D.G. (1988), "Experimental Investigations of Route & Departure Time Dynamics of Urban Commuters", *TRR*.
6. Mannering, F., Jones, B., Sebranke, B. (1989), "Generation & Assessment of Incident Mgmt. Strategies Vol 4: Seattle Area Incident Mgmt: Assessment & Recommendations". Technical Report.

Contact-free IC Cards for New Railway Ticket Systems

Shigeo Miki[1], Noboru Nagai[2], Hiroshi Matsubara[2]

Hideaki Yoshino[2], Koichi Goto[2]

Abstract

Two types of smart cards which can communicate with an automatic gate machine without contact are developed. The first one uses medium wave radio as communication media. The microprocessor in the card shares the logic with the gate program whether it can go through a gate, thus giving an open ended ticket without changing the program of gate machine. Another type uses microwave for communication. The card is suitable for an urban area season ticket where the large number of commuters go through the gate every day.

The contact-freeness brings the new relation between passengers and railway company. Thus a new ticket system named *customer-based* system could be constructed, whereas the conventional system may be called *trip-based* system.

Introduction

Recently, railway systems adopting automatic fare collection (AFC) system are increasing. The trend covers the old railways like New York CTA [Leas 90] as well as newly constructed railways like Hong Kong, Singapore [Rich 90] or Madrid. In Japan, major private and municipal railways except Tokyo area adopt AFC for some ten years. JR East Railway Company, the largest one of privatised and divided companies from old Japanese National Railways, began to introduce AFC system to the Tokyo Metropolitan Area and other railway companies decided to follow it.

All of these systems use magnetically strip or coated tickets for information storage media. But the system is not so comfortable for passengers, especially for the

---

[1]Chief, Miki Laboratory, Railway Technical Research Institute, Tokyo, Japan 185
[2]Same

users of season tickets who dominate the passengers at urban areas in Japan. They must pick their ticket out of the case, insert it to the slit of the gate, pick it up from the ejector, and back it to the case again.
Recently, IC(integrated circuit) or smart cards gradually widen their application area. In transportation, a few companies begin to use or test them as bus tickets[Slev 90] and others[Sutt 90]. However, in spite of their inherent superiority over the magnetic cards, the cost bars their prevails. Moreover, the speed is not enough for railway gate. Particularly in Japan, where eighty passengers go though the manned gate every minute at morning rush hours, it is vital for practical use.
If the ticket can communicate with the automatic gate speedy and without contact, the above inconvenience is resolved and the railway company can fully develop the advantage of smart cards. At least one company other than us are testing contact free cards[Grim 90]. In this paper, we report the revised versions named *muCard*s of the card reported in [Goto 89] and discuss the new railway ticket systems using them.

Contact-free IC Cards

The objective of contact free ticket is to offer a convenient manner for the passengers. However, the unlimited freedom is not realistic for the system. So we impose passengers to *show* the ticket to the machine just as they do at the manned gate. There are two major reasons for this: The first is for reliability. Nearer the distance, the more increases the reliability. The second reason has a psychological sense. Although it is convenient if a passenger can go through the gate without special action especially when he carries large baggage, it may be not so comfortable if he feels "body checked" by the machine. It is important for the passenger to show his will to go though the gate by putting his ticket near the sensor. Thus the first target of communication distance is settled to 30cm and process time to 0.2s to cover the normal men's walking speed of 1.25m/s.
We have developed two types of cards and the tentative specification is shown in Table 1. The first two columns are for medium wave card. Their fundamental functions are basically the same, but the thickness and speed is improved for one on the second column.
Medium wave card contains a microprocessor in it. It receives information about gate, date, time and so on from the gate, decides by itself whether it can go through the

gate and returns an appropriate response to the gate. Fare information is given by the gate and deducted from the card if it has the stored fare function. This scheme reduces not only the load of gates, but also the transmission information between

Table 1. Tentative Specification of *muCards*

item		1	2	3
radio frequency	Hz	medium wave 200k	200k,400k	microwave 2.45G
size weight	mm g	54×86×4 25	54×86×1.4 11	54×86×1 8
CPU			yes	no
memory byte writable area			256 any	128 specific 32
communication distance speed time	cm kbps s	30 4.8–9.6 0.2–0.4	30 12.8 0.15–0.25	30 60–70 0.1

them, hence the high speed. Moreover, it brings an important feature that new ticket can be easily planned without changing the program of gate machines.

Another type uses a microwave technology. The card of this type contains a simple circuitry for high speed reading and writing of data. The validity and fare are calculated in the gate. One of important features is its price. If the cards are consumed as many as Japanese season tickets which amount 20-30 million a year in Tokyo area, the cost would be about one US$.

Thus the two cards have their merits. The microwave card is suitable for large scale application like season ticket or stored fare ticket. A new type of combined ticket of above two will offer an elegant system for both passenger and railway. An experimental prototype of the scheme have been developed. The medium wave card may be used for higher grade service with additional functions.

Experimental Results

Figure 1 shows relation between success rate and the moving speed of the card, the data of medium wave card being for one on the first column of Table 1. Microwave card clears the target speed, whereas the medium wave one is a little short. Recently developed thin *muCard* on the second column seems to clear it and is now under test.

Communication error of medium wave card is below 5×10^{-5} settled for magnetic ticket in Japan at the communication distance of 20cm and card speed of 1.1m/s. Although the first version of microwave card do not clear it, the possible causes were detected during the test and a new version is expected to clear the value.

Both proved the possibility of practical use of *muCard* to the railway ticket system and similar applications.

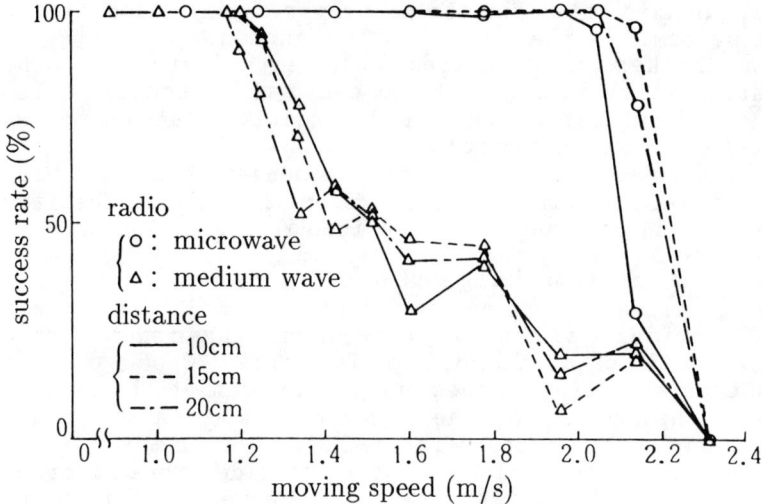

Figure 1 Velocity Characteristics of *muCards*

Advantages of *muCards*

The contact-freeness of *muCards* not only offers the passenger the convenient way of passing gate, it brings some advantages to the railway company.

1. forgery IC circuit is very difficult to fabricate for ordinary people. Moreover, a company can freely introduce a security code to the system.

2. fraud Card information is not only read by the gate but also altered. Thus the incomplete use of the card is easily detected. The same countermeasure against fraud is of course possible with the magnetic system. But the commuter is apt to go to the manned gate because of the troublesomeness of the magnetic season ticket and the railway cannot compel him to go though the automatic gate. It means that the perfect check is in reality impossible.

If a passenger can easily go through an automatic gate as a manned one, the railway could ask him to go there. Even if he refuses and goes though the manned gate, a checker without door mechanism can inspect his card automatically. It will be inexpensive and free from maintenance compared to automatic gate because all parts are electronic.

3. system integrity In an open ended railway the AFC is difficult to adopt because it is impossible to install gate machines to all the stations. Such a system must

leave manned gates for the passenger coming from outside the system. So the cost is not so reduced. The checker just described above can be installed to the station which is too small to equip the automatic gate. Thus the "automated" area will be greatly widened compared to expensive magnetic gate.
4. new railway systems   As the passenger never part with the ticket in normal use, it become his own. The fare system will be changed as described in the following.

New Ticket Systems Using *muCards*

Most railway stands on the ticket system that the ticket is the evidence of contract between passenger and the company and the payment is done on its basis. Once a passenger ends his trip and goes out of the station, he is also out of the system until he buys a ticket again as another passenger. The pattern is the hotbed of frauds. If he had his own card, and the history of rides were recorded on it, he would be very reluctant to cheat. Moreover, the railway company could greatly reduce its duty in the rear. If the post payment system is adopted, the company can deduct his fare on his record. In the prepaid system, passenger may be refunded or deducted when he adds credit to his card. This means the fundamental change of ticket system because the fare is not decided by the ticket only. In contrast to the conventional *trip-based* system, we may call the new system *customer-based*. The contact free card will be the most suitable for these systems.

Acknowledgements

The development of the cards are conducted under the RTRI research theme with the request of East Japan Passenger Railway Company. The authors thank to Mitsubishi Electric Corporation and Sony Corporation for manufacturing *muCard*s of each type.

References

[Goto 89] Goto, K. et al., *Development of a Contact-free IC Card for Railway Ticket Systems*, IFAC CCCT89, 1989.9
[Grim 90] Grimsey, C., *The London Transport Perspective*, International Conf. on Automatic Fare Collection in Public Transport, 1990.11.
[Leas 90] Leas, J.W., *North American Experiences with Automatic Fare Collection*, ibid.
[Rich 90] Richards, M.C., *Advanced Magnetic Ticketing Systems and Their Application in Singapore.* ibid.
[Slev 90] Slevin, R. *The Milton Keynes Experience with Smartcards in UK*, ibid.
[Sutt 90] Sutton, D., *Single Smartcards for a Multiservice and Multioperator System: The case of La Plagne, France*, ibid.

FAULT-TOLERANT COMPUTING ARCHITECTURE FOR MF 88 AND MP 89 ROLLING STOCK

By Mr Henri BORDENAVE, Ingénieur Département Etudes, RATP

INTRODUCTION

For several years now, the French and foreign railway industries have witnessed the development of a new generation of on-board computer systems for rolling stock. It mainly features distributed computing on the train : the different functions processed by this computing architecture are resident on several computers geographically located along the train, and linked together via a powerful communication link : the Local Area Network (LAN).

This article will describe the computing architectures as implemented on board RATP's rolling stock.

WHY A COMPUTING ARCHITECTURE

Historically, the computing architectures are derived from several processes :

- Deep thinking about the system brought out the need to conceive a train system as a consistent whole, and not as a mixture of equipment sets with potential interfacing.

- Optimizing the train cabling : with increasing connections between equipment sets, which resulted in raised cabling costs on the one hand and diminished reliability on the other hand, solutions were sought for so as to rationalize every solution regarding communications between on-board equipment sets.

- Searching for Fault Tolerance. Under certain conditions, computer techniques allow us to build fault-tolerant systems. It is therefore interesting to take advantage of this possibility to improve train availability.

- Improving the maintenance procedures, accordint to two main lines : first a better information transmission towards the operating or maintenance staff, then the possibility to establish synthesis diagnoses on the whole train, from the primary diagnoses as drawn up by the subsystems.

- Taking all or part of these objectives into account led to systems made of distributed computers (taking the notion of train system into account), possibly made redundant (fault-tolerant), linked together by at least one LAN (with optimized cabling, great masses of information exchanged because of function distribution and computer-aided maintenance).

DESCRIPTION OF MF 88 AND MP 89 ARCHITECTURES

The first RATP rolling stocks equipped with this architecture are the MF 88, a 3-coach conventional rolling stock (made by ANF and FAIVELEY for the electronic part), to be operated on line n° 7bis, and the MP 89, a rubber-tyred rolling stock (made by GEC ALSTHOM), which will be operated on the new automatic Meteor line and on man-operated lines n° 1 then n° 11.

MF 88 Rolling Stock is made of three coaches, whereas MP 89 Rolling Stock is made of six, with similar architectures.

The MF 88 architecture is made of eight computers distributed along the train and applying the hot redundancy principle : each function is processed at a time by 2 computers, that receive the same imput and do the same processing ; however, the output from one computer does go towards the controlled process, whereas the output from the other computer is either inactive or active but not taken into account by the controlled process (as is with the smart controlled process).

The eight computers can be divided into two types :

- The computers which carry out the "train" functions, namely which are about the general functions of the train set and its ressources : these are the two drive computers (ORC), with mutual redundancy. The active drive computer is the train master, the other computer being in hot redundancy.

- The computers carrying out the "coach" functions (OR1M, OR2M, OR1R, OR2R), which do pilot in each coach the units which carry out the different train functions : Traction, Braking, Energy conversion, etc. These computers being in hot redundancy two by two are controlled by the master drive computer, which sends remote controls to them and to which they report via remote monotoring.

FAULT-TOLERANT COMPUTING ARCHITECTURE

The same function is distributed along the whole train, on various computers. The managing function of doors, for instance, cans be divided into : Management of the general control of the train doors in the drive computer, Management of the coach doors in OR1R and OR2R, Door management in the microcontroller located at right angles to the door.

THE NETWORK

Type of network

The link between these computers is provided by a redundant network. On the MF 88, it is of the type FACTOR made by the APTOR Corporation. Based on Ethernet (IEEE 802.3 Standard) at a rate of 10MBPS, it has the additional advantages of determinism (CSMA/DCR) and of medium redundancy. This network makes up the cabling for the whole train system (1).

On the MP 89, the network in use is the TORNAD* network by GEC ALSTHOM. Based on the MAP protocol and the Token Bus (IEEE 802.4 Standard), this network also has the advantages of medium redundancy and of determinism.

The services offered by these two networks are globally those of the ISO layer n° 4 (ISO 8073 class 4) for traffics as "long and rather slow messages", linked to a direct access to the MAC layer for short messages and with real-time features described as being critical.

The networks characterized as purely "master-slave" have been abandoned because their architecture did not allow them to easily process the huge traffic between dual computers.

Type of traffic

Typically, the communication flow is carried ou by either of the following methods :

---

(1) A score of train lines are kept however for signals entering sefaty functions, according to classical techynology.

* Either with messages emitted in a cyclical way without network acknowledge, the acknowledge being made at the application level. The lost message is tolerated by the repetitive side of the cyclical traffic. In nominal state, this type of traffic is easy to carry out ; it does not demand any great network performance (MAC layer access), it is fast and deterministic. Yet the use of the network passband is very bad since a major part of the traffic is only made of repetitions, that is to say is useless. In addition, the application is loaded with additional work.

* Or with messages emitted on events. The network then has to secure the safe arrival of the message to the receiver. This supposes that acknowledge and flow control mechanisms are implemented by the network. The application is then unloaded from this work and the network passband is then optimized **on average** and on average only : the avalanche cases proper to this type of traffic have to be taken into account.

In practice, the increased power that has to be integrated at the board coupling level in this case grows rapidly and is located at the level of an MC68020 type processor. However the traffic processed this way is more universal in type and number of messages.

The choice between the above two solutions typically depends on application needs, namely :

- the response times in Traction/Braking as regards the network speed (presence of an ATO, for instance),

- the size of the information needed for maintenance, as regards the message size and the need to foresee a message fragmentation.

Experiments with Formal Specifications on MAGGALY

P. Dauchy
Université de Paris-Sud
Laboratoire de Recherche en Informatique
91405 Orsay FRANCE

P. Ozello
Institut National de Recherche sur les Transports et leur Sécurité
2, Avenue du Général Malleret-Joinville
94110 Arcueil FRANCE

Abstract

Until lately, safety functions for automatic train driving systems have mostly been implemented with classical techniques such as relays, way circuitry and, more recently, electronic devices. By now, microprocessors are being massively introduced into very high safety level transportation systems. While (in the French systems) hardware safety is ensured by the use of a coded monoprocessor, software safety can be reached by rigorous development and validation methods.

We present here an experiment on the development and use of formal specifications on the on board software of Lyon's unmanned subway MAGGALY, using the PLUSS modular algebraic specification language. This experiment was done in parallel with and independently of the actual industrial development. Our specification is executable and has been used as a prototype of the system, thus allowing a validation of the specification against the user's needs. It is also used to generate test data sets for the system. Logic programming with constraints is used to derive relevant testing values from the axioms of the specification.

Key words: software validation, algebraic specifications, train protection systems.

1 Presentation of the MAGGALY subway

MAGGALY (Métro Automatique à Grand Gabarit de l'Agglomération LYonnaise), an automatic, unmanned subway, will run by the end of 1992 on Lyon's line D. This 13 km. line serves

13 stations with a flow of 20000 passengers per hour in each direction. The delay between two trains is 90 seconds.

In order to save energy and reduce wear, one-element trains are used in off-peak hours and two-elements ones during peak hours (the economic advantages of this solution are discussed in [DHG 85] and [Luc 88]). Hence the need of an automatic coupling/splitting function, and of a moving block method in order to limit the duration of couplings. This moving block notion has been generalized to the whole line D. Anticollision between elements is not done by passive way circuitry detection, as in classical railway signalling, but by computing in real time the exact distance between a train and its target. It is then compared to the stopping distance, which is a function of the element speed and of the slope of the track. This is implemented as follows :

- each element computes its position and direction and sends them to the ground equipment;
- from this information, the ground equipment computes the order of the elements, their directions and precise positions, and sends to each element its target number and position;
- each element computes the distance separating it from its target; when it is less than the safety distance computed from its speed and the track slope, an emergency stop is triggered.

The safety of MAGGALY automatic driving is ensured in three ways :

1. Devices such as optical gates and phonic wheels are failsafe, i.e. : no single failure can be hazardous; for every hazardous succession of two failures, the first one must be detected; and failures occur independently of each other.

2. A coded monoprocessor ([For 89]) detects errors such as hardware failures, parasites and data transmission defects. The probability of an undetected error is guaranteed to be less than a given threshold.

3. Software safety is ensured by a rigorous development method, the use of a quality plan, modelling and validation of the specifications, an analysis of the software errors effects, and a validation by testing (unit, integration, functional and on-site). Furthermore, the functional testing results are checked by independent experts.

2 Algebraic specifications

A specification expresses the behaviour of a program or system without dealing with implementation details. An *algebraic* specification defines a class of *algebras* or *models*, each one being a number of operations on some sets of values. Each model corresponds to a particular implementation of the specification.

An algebraic specification is composed of:

- A *signature*, i.e. a set of names of *sorts* (or types) and operations, each of these having a given arity. Given a signature Σ, we call Σ-*algebra* any algebra whose names of sets and operations are those of Σ. A Σ-*term* or *term* is any correct composition of sorted variables and operations of Σ.

- A number of *axioms*, which are equations between Σ-terms or conditional formulae of the form $<$ conjunction of equations $> \Rightarrow <$ equation $>$. The axioms express the required properties of the system. A Σ-algebra *implements* the specification if all axioms are valid for any assignment of the variables involved.

The algebraic specification language used here is named PLUSS (Proposition of a Language Usable for Structured Specifications), and described in [Gau 85] and [Bid 89].

Examples of axioms will be shown later.

3 Presentation of the specification

This specification has been developed from the manufacturer's informal specifications ([Gui 90]). The organisation of the specification is shown on the graph of dependencies in Fig. 1, where the nodes represent the specification modules and the arrows the *use* of a module by another one (a module *using* another one may use any operation defined in that module).

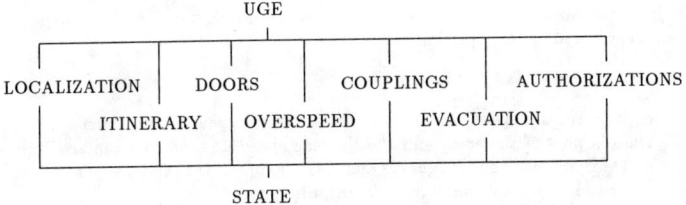

Figure 1: Graph of the specification modules

Due to the large number of train parameters, we chose to "pack" them in a specific data type State described in the module STATE. It is only a "record" of all parameters involved. Each of the safety functions takes as argument a value of type State and updates the relevant parts of the state. This state is itself structured in order to express which parameters are accessible to a given safety function.

For the sake of brevity, the modules defining the basic data types such as speed, abscissa... which compose the state are omitted here.

The module UGE describes the general behaviour of the system, which is the cyclic execution of three functions on an object representing the state of the train. The functions performed during a cycle are the following: input describes the reading and validation of the data transmitted by the interfaces, and especially from the ground-based software; processing groups the different operations performed on this data, described in the modules LOCALIZATION, ITINERARY, DOORS, OVERSPEED, COUPLINGS, EVACUATION and AUTHORIZATIONS; and output describes the sending of the appropriate signals to the ground-based software and to the train devices.

The safety functions performed by the system are the following :

- localization determines the position, speed and direction of the train;
- itinerary elaborates the train itinerary and target;
- doors checks that the doors only open in safe conditions;
- overspeed ensures the respect of the speed limits in order to avoid collision with other trains and to protect the interlockings;
- couplings checks the couplings and prevents having too long trains;
- evacuation deals with the passengers' evacuation demands;
- authorizations deals with the alarms and authorizations computed by the previous safety functions.

4 Experiences performed on this specification

4.1 Symbolic evaluation

Since the axioms are positive conditional formulae, they can be used as rewriting rules. Given a specific value of the state of the train, an expression including this value can then be rewritten, replacing any occurrence of a defined operation by its definition (the right-hand part of the appropriate axiom).

For example, given a value s of type State, we can use the axiom whch defines the overspeed alarm function al-overspeed by the equation

```
al-overspeed (s) = gt(speed (s), limspd (s))
```

in order to rewrite the term al-overspeed (s) as the term gt(speed (s), limspd (s)). The speed of the train is part of the state and the limiting speed limspd (s) can similarly be rewritten as min (target-spd (s), imposed-spd (s), stop-point-spd (s), way-spd (s)), meaning that limspd (s) is the smallest of four limiting speeds.

This rewriting is repeated until we obtain a single value giving the result of the operation on the particular state chosen. In the example shown, successively rewriting the operations into their definitions will lead to a boolean value giving the value of the overspeed alarm.

We thus have a prototype of the system. It can be used to increase our confidence in the validity of the specification or to help validate a program which implements it by comparing the program results with the prototype results.

4.2 Test data generation

A test is defined as a pair of terms without variables. Its execution is the evaluation of both terms by the program being tested; if the results are the same, the test is successful. We wish to test whether the program satisfies the axioms of the specification. Therefore, our tests are ground instances of these axioms, i.e. all variables occurring in the axioms are replaced by terms (without variables) of the appropriate type. Furthermore, to test a conditional axiom, the values instantiating the variables must satisfy the premises.

This theory of program testing is described in [BGM 91]; we will not describe it further, but only illustrate it on an example.

For instance, suppose we wish to test the following axiom :

```
doors-state (s) = wait-open-doors & in-siding (s) = false
 & closed-way (s) = true & closed-pf (e) = false
 & and (localized (s),
        and (in-station (s), gt (spd-open-doors, speed (s)))) = false
=> next-doors-state (s) = emergency-stop
```

We must find a value S of type State such that in S,

- the doors state is "wait until a door is found open";
- the train is not in the siding part of the way;
- the doors are closed on the way side and open on the platform side;
- the train does not know its position, or it is not located in a station (with all doors facing the platform), or the speed is too high for the doors to be open.

This test will be applied to a program by computing the values of next-doors-state (S) and emergency-stop and comparing them.

The use of logic programming allows us to generate adequate values for the variables in an axiom. Clearly, one test is not enough; for instance, if we consider the truth table of the operation and, we see that the equation and (x, and (y, z)) = false has seven solutions. Similarly, the equation in-station (s) = false has more than one solution (there are many possible ways for a train not to be correctly located in a station). The equations appearing in the preconditions of the axioms are thus expanded into different situations which reflect the axioms defining them. This way, many possible scenarii are generated. However, such decomposition must be stopped at a given point, so as to limit the number of tests generated. We define special directives in order to express which decompositions must be made, depending on the operation concerned and on its arguments. The system used can be very accurately tuned and allows the user to focus the testing process on a particular aspect of the specification.

4.3 Further work

Modula-2 programs are currently being developed from our specification and also directly from the manufacturer's informal specifications. We are currently applying the test data sets we have generated to them, in order to evaluate their quality. In particular, we intend to do mutation testing, i.e. to introduce errors in the programs and check whether our test data sets will detect these modifications.

We also intend to perform proofs on these programs, using the axioms of the specification as pre- and post-conditions.

References

[BGM 91] G. Bernot, M.-C. Gaudel, B. Marre : Software testing based on formal specifications : a theory and a tool, Software Engineering Journal, March 1991.
Also : LRI internal report n°581, June 1990

[Bid 90] M. Bidoit : Pluss, un langage pour le développement de spécifications algébriques modulaires, Thèse d'État, Université de Paris-Sud, May 1989.

[DHG 85] Yves David, Jean-Jacques Herry, Jean-François Gobard : Cost Reductions for Unmanned Transports Systems, ASCE, March 1985.

[For 89] P. Forin : Vital coded microprocessor principles and application for various transit systems, IFAC CCCT'89.

[Gau 85] Gaudel, M.-C. : Toward Structured Algebraic Specifications, Esprit '85 Status Report, North Holland, 493–510.

[Gui 90] M.-C. Guillaumin : Spécification de l'UGE sécuritaire (éd. 10), June 1990.

[Luc 88] E. Luca : Bilan de la composition variable pour les rames du métro de Lyon, Recherche Transports Sécurité 18-19, September 1988.

SAFETY MICROPROCESSORS IN GUIDED GROUND TRANSPORT

M.El Koursi* & A.Stuparu *

ABSTRACT

The use of microprocessors performing safety tasks such as Automatic Train Protection systems made more difficult their safety-worthiness assesment. In this paper we will first examine how the microprocessors have been used in safety applications in land transport field and we will proceed to a comparative examination of different basic architectures and present according to significant criteria a qualitative analysis.

INTRODUCTION

In ground transport, microprocessors are being used since only a relatively short time in safety functions. The first fully automatic system using a safety microprocessor architecture was Vancouver "Skytrain" put into service in 1985. However, some applications in the railway signalling field are older due to the fact that this type of equipment has given rise to a considerable amount of research and development since about ten years.

In safety operation of microprocessors, the **information redundancy** and the **hardware redundancy** concepts have been used in various forms. In this contribution, we will examine how these two concepts can be used in specific safety applications to ground transport(ELK.90) and proceed to a comparative examination of different basic architectures.

MONOPROCESSOR ARCHITECTURES

These architectures are based on the information redundancy concept. This redundancy can be integrated into the software or installed in the very processor. We can classify them into three categories:
- the information redundancy by coding, which consists in giving to the processed information a certain redundancy in form of code which processed separately and which allows the checking of the result through a "signature". This is the case of the **coded**

---

* Researchers Engineers, INRETS, 2 Avenue Général Malleret-joinville-F 94114, ARCUEIL, France

monoprocessor(fig 1.1) used in France to be found in the train protection of line A of the RER express metro in Paris put into service in 1988, of the POMA system in Laon put into service in 1989 and the MAGGALY(Metro Automatique à Grand Gabarit de Lyon) system, which will be put into service in 1991. We can also find it in the fixed automatic pilot of de VAL(Véhicule Automatique Leger) in Chicago designed for a single track operation (FOR.89).

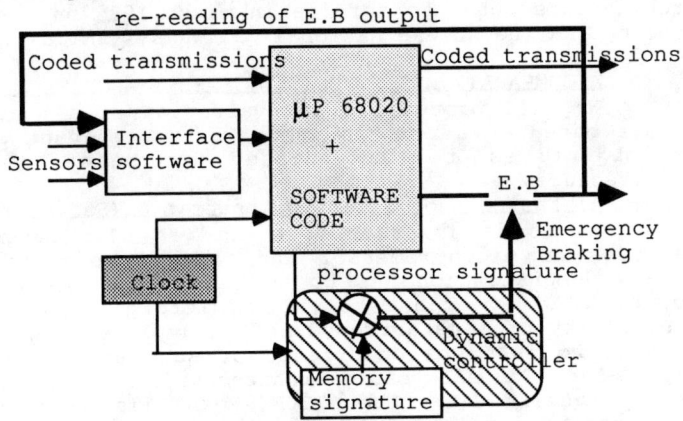

Fig1.1:Schematic diagram of the coded monoprocessor

- the heterogeneous time redundancy consists in making run alternatively two different programmes by means of the same microprocessor(VOG.88). After each step of the algorithm, the results given by the two programmes are compared mutually in order to verify their concordance. This is the system developed by LM-ERICSSON to be found in computer based interlocking in Sweden, Norway and Finland.

- the self-testing microprocessors are capable of detecting immediately their own faults. These specialized processors are based on the so-called Self-Checking circuits. A processor type is being developed by French Company CSEE for the MAPS" Micro Automate Programmable de Sécurité".

DMR(Dual Modular Redundancy) architectures

These architectures contain two distinct processors, generally homogenous to decrease the maintenance costs, accompanied by their own environment equipment. These architectures differ by the nature and the safe design technologies of the comparators. Into these architectures are often integrated functional tests for protecting them against latent failures. This is the case for instance of the PUSH and SELTRAC system in Berlin, Vancouver and Detroit system. They are also used in safety functions on the Metromover in Miami and on the on-board automatic devices of the Docklands line.

TMR(Triple Modular Redundancy) architectures

These architectures are based on the redundancy of three microprocessors associated with a system of comparison and majoratory vote.

We can mention the SSI(Solid State Interlocking) system installed in the British computer based interlocking designed by British Railway(CRI.87) and the SMILE" Safe Multiprocessor for Interlocking Equipment" architecture put into service in march 1985 on the Keihin line by JNR" Japaneese National Railways"(AKI.87).

A COMPARATIVE SYNTHESIS OF SAFETY ARCHITECTURES

We will proceed to an evaluation of a qualitative nature based on a certain number of significant criteria or objectives to be reached by means of these architectures.

1) Easiness of Safety Demonstration(Saf.Dem)

As it is impossible to analyse all the possible failures of microprocessor and their consequences, the safety demonstration has to be based on a probabilistic approach which consists in admitting a priori the possibility of a catastrophe and in implementing a set of methods to make the occurrence of such events extremely improbable. For each basic architecture, we can define the Safety factors $S(t)$, the MTBUF (Mean Time Between Unsafe Failures) and the MTTF (Mean Time To Failures) expressions :

$$S(t) = 1 - P_{nd} + P_{nd}.e^{-\lambda_{eq}.t}$$

$$MTBUF = \frac{1}{\lambda_{eq}.P_{nd}} \qquad MTTF = \frac{1}{\lambda_{eq}}$$

P_{nd} : non detection Probability

λ_{eq} : architecture equivalent rate of failures.

When evaluating the architecture safety level, the most difficult is to determine the P_{nd} which depends of various parameters. For this criterion, we can distinguish between three classes:

-- **Mathematically verifiable or demonstrable safety**

This is the case of the so-called architectures with "signatures" or self-testing inside the processor. These architectures are based on coding techniques allowing to demonstrate the safety level through probabilistic calculations. In the coded microprocessor, the retained code is of the following form:
$X = A.x$ where X is a coded operand
x is the uncoded value of operand
A is a randomly selected prime number.

This code allows the detection of operation error with $1 - 1/A$ probability. For a 32 bit code length the P_{nd} key is $2.3 \ 10^{-10}$. The MTBUF is $4.3 \ 10^{14}$ hours (under hypothesis $\lambda_{eq} \# 10^{-4}$ failures/hr).

--Safety based on functional tests and hardware redundancy
In the case of systems with hardware redundancy, the safety levels can be reached theoretically as a function of the overall testing coverage rate and of the type and frequency of the control and of the comparison type between the redundant systems. The MTBUF is generally about 10^{11} hr for the DMR, TMR and DDMR architectures.

-- Very difficult to demonstrate safety
This is the case of systems based on software redundancy. Inspite of the precaution we may take, it is practically impossible to prove that two distinct softwares do not have any common-mode failure. In this case, the confidence that we can have in the system is based more on an intimate conviction than on a quantified assessment.

2) Fault Tolerant(F.T)
Only the multiprocessor architectures with more than two microprocessors such as TMR and DDMR(Dual Duplex Modular redundancy) structures meet completely this requirement. The DMR can meet this criterion at least during short periods and in a degraded way if the processors have self diagnosing programmes allowing to determine which of them has a failures.

3) Protection against Software Errors(Prot.Sof.E)
The systems with two softwares developed in diversity, and coded monoprocessors in which all the operations on variables are doubled by independent operations on the codes present a certain degree of protection against software errors. Moreover in these systems, a compiler defect has little probability of producing the same effect on the two operating channels.

4) Programming Easiness and Costs(Prog.Eas)
The programming effort in redundant software systems is practically double, which represents an important supplementary cost. The systems with functionnal tests require in addition to the development of the application software, the installation of the testing programme. The coded microprocessor requires a tool for predetermining the signature.

5) Hardware Costs(Hard.Cost)
This criterion seems rather sensitive when comparing monoprocessor and multiprocessor architectures. However, the situation is more complex, and finally it is rather difficult to make a statement on this criterion. Coded monoprocessor systems contain decoding components generally realised with a safety design and therefore expensive. The self-testing systems are more specific and made in small series. More generally monoprocessors are often doubled for fault tolerance reason.

Finally, we can set up the table 2 in which the different criteria are estimated in a rough way with qualitative appreciation (+++: well, ++: favourable, +: average, - : unfavourable).

ARCHITCTURES	F.T	Saf Dem	Prot Sof.E	Prog Eas.	Hard Cost	Performance
Monoprocessors						
Self-testing	—	++	—	+++	++	+
Soft-diversity	—	—	+++	—	+++	+
Coded µP	—	+++	++	++	++	+
Multiprocessors						
Dual (DMR)	+	++	—	++	+	++
Triple (TMR)	+++	++	—	++	—	++
DDMR	++	++	+++	—	—	++

Table 2 qualitative appreciation

CONCLUSION

Almost all the described architectures allow to achieve a very high safety level by taking the appropriate precautions. From the safety point of view, monoprocessor architectures may offer a good solution, in so far as their performances in time and operating capacity are sufficient to meet the list of specifications for the application considered. The safety demonstration of redundant architectures with Dual, Triple or even quadriple redundancy is not very easy taking into account the common-mode failures (manufacturing faults, software and compiler faults,...). These architectures are particularly well adapted for fixed equipments, because they allow on-line maintenance, in so far as the perfect independence of the processors is insured. When the choice of a digital solution is justified, it is important to choose an architecture whose safety is easy to demonstrate.

References

[AKI.87] K. AKITA " Solid-State Interlocking railway signalling, SMILE" IEE, London, september 1987.

[CRI.87] A.H CRIBBENS" Solid-State Interlocking (SSI):an integrated electronic signalling system for mainline railways"IEEE proceeding,vol134,Pt,B,N°3,may 1987.

[ELK.90] M.EL KOURSI and A.STUPARU " étude comparative des architectures microprogrammées utilisées dans les applications de sécurité", Rapport INRETS N°134 december 1990.

[FOR.89] P. FORIN "vital coded microprocessor principle and application for various transit systems" IFAC-CCCT september 1989, Paris.

[VOG.88] U. VOGES(ed) "Software diversity in computerized control systems" Dependable and Fault-tolerant systems, vol2, Spring-verlag wien New York, 1988.

SIMULATION OF A VEHICLE PLATOON CONTROL SYSTEM FOR AUTOMATIC HIGHWAY USING THE FUZZY CONTROL CONCEPT

Sijiu Liu[1] & Andrew Frank[2]

Abstract

This paper discusses the concept of the longitudinal control of vehicle platoons on the highways. The paper discusses the simulation of vehicles with non-linear properties, such as air drag, saturation (acceleration and brakes) and pure time delays in the engine and brake systems, as well as a mix of vehicle sizes with varying performance on the highway. A non-linear controller is developed by using a fuzzy logic algorithm concept. Tests of a computer simulation for six vehicles in a platoon of three different vehicle sizes and for five different driving situations including emergency stop show the concept is viable and can be extended to more vehicles in a platoon.

Introduction

The main objective of an automatic highway is to improve: highway efficiency by 2 times, safety and convenience. An approach, termed "vehicle following", is often used to implement a longitudinal control on a vehicle platoon. This approach requires communications between successive vehicles and needs sensors to determine space information with a front neighbor vehicle. In a platoon, this system is actually a non-linear control system, even though some past research has presented study results which use only a linear controller.

In 1974, a method called "the fuzzy controller" was proposed by E.H.Mamdani[2]. Now a special fuzzy processor chip has entered the market which could help propagate this technique. The purpose of this paper is to show how to use the fuzzy logic algorithm concept to develop a controller for vehicle platoon control on the highway.

The Vehicle Modeling and the Closed Loop System Structure

The longitudinal dynamic model of a vehicle platoon system referenced in [1] is represented in the block diagram of Fig. 1. The basic longitudinal dynamic process of a vehicle is a mass-damped system with damping from air drag and the friction forces.

Additional dynamic processes comes from the engine and braking systems. For the model shown in Fig. 1, both of them receive a control signal from the controller and create an output force to push or to stop the vehicles motion. Because this system works in a lower bandwidth relative to the engines working process it can be considered as a first-order element with a pure time delay. Similarly, the braking process has a dynamic structure, only with a smaller time constant with a first-

---

[1]Research Assistant, Department of Mechanical, Aeronautical and Materials Engineering, University of California, Davis, CA 95616.

[2]Professor, Department of Mechanical, Aeronautical and Materials Engineering, University of California, Davis, CA 95616.

order lag and a larger pure time delay because of the mechanical structure.

One more non-linear effect is the torque or force saturation caused by the limit of both the engine and braking systems. As described above, in the normal case, the driving power needed is much less than the maximum engine ability. The extra power is used for maneuvers such as acceleration and going uphill. Usually the limit of the driving force from the engine is about two to four times (depending on vehicle size) the value for level no wind driving at 25 m/s. For the braking case, the standard mechanisms can produce the maximum average deceleration of about 0.4g to 0.6g (about $4m/s^2$ to $6m/s^2$) based on tests, which means it has the value about 4000 N. to 6000 N. for a vehicle of 1000 kg.

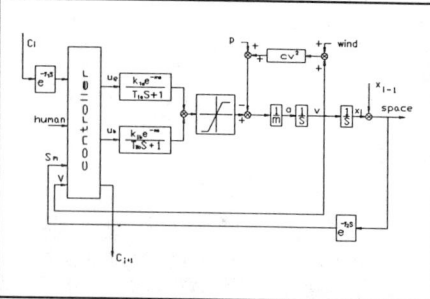

Figure 1. Block diagram of each vehicle dynamics with assumed sensors and communication system.

The control law for each vehicle is based on the sensor signals, the vehicle size and response capability. Generally it is assumed that larger vehicles have slower response. This was presented in [1]. In the normal case, (which means no hills or valleys, wind, acceleration and deceleration), the driving force should be equal to the opposing force. A mid size car is assumed to have a mass of 1000 kg and the drag coefficient of 1.19 $N*s^2/m^2$. For a normal running speed of 25 m/s (about 55 mph), the driving force is 739 N.(147 lb.). The equivalent time constant (for a first-order lag after linearization) of the acceleration/speed control is about 16 seconds.

The term multi-mode controller means the use of different control strategies according to different conditions. Here, the basic condition considered is the space between the vehicles and the relative velocity. For example, if the space is near the desired space and the relative velocity is near zero, keeping the relative speed between vehicles zero can maintain this space. A linear speed regulator (cruise controller) is assumed available. When the spacing becomes larger or smaller then desired, the vehicle following system should adjust its velocity accordingly. However, if the space is too small or if the speed of the following vehicle is too fast relative to the front one, an emergency situation should occur. The vehicle should use the maximum braking power available and a special control algorithm to avoid a collision. On the other hand, when the space between vehicles is too long then the following vehicle should not consider a dangerous situation, and a standard speed control loop with an independent desired speed reference is a candidate for control. Thus the closing speed and the following space are the basic feedback variables, but a selection of which control algorithm to use and under what conditions is necessary.

It may also be necessary to receive a desired space and speed issued by the driver himself for control. For example, if a vehicle wishes to enter the platoon, the desired space of an adjacent controlled vehicle needs to increase its normal value to allow the extra vehicle to enter (a lane change). The amount of increase will depend on the vehicle size and reference space for that vehicle. After the entering maneuver is finished, the desired space should be maintained.

Beside the above, getting information from the front vehicle is very important. For example, it is too late to wait for the space to be close enough to determine braking. So a better way is to let the front vehicle pass a signal to the following one simultaneously as it applies its brakes. Similarly, each vehicle needs to pass the reference speed signal to the following vehicle to improve platoon control performance. Actually, this information can be quantized to several levels to simplify a communication device. This is similar to the current vehicle's use of a brake light to inform following vehicle drivers that braking is beginning to happen. Thus besides giving a signal to its own engine and brake system, the controller also needs to give a signal to the following vehicle.

In summary, the controller needs to reference the desired space between vehicles and the speed of the vehicle, sense the actual vehicle speed and measure space for feedback, receive a feedforward signal from a front vehicle, output the control signal for the engine and brake system, also send a synchronized speed signal to the following vehicle. If no front vehicle exists the leading vehicle should get it's reference speed from a roadside communication system which transmits the speed limit for that specific road.

Controller

1. The table of control rules

The criteria conditions for multi-mode control are based on the space between vehicles and the relative speed between vehicles.

On the phase-plane shown in Fig 2, the control modes are indicated with certain regions which are separated by boundaries. The region I is a linear control zone. The system should work in this region as the gross transient process is completed. Region II is a deceleration zone, running by coasting or by engine braking. The region III is the emergency zone using maximum braking to stop the vehicle as quickly as possible. Region IV is an acceleration zone and region V is an individual operational zone which does not consider a front vehicle because the distance is too long.

Here, suppose the phase plane is divided into a number of small blocks which separates the space axis into columns and the speed axis into rows. It should be noticed that the quantization is not limited by a linear relation. Furthermore, this relation can be described by a control rule table T. The variable T(ie,ic) are the elements of T and represents the controller mode under the condition that the quantized values of space error and its change rate are ie and ic respectively.

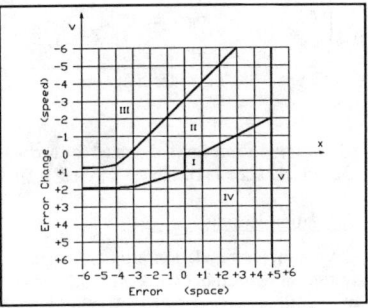

Figure 2. phase plane of closing speed-space

In this way, the control process can be summarized as follow:
1) get space error and its rate by $e = x_d - x_m$ and $c_t = (e_t - e_{t-1})/t_s$, where x_d and x_m are desired space and measured space, the t_s is the sample period.
2) quantize e_t and c_t to ie and ic, both ie and ic are integer values.
3) using ie and ic as the index, search the control rule table T to get the control mode index iu = T(ie,ic).
4) find the actual value of controller output by executing the control routine which would result from the selected index iu.

It should be noticed that the total above control algorithm is based on a control rule table T which needs to be set up using human experience and could be modified by the driver thus retaining driver control.

Setting up this table is a problem to be solved. It is not easy to answer questions such as what value of controller output should be chosen if, for example, ie=3 and ic=2. However, it is easy to describe a person's experience by a sentence such as if an error is small and an error change rate is small then the controller output should be small.

2. Fuzzy Control Algorithm

Generally, fuzzy set theory is not simple, but only some subsets of the total concept are involved with developing a control algorithm. The development is to be published in another paper.

3. A vehicle platoon controller with a fuzzy algorithm

The vehicle platoon controller scheme is shown in Fig. 3. There is a speed regulator designed to keep the actual speed of the following vehicle to a reference speed Vr and to reject disturbances such as road grade or wind conditions. The reference speed V_r consists of two parts. One is a synchronizing signal which comes from the front vehicle. The other is the output of a space controller

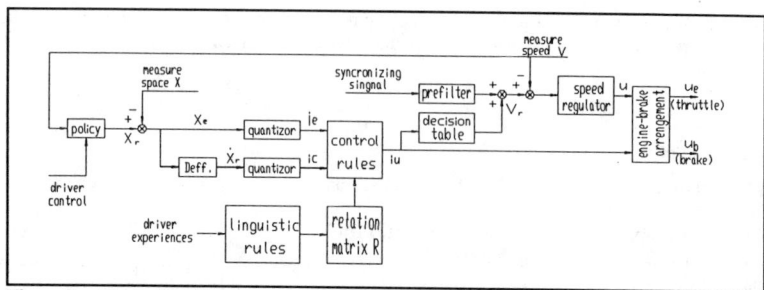

Figure 3 Controller diagram

whose value represents the need for acceleration or deceleration. The desired space X_r depends on the vehicle speed [1]. However, their relation can be adjusted by the driver or human operator. [3]

Simulation Results

A computer simulation program representing the vehicle dynamics, control delays, engine/braking systems is written in Microsoft C 5.1 and runs in an IBM PS/2 Model-70/386, to reveal the properties of longitudinal control of a platoon of vehicles on a highway. Six vehicles of three different size (A, B, C) are selected and placed in a platoon in a certain order. The order is A1,B1,C1,A2,B2,C2. The plant (vehicles) and control system parameters are represented in Table 1.

Tdb(s) vehicle	M(kg)	C2(N.*s²/m²)	T1(s)	Maxe(N.)	Tde(s)	T1b(s)	Maxb(N.)	T2(s)
type A	750	1.00	1.5	1500	0.3	0.7	-3700	0.5
type B	1000	1.19	1.8	1800	0.4	0.8	-4300	0.7
type C	2000	1.50	2.0	2200	0.5	1.0	-7000	0.8

Table 1. The plant and controller parameters

The control system simulation was tested in five situations.

1. The system response to rejecting a disturbance from the road environment such as a sudden uphill grade of 6 degrees shows a response which reaches steady state in about 30 sec.
2. The system response to a speed limit change on the highway such as from 55 to 33 MPH (25 m/sec to 15 m/sec)is adequate. It is supposed that the speed limit is posted on the highway and the first vehicle in the platoon senses the speed limit as it passes the post, this may be done via a radio transmitter or other communication system from the road to the vehicles. Due to the use of a synchronizing signal and a speed filter, speed following is good even for the different size vehicles in the platoon with different performance values.
3. As one vehicle leaves the platoon, the system keeps the space between the vehicles, the following space suddenly doubles for one vehicle. So the following vehicle should speed up to close the gap or space between vehicles, then it needs to slow down to the original speed as the space approaches the desired spacing. All vehicles following must also do the same. The simulation shows this response is appropriate.
4. Figure 4 shows the system response to an emergency. This time the vehicles should be stopped as soon as possible. The brake decision depends mainly on the rate of relative speed in the case that space is less than the standard safety range. Note the time required to come to a stop is not the same but collisions are avoided. One vehicle comes to a stop with very little tolerance. The algorithms for following control could be further tuned.
5. Figure 5 shows the process of merging. When a vehicle in another lane adjacent to the platoon wants to enter the platoon, the vehicle closest behind the entering vehicle should reduce its speed

and double the space to let the vehicle enter. This can be initiated by using a turn signal as people now do to change lanes. After this, the vehicle adjusts the speed to return to the normal space.

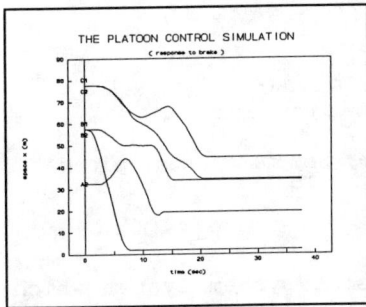

Figure 4. space response to brake

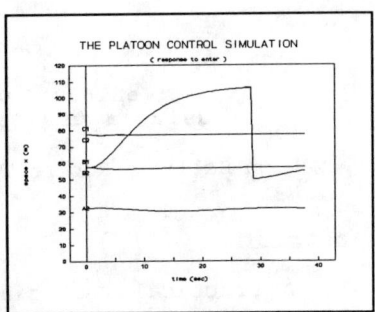

Figure 5. space response to an entering vehicle

Conclusion

The above discussions have shown that a fuzzy control algorithm can be applied to the design of a longitudinal controller for a vehicle platoon on the highway. From analyzing the computer simulation, it is found that the speed fluctuations which occur due to hills on freeways are not large because of the speed control loop. Also by means of the speed synchronizing signal and pre-filter, an intentional change of speed has little effect on spacing. Thus for a general space control algorithm, it is not necessary to pursue quick response, rather a smooth response is more important based on human factors and values. But because of the difference in braking ability between the different size of vehicles, a space large enough to avoid a collision with the front vehicle is always needed. The space between consecutive vehicles is easy to control, mostly because of the synchronizing signal used to sense what is occurring in vehicles at the front of the platoon. The action of this signal is similar to the use of the brake light by a driver in current vehicles. Thus, the sensing and spacing algorithms proposed seem appropriate. The fuzzy algorithms select the control policies for the various conditions which may occur.

This study shows that the fuzzy control concept can be used to successfully implement control policies which, to a degree, reflect human experience. More research is necessary to verify the robustness of the policy and further tune the fuzzy concept.

Reference

[1] A.A. Frank, S.J. Liu and S.C. Liang, Longitudinal control concepts for automated automobiles and trucks operating on a cooperative highway, SAE conference, Toronto Canada, 1989.
[2] E.H. Mamdani, Application of Fuzzy Algorithms for Control of Simple Dynamic Plant, Proc. IEEE. vol.121, pp. 1585-1588, 1974.
[3] J. Glimm and R.E. Fenton, An Accident-Severity Analysis for a Uniform-Spacing Headway Policy, IEEE Transaction Vehicular Tech. Feb. 1980.

Vehicle Lateral Guidance Using a DSP Based Vision System

Mahlon Heller[1], Robert Trahms[2], Sompol Chatusripitak[3]

Abstract

A practical real-time vision system system which uses a digital signal processing chip to obtain a tracking error for lateral guidance of vehicles at highway speeds is presented. Minor changes are made to the highway infrastructure which reduces the complexity of the vision system. Longitudinal vehicle control is not addressed.

Introduction

The Lateral Guidance System(LGS) is a device which limits the lateral movement of a vehicle to a narrow tolerance of displacement from a reference. The target lateral displacement is ±1 inch. Significant benefits to transportation systems are expected as a result of utilizing the LGS. The most prominent benefit may be the increase in safety to drivers by minimizing the probability of collisions caused by erratic movement of vehicles (Bender 1991). At the mature stage of the LGS technology, it is envisioned that commuter stress will be reduced as well as allowing effective use of commute time. The LGS will certainly increase the capacity of the transportation facility by making it possible for vehicles to be operated safely on narrower lanes. Conventional 12-foot traffic lanes may be restriped to 10-foot lanes, creating room for new lanes without adding new pavement. In addition, in the next decade LGS may be applied in mass transportation. It can offer unprecedented flexibility to transit systems by simply

---
[1]Prof., Dept. of Elect. Engr., Calif. State Univ., Sacramento, 6000 J Street, Sacramento, CA 95819-6019
[2]Grad. Student, Dept.of Elect. Engr.,Calif. State Univ., Sacramento, 6000 J Street, Sacramento, CA 95819-6019
[3]Sr. Elect. Engr., Calif. Dept. of Transp. New Technology Development, 5900 Folsom Blvd., Sacramento, CA 95819

rearranging the references on the road in lieu of reconstructing rails or guideways.

In the near term, incremental benefits of the technology can be exploited in related areas such as the automated highway maintenance and construction technology. In California, work is under way to design and construct prototype automated systems to perform pavement crack sealing, reflective marker installation, and paint striping. All of these machines can immediately take advantage of LGS.

Why a Vision System?

A number of technologies have been considered for LGS applications, each of which offers advantages under certain circumstances. Vision is perhaps the most flexible. The major disadvantage of current vision systems for LGS is their high cost and complexity as most vision-based LGS's do not assume changes in the highway infrastructure (Kuan, Phipps, Hsueh 1988). This project reduces the cost and complexity of a vision-based LGS by painting an ultra-violet(UV) reflective, clear reference stripe down the center of each vehicle lane (Ishikawa, Kuwamoto, Ozawa 1988). When illuminated with UV, the stripe becomes pronounced in a video frame.

Vision Development System

The image data gathering system consists of a 0.02 LUX, 512x512 pixel, CCD Camera with auto iris mounted in the automobile, a Super VHS video recorder, video monitor, and a 12 Volt dc to 115 Volt ac convertor. The camera will be mounted in front of the radiator pointing downward along with a UV source to illuminate the reference stripe. This location permits the camera to always see the reference stripe regardless of the location of the vehicle in front. Furthermore, glare from the pavement is eliminated.

After data have been collected while manually steering the vehicle, the VCR is connected to a frame grabber which is installed in a 25MHz, 486PC system equipped with a 200Mbyte hard-disk. Application image processing software installed on the 486PC is used to obtain attributes of the digitized frames and to apply image processing techniques. The objective is to develop real-time software algorithms which will identify the location of the reference stripe to be tracked relative to the center of the vehicle and as a result produce a lateral displacement relative to the reference. Real-time is defined as "production of lateral displacement every frame-time" where a frame time is 1/30 second.

On-Vehicle Real-Time Vision System

A block diagram of the real-time vision system installed in the vehicle to provide a lateral displacement signal to a vehicle steering controller is illustrated in Figure 1.

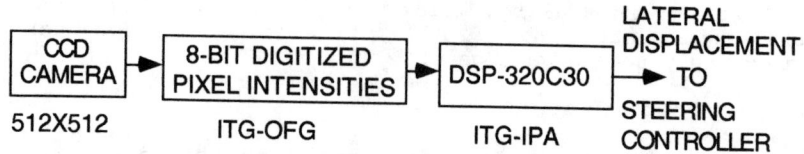

Figure 1. On-Vehicle Real-Time Vision System

The video scan lines are fed to an Imaging Technology Inc. Overlay frame grabber (ITG-OFG) mounted in the 486PC. The important point is that the ITG-OFG allows us to pass the 8-bit digitized pixel intensities directly on to a Texas Instruments 320C30 DSP mounted on a board developed by ITG called an Image Processing Accelerator (ITG-IPA). Interpixel processing is performed without waiting for a complete digitized field or frame. At the end of a frame time, a lateral displacement digital value is sent on to the vehicle steering controller. Lateral displacement, time, and other parameters that are the results of the DSP real-time software algorithms are stored in DSP memory to aid in debugging and evaluation of the algorithms used. Actual lateral displacement will be computed off-line using the recorded image frames. This actual value will be compared to the lateral displacement computed real-time by the on-vehicle DSP.

The ITG-IPA, also slotted in the 486PC, performs 32-bit fixed and floating-point operations and is rated at 33 MFLOPS. The ITG-IPA we are using has 2Mbytes of on-board ram. The DSP chip pipelines instructions and can perform two multiplies or adds simultaneously.

Vision Software Algorithms

The first step is to process a selected portion of the camera image as the entire image is not necessary for line detection (Dods 1982). The digitized video frame consists of two interlaced fields of 262.5 rows by 512 pixels. We selected a rectangular window in the first field. We will refer to this window as a *subfield*. The size of the selected subfield for processing is 25 vertical pixels by 100 horizontal pixels. The processing steps are illustrated in Figure 2.

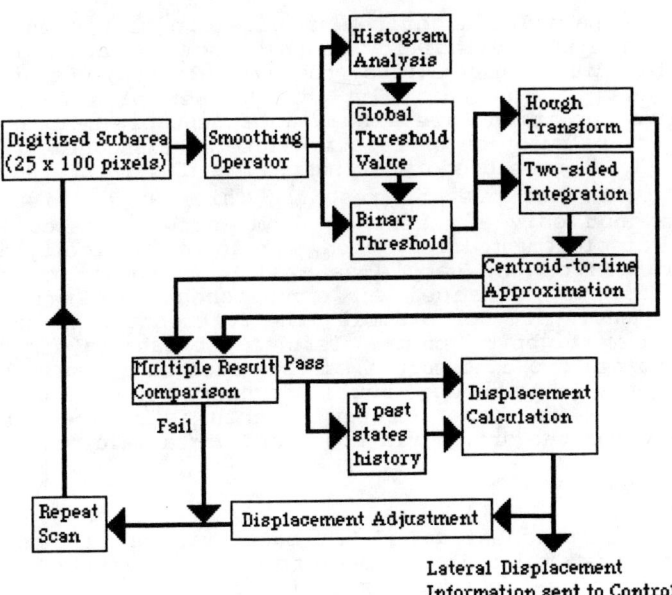

Figure 2. Vision Software Processing

While receiving the new digitized subarea using a defined coordinate origin, the process applies a 3x3 smoothing operator to the incoming pixels. The smoothed area is then processed by a histogram analysis routine, which determines the global threshold for the conversion of the next subarea sample to a binary image. Simultaneously, the subarea pixels are converted to a binary image as they are input, by using the global threshold from *the previous subarea histogram analysis*. This allows thresholding on a real-time basis. This binary image is passed to a series of algorithms that estimate the slope and position of the line segment.

One such algorithm is the Hough transform that can extrapolate line information such as slope and position. Another formulated algorithm in parallel with the Hough is what we call Slice Two-Sided Integration(STSI). STSI sums pixel intensities from both directions of a horizontal slice of the subarea until integration paths intersect. This intersection determines that the line center resides at this intersection point and identifies it as a centroid. Once a series of centroids has been determined from various slices, an approximation to the geometric line connecting the centroids is made.

These parallel algorithm results are then compared to determine the reliability of the incoming data. If a detectable discrepancy exists, the sample is rejected from the process. Alternatively, if the set of resulting slopes and positions are within an acceptable margin of error, the calculation of relative lateral displacement is commenced. A past history of sample results is stored to compare with the present result. This sample history provides not only a reference from which to calculate lateral displacement information, but lateral velocity and acceleration of the sample as well. In addition, the sample history provides an error check for erratic displacement. If, for example, the reference line moves from extreme right to extreme left in the subarea, or the slope between two consecutive samples changes by more than 20 degrees, a watchdog routine checks the next few samples. If the same erratic pattern is noticed, the process warns the driver via sound and visual alarms.

Conclusion

A description of a LGS DSP based vision system has been presented which in the near term can be utilized for lateral guidance of vehicles performing various highway maintenance tasks and which within the decade can be used to guide transit and private vehicles. Data taken under fog and rain conditions indicates that the proposed system will operate under these prevailing environmental conditions in metropolitan California cities. Techniques and materials are under investigation for application of vision under snowy conditions. The system will be integrated with a vehicle steering controller presently under development.

Appendix. References

Bender, J.G. (1991). "An overview of Systems Studies of Automated Highway systems." *IEEE Trans. Vehicular Technology*, 40(5), 82-99.

Dods, J.S. (1982). "The ARRB Lateral Position Indicator." *Technical Manual ATM No. 15*. Australian Road Research Board, Victoria, Australia.

Ishikawa, S., Kuwamoto, H., Ozawa, S. (1988). "Visual Navigation of an Autonomous Vehicle Using White Line Recognition." *IEEE Trans. Pattern Anal. Machine Intell.*, 10(5), 743-748.

Kuan D., Phipps, G., Hsueh, A. (1988). "Autonomous Robotic Vehicle Road Following." *IEEE Trans. Pattern Anal. Machine Intell.*, 10(5), 648-658.

Observation and Modeling of Mobile Environment

Hannu Hakala[1], Jari Kaikkonen[1], Pentti Mattila[1]

ABSTRACT

Perception of the environment in a mobile robot is a difficult task. Other moving objects and various changes in the environment and motion of the robot must be handled together.

Control of mobile movements in a known or especially in an unknown environment involves knowledge of exact details of other objects in surrounding areas. These details may include information such as instant position, speed and direction of movement, dimensions and collision possibility.

The observation of environment is done by using various sensors, which can give information about distances and directions of other objects, which can be either stationary or moving. Depending on the accuracy demands of the application the number and complexity of sensors may vary a lot.

Sensor signals are preprocessed to minimize the amount of data, that is transferred ahead to main processing units. To make this processing simple and easy data from different sensors must be presented similarly. For each transducer type there is an own filter to produce a common data representation. This representation is done in such a way that adding and removing of sensors is possible without modifying other parts of the system.

The modeling of the mobile environment is based on the combined information from various sources. These sources include a priori data, e.g. maps of the area where mobiles are used, measured distances, and data of other objects. This environment model could be described as a dynamic map of the mobile environment.

Environment observation

The observation of environment is done by using various sensors. They can give information about distances and directions of other objects, which can be stationary like walls or edges or which may be moving such as other mobile robots or for example doors. Additional properties of targets can be searched and detected with other tranducers, like light sensitive, infrared, smoke detection, and acceleration sensors.

Depending on the accuracy demands of the application the number and complexity of sensors may vary a lot. The amount of data is huge and must be processed fast and efficiently to make real time use possible. To ensure fast enough responses only a part

---

Machine Automation Laboratory, Technical Research Centre of Finland, P.O.Box 192, SF-22101 Tampere, Finland, Tel. +358-31-163626, Fax +358-31-163494
email: Hannu.Hakala@kau.vtt.fi, Jari.Kaikkonen@kau.vtt.fi, Pentti.Mattila@kau.vtt.fi

of the sensor data is allowed to go through to processing stages. The selection must be made properly and only the most necessary data should be processed.

Sensor signals are preprocessed to minimize the amount of data, which is transferred ahead to main processing units. To make this processing simple and easy data from different sensors must be presented similarly. For each transducer type there is an own filter to produce a common data representation. This representation is done in such a way that adding and removing of sensors is possible without modifying other parts of the system.

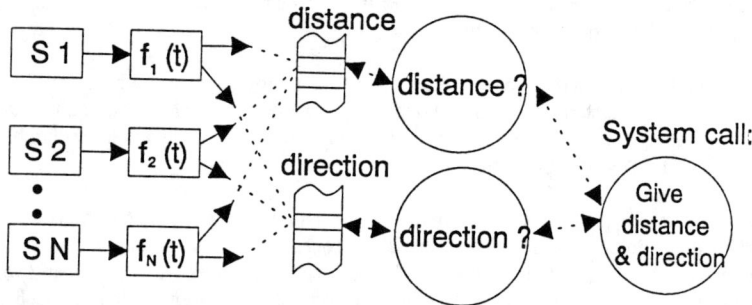

Figure 1. Common data representation for sensor signal

Combining sensor data

The modeling of mobile environment is based on the combined information from various sources. These sources may include maps of the area where mobiles are used, measured distances of that area, data of objects which are in that area and the sensor data. This environment model could be described as a dynamic map of the mobile environment.

Environmental data is updated according to the changes in the surroundings. Updating will be done if the checking of the changes through various sensors gives the same result or sensor readings are multiple times enough accurately same.

Additional information can be given by external systems through communication links. This external information exchange makes it possible to test system functions quite efficiently during prototyping.

Difficulties in modeling

Modeling of a mobile environment is not a very difficult task - in principle and when the environment is at least partly structural. A model can be created easily by forming a data base and feeding in all available information of the environment. This information will include locations of different objects in the environment (walls, ceilings, furniture, trees, rocks, other mobiles), information about properties of objects, information about temperature, weight limits, lighting conditions etc. . Feeding in all this enormous amout of data is possible but it consumes a lot of time.

Difficulties will start when a model must be embedded into a real mobile robot, where there are quite hard limits to the size of the data base and processing capacity. Additionally, when a mobile robot should be able to act in real time, the needed data must always be available quickly. Perhaps the most difficult part in mobile environment modeling is to determine what data should be stored in the model and in which form so that basic requirements could be fulfilled.

More difficulties will arise when operating environment of the mobile is not structural. In such cases, e.g. in forests, it is very difficult to generate a usable environment model in advance so it must be generated with sensors during action. This will cause more problems to data representation because it is difficult to find out if a sensor reading is correct and wheather it should be combined with a previously determined data.

Environment model types

In our approach two kind of environment models are used: accurate model and rough model. Criterias for creating these models are different but they are used almost by the same way. Interfaces to these models are designed so that it is possible to use same data base search calls in both cases.

When using environment model for simulation purposes the size of the model is not a limiting factor. Almost always there is enough disk space in use for the model. Timing is not as important as it is in case of a real mobile robot and it is easy to increase processing capacity if needed.

A rough environment model is needed in the mobile robot. Accuracy and complexity of the model depends much on the application. In general it should be as simple as possible. In some applications there is no need for an environment model at all. In those cases a very simple environment model is formed with sensors or the actions of the mobile are planned on behavioural bases and are taken according to sensor perception. The capabilities of these mobiles may be limited when compared to systems where environment model is used. On the other hand this kind of an approach is very useful when the mobile is used in an unstructured and unknown environment and when the mobile robot have to response to unpredicted situations.

In all these cases it is essential that data is organized properly and that each moment all the necessary data can be found easily and fast.

Data representation in models

It is important that data representation is carefully fitted to each application optimally. With suitable data representation it is possible to limit the size of the model significantly and to simplify software structure. [1]

Sensors and actuators are connected to environment models with multilevel data interfaces. This allows them to be changed flexibly and enables their replacement with simulated sensors which is very useful feature when simulating system operations.

Interface levels can vary from raw sensor data to strongly processed data which will give e.g. description of local environment objects. Processing can be done either by main processor or by special sensor processor located near or in the sensor block.

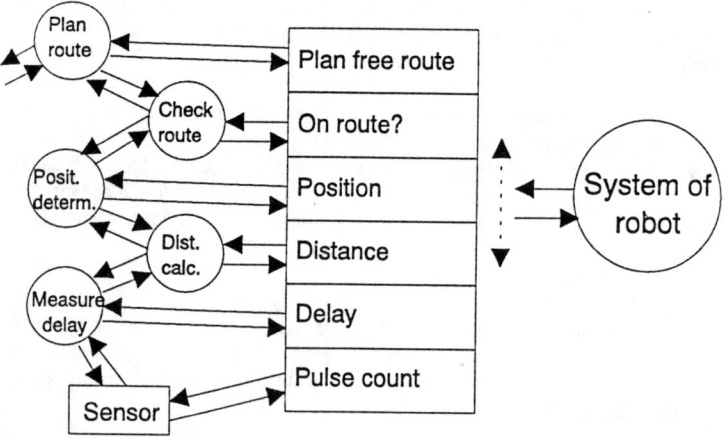

Figure 2. Multilevel data interface for sensors and actuators

Data utilization

It is not enough that the data can be stored in the model. It must be used efficiently, too. Data utilization can be done sophisticated by defining a set of functions which perform special data base operations. These functions can be defined so that they are not depending on realization of the environment model.

As an example there could be a function which returns information of mobile robot's local environment: objects, obstacles and free routes. The realization of this function is reasonable to do so that the answer includes same basic information of the local environment even if there were no environment model at all. In this case the knowledge would be obtained by using sensors to figure out the structures of the environment and returned information would include estimation about correctness of delivered information. In some applications local environment information could be obtained from e.g. commercial Geographical Information System (GIS) through a radio link.

Data updates

After enormous amount of data is fed into environment model one potential problem will arise: how to update the model? This is quite a difficult and heavy task. If a normal office environment is considered there is a lot of information which will change even daily. Chairs may be moved from place to another. One potential problem is all the people who move around the office and will not be found from standard place (except in government offices, perhaps).

Updating can be performed in two ways. Basic environment data should be checked regularily. Additionally, the mobile robot itself must be able to update its model during actions. Special rules must be made according to which mobile robot can compare the model and the results based on measurements and decide wheather update actions are necessary. A rule may tell that an object called chair has such properties that it can easily be moved - so a change to environment model can be done easily and with quite a low error probability.

If application allows using GIS data bases as environment models or as parts of the environment model updating situation is easy because they are regularly updated by various official authorities.

Physical realization

We have created an environment model of our laboratory for test purposes. We measured all the rooms and corridors of the laboratory (some 650 square meters) to obtain an accurate map. Measuring accuracy was 0.5 cm (0.2"). Additionally, we fed in information about furniture and their location in laboratory. Data was fed in mainly graphically and all the information was fixed to geometrical points.

Preprocessing was done on PC/386 workstation with ANVIL 1000MD software. Postprocessing was done with PATRAN software on SUN 4/60 GX unix workstation. A part of geometrical information was transferred from the neutral file format of PATRAN into format of a JC sensor simulation software.[2] JC software was used to visualize the results. Additionally, neutral file format was converted to Ingres data base format. Ingres will be used as a model control tool.

Conclusions

Methods for observing and modeling the environment of a mobile robot have been developed. These methods can be used both in simulations and in real use and they can used to estimate the capabilities and functionality of a mobile robot.

Three potential problem sources exist: model creation takes a lot of time, sensor perception includes a lot of error sources and today there are very few tools which are usable.

References

[1] H. Hakala and J. Kaikkonen, New Approach to Data Representation in Mobile Robot Models, in Proc. IEEE Conf. Systems Eng., Pittsburg, Aug.1990, pp 56-59.

[2] Jari Ojala, Kenji Inoue, Ken Sasaki, and Masaharu Takano, Development of an Intelligent Wheelchair Using Computer Graphics Animation and Simulation, presented in First Eurographics Workshop on Animation and Simulation, Zürich, Oct. 1990.

MEASURING HIGHWAY INVENTORY FEATURES USING STEREOSCOPIC IMAGING SYSTEM

Hosin Lee[1], Michael A. Weissman[2] & Joseph P. Powell[3]

Abstract

The new stereoscopic measurement approach is discussed in this paper. The objective of this paper is not intended to present another new videologging system. Instead, this paper is to present a stereoscopic measuring capability which can be easily added to the existing videolog system with minor modification. Preliminary analysis results show that this stereoscopic measurement technique can be applied for measuring roadside features with a reasonable accuracy.

Introduction

Photologging of roadways has been conducted by many cities, counties, and states, including the state of Washington since the mid-1970's. The general objective of photolog system is to allow access to visual information about any section of roadway in the office rather than requiring an expensive trip to the road site.

Recently, video technology has been introduced to such applications, attempting to replace the traditional 16 and 35mm film format used in photologging. Both Wisconsin and Connecticut Departments of Transportation have demonstrated successfully the use of optical laser disc storage and retrieval. Furthermore, the concept of direct image capture without the film conversion to laser disc in videologging process has been demonstrated by the Iowa Department of Transportation (McCall 1990). These electronically stored videolog images on laser disc can provide the capability for instantaneous retrieval of images and linking the images with other computerized highway data files in pavement management systems (PMS) or geographic information systems (GIS).

[1]Assistant Professor, Civil & Environmental Engineering, Washington State University, Pullman, WA 99164-2910
[2]President, Microscience, Inc., 31101 18th Avenue South, Federal Way, WA 98003.
[3]Vice President, Engineering, Pavedex, Inc., E. 9514 Montgomery Ave., Suite #26, Spokane, WA 99206.

As videologging procedure is becoming more and more popular, additional use of the videologged images is being explored in this paper. That is the stereoscopic measurements from the videologged images of highway inventory features including size and location of various traffic signs, highway geometric, bridge clearances, curb heights, roadside slopes, and so on. The concept of measuring necessary highway inventory features is demonstrated using a small sample of stereoscopic measurements.

Stereoscopic Imaging System

The photographic reproduction of stereo pairs has been available almost as long as photographs themselves, and some very important applications have been developed in engineering area. However, the general usage of stereoscopic imagery has been restricted because stereo pairs of photographs are difficult to view (Weissman 1988). Recent advances in digital image processing is opening up the potential for the application of stereoscopic techniques in many new areas such as highway inventory videologging as discussed in this paper.

In a 2-D digital image of a surface, lengths are measured by converting pixel size to physical distances. However, in a two-dimensional image of a 3-D scene, lengths cannot be determined because the conversion factor depends on the distance that subject is away from the camera. The stereo system gives us the ability to determine that distance, via triangulation.

A typical highway inventory measurement system would have two cameras mounted so that the separation distance of their focal points and the angles of their lines of sights are known. This leads to a mathematical relationship between the pixel positions in the two images and physical x-y-z locations.

Microscience has developed software, called SIS (Weissman), which provides the capture of video images, image enhancement, storage and recall of stereo pairs, and alignment techniques needed for basic viewing of the 3-D scene. Once images are aligned using SIS, range equations are applied to determine distance from the cameras (Weissman).

Pavedex Highway Inventory System

The state of Washington recently contacted Pavedex, Inc. of Spokane, Washington to conduct a videolog demonstration. This demonstration project was divided into four phases, and efforts made for the phase 4 are discussed in this paper. The scope of the overall videologging demonstration project is summarized below.

Phase 1: Video Image Data Acquisition:

Approximately 75 miles of roadway selected by WSDOT were inventoried by the Pavedex survey vehicle. The basic highway features evaluated by Pavedex includes approximately 30 miles of pavement marking for inventory, 25 miles of perspective view, and 5 miles of shoulder/drainage view.

Phase 2: Image Database
Video images of all signs, all pavement markings, and one video image every 50 feet for each two-mile pavement section were stored on optical disc and the random retrieval of the images by accessing an image database was demonstrated.

Phase 3: Computer-Assisted Video Inventory
A roadsign inventory was collected for 30 miles of roadway. Each video frame is digitally encoded with a project-segment code and the distance from the beginning of the project-segment. An operator, viewing the video tape, enters the type and condition of various traffic signs.

Phase 4: Stereoscopic Image Measurement
A stereo measurement system was demonstrated that enables 3-D measurement of various elements from a video image. Heights and Locations of sample street signs were measured for the demonstration of the system accuracy.

Video Camera Calibration

Most video cameras have been reported to exhibit relatively large radial and decentering distortion. Magnitude of these effect may be in excess of the pixel size at the sensor plane (Fryer 1989). Video camera lens distortion causes an incoming ray of light to be deflected from its original direction. Therefore, calibration of the video camera is essential in any photogrammetric measurement system (Moffitt 1980).

For this demonstration project, the calibration of the video camera was completed by using a cubic frame in the field of view. The video camera lens was focused at infinity during calibration process because most video camera lenses have been constructed with their minimum distortions occurring at a principal distance equal to their focal length (Fryer 1989).

The corrections for systematic error can be determined by this calibration method by means of additional measurements in a triangulation routine. This resectioning method is known to result in much better refined imagery for subsequent use and thus in better accuracy of the measurements (Gruen 1989).

The size of cubic frame was 96 x 72.25 x 144.5 inches, and the frame was located about 20 feet away from the video camera. The pixel resolution to be used for measuring distance using the video images is shown in the Table 1. Eight measurements were made from the cubic frame for the calibration of the video camera which were used to identify the relationship between pixel positions and the physical distances as shown in the Table 2.

Table 1. Pixel resolution for current system configuration.

Distance from Camera	X distance per pixel	Y distance per pixel	Z distance per pixel
20 feet	0.46 inch	0.74 inch	1.33 inch

Table 2. Measurements made on eight corners of the cubic frame.

Measure No.	Measured by Video			Measured by Tape		
	x	y	z	x	y	z
1	0.0	0.0	0.0	0.0	0.0	0.0
2	96.0	0.0	0.0	96.0	0.0	0.0
3	96.0	72.3	0.0	96.0	72.3	0.0
4	0.0	72.3	0.0	0.0	72.3	0.0
5	0.3	0.8	140.6	0.0	0.0	144.5
6	95.9	-0.4	145.3	96.0	0.0	145.0
7	96.0	72.2	145.2	96.0	72.3	144.5
8	0.5	71.8	140.7	0.0	72.3	144.5

3-D Measurements of Highway Inventory Features

Measurements were made from the videotape where pictures were taken using Pavedex videologging vehicle. Image pairs were digitized from the video tape. After the operator selected points in the two images of the pair using a mouse, the distance between the selected points on the screen was computed. The measurement of the dimensions and height of highway inventory features was attempted using stereoscopic 3-D Imaging system. The measurements were computed using the Microscience software, and compared against the physical measurements made by the use of the surveying tape.

The measurements were computed using Microscience SIS software, and compared against the physical measurements made by the use of the surveying tape. The five measurement results are shown in the following table 3.

Table 3. Preliminary 3-D Measurement Results.

Measurement Object	Measured by Video	Measured by Tape
Concrete Post	67.1 in	14.0 in
Concrete Post	166.5 in	150.0 in
Concrete Post	151.6 in	164.0 in
Height of Post	27.6 in	28.5 in
Sign Post	64.8 in	80.2 in

The accuracy of the stereoscopic measuring system seems somewhat questionable, and it could have been affected by (El-Hakim 1989): 1) video camera resolution, 2) video camera orientation, and 3) quality of the calibration. The camera set-up could have been better designed in order to capture the images for stereo measurement. Optimum camera set-up configuration would definitely improve the precision and repeatability in addition to the accuracy of the system.

Summary and Conclusion

As part of photologging demonstration project conducted by Pavedex for the Washington Department of Transportation, 3-D measurements of highway inventory features were made such as the size and height of the traffic sign. Other measurements can be made on bridge clearance, curb height, roadside slope, stopping sight distance, etc.

The concept of measuring necessary highway inventory features is demonstrated in this paper using a small sample of stereoscopic measurements. Based on the limited set of data, the stereoscopic measurement system seems to show fairly consistent results except one significantly different measurement. However, the accuracy of the system can be improved by increasing the video camera resolution and the quality of the calibration procedure

It is believed that more precise measurements can be made from video images using a stereoscopic image measurement system through optimum camera set-up configuration. Continued development of the stereoscopic measurement techniques with more sample test measurements will be required to improve video camera optics and calibration techniques.

References

1. El-Hakim, S. F. (1989). "A Hierarchical Approach to Stereo Vision," Photogrammetric Engineering and Remote Sensing, Vol. 55, No. 4, April 1989, pp. 443-448.
2. Fryer, J. G. and Mason, S. O.(1989). "Rapid Lens Calibration of a Video Camera," Photogrammetric Engineering and Remote Sensing, Vol. 55, No. 4, April 1989, pp. 437-442.
3. Gruen, A. W. (1989). "Digital Photogrammetric Processing Systems: Current Status and Prospects," Photogrammetric Engineering and Remote Sensing, Vol. 55, No. 5, May 1989, pp. 581-586.
4. McCall, B. and Whited, J. (1990). "Video Imagery Systems for Highway Applications," Report No. FHWA-DP-90-085-004, FHWA, May 1990.
5. Moffitt, F. H. and Mikhail, E. M. (1980). "Photogrammetry," Third Ed., Harper & Row, Inc., New York, 1980.
6. Weissman, M. A., "Using Stereoscopic Imaging," Advanced Imaging, pp. 24-27.
7. Weissman, M. A. (1988) "Stereoscopic 3D Imaging," ESD: The Electronic System Design Magazine, April 1988, pp. 59-64.

Subsurface Pavement Structure Inventory
Using Ground Penetrating Radar
and a Bore Hole Camera

Mr. M.Inagaki[1] Dr. H.Tada[2] Dr. A.Kasahara[3]
Mr. H.Tomita[4] Mr. T.McGregor[5]

Abstract

The Pavement Management System (PMS) is now a necessary pavement evaluation tool. The PMS data base has been historically fed with surface inspection data, but increasing demands for accurate predictions require subsurface pavement condition data. A highly effective subsurface inspection technique marries Ground Penetrating Radar (GPR) with Bore Hole Camera (BHC) video technology.

Introduction

Surface condition surveys provide data base information to the PMS which target rut, crack, and profile measurements. Falling Weight Deflectometer (FWD) measurements indicate the load bearing capacity of the combined subsurface strata without determining the composition of these strata. Historical subsurface structure data generally is either not accurate or does not exist. In order to predict the life of a pavement and eventual failure, the current condition of the pavement must be determined.

----

[1] Director of Analysis, Geo Search, Co., Ltd. 22-19 Unane 1-Chome, Setagaya-Ku, Tokyo, Japan.
[2] Chief Director, Road Management Technology Center 2-15-15 Ningyo-Cho, Nihonbashi, Chuo-Ku, Tokyo, Japan. (Affiliated with Ministry of Construction)
[3] Professor, Department of Civil Engineering, Hokkaido Institute of Technology, 419 Teine Maeda, Teine-Ku, Sapporo, Hokkaido, Japan.
[4] President, Geo Search Co., Ltd.
[5] President, Subsurface Imaging, Inc., 1400 Hermann Drive, Suite 4-C, Houston, Texas 77004

Surface inspection and FWD inspection are useful and necessary but do not provide enough data. Additional data acquired by a GPR system which incorporates a BHC effectively solves this insufficient data problem.

Ground Penetrating Radar

GPR is widely used in environmental and civil applications, but the speed capability of the traditional "bow-tie" antenna GPR systems has been too low (2-3 km/hr) for them to operate safely in a high vehicle traffic environment. The introduction of the horn-type antenna GPR system has allowed higher frequencies to be employed which enable the system to gather data at a more rapid rate (40-60 km/hr) providing higher shallow depth resolution representations of the target pavement.

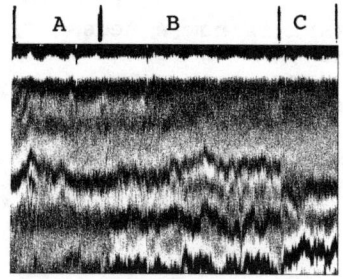

Figure 1. GPR trace of pavement structure showing segmentation.

Figure 2. GPR data acquisition vehicle in operation.

The Bore Hole Camera

GPR systems are limited in their ability to determine thickness of substructure strata. In order to calibrate the GPR system, ground truth data is required. Ground truth data is usually obtained by taking a standard 4" (100mm) internal diameter core sample. Taking such a core and repairing the pavement afterward is time consuming, destructive to the pavement and also disturbs the in situ nature of the target substructure; voids and loose sands collapse inward as the core is removed making them invisible to the pavement engineer. A system called a "DOUROSCOPE" after the Japanese word "douro" which means "road" or "highway," which has been developed by the company Geo Search, Co., Ltd., of Tokyo, Japan, that overcomes this ground truth problem.

The DOUROSCOPE system is quick (a data acquisition sequence takes 40 minutes start to finish), the data is accurate and easy to read, and repair of the bore hole is easy and minimally destructive to the overall pavement structure.

Figure 3. Operator boring hole prior to insertion of DOUROSCOPE data acquisition probe.

Work flow: A 25mm internal diameter hole is bored 150cm into the pavement structure, then a video camera mounted in a long probe is inserted into the bore hole. The control unit is switched on and a video record is taken of the internal surface of the bore hole as the feeding device withdraws the probe. When the camera lens of the probe reaches the surface, the control unit is switched off and the hole is repaired with a fast cure concrete. The control unit produces a VHS taped record of the down-hole probe travel that is flattened to a two dimensional image that scrolls on the monitor screen during playback. In post-processing, video capture of target frames and paste-up to a presentation format are performed.

The DOUROSCOPE record as shown in Figure 4. is produced from the VHS tape record of the penetration by frame capture photography with three frames captured per photograph. The captures are then placed on a presentation board and a color copy is taken for presentation to the pavement engineer. The depth notations to the left of the photo record of the DOUROSCOPE penetrations are inserted by an on-board title generator synced to a pulser on the feeding device roller, thus providing highly accurate depth gradations on the visual record. The data acquired by the GPR system requires dielectric constant information for calibration which can be determined from visual identification of material composition from the DOUROSCOPE record which provides a high degree of articulation.

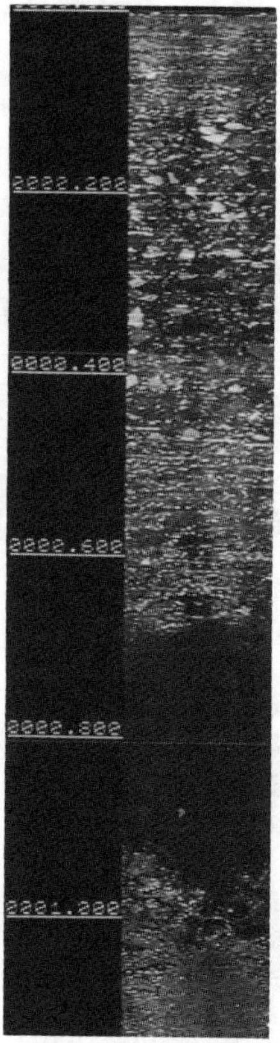

Wearing course (fine asphalt)

Binder course (rough asphalt)

Base material (cement mixed gravel)

Void

Sub base (crushed stone)

Sub grade (sand)

Figure 4.

Data Analysis

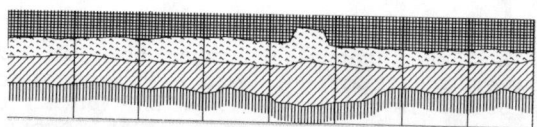

Figure 5. Subsurface map produced using automatic graphic software.

It is essential that the data report be easy to read and contain a clear record of the pavement substructure. The above record was produced by processing gray scan chart records of GPR data with a digitizing pad and stylus, and then data is inserted into a graphics program that produces easily readable patterns of strata configuration, size, and depth.

Other Applications: Void Detection

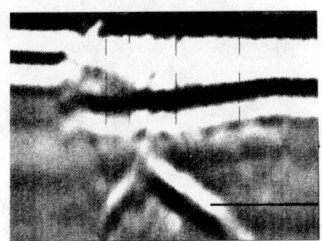

Void Signal

Figure 6. GPR data showing subsurface void.

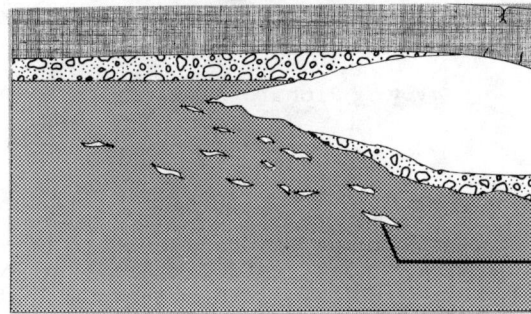

Void

Loose Sand

Figure 7. Subsurface void in actual and formative stage.

A loose sand condition which can be detected by the DOUROSCOPE has a close relationship to the formation of void areas. Loose sand areas can be the result of water migration due to sewer or water main failures, changing migration patterns of ground water, or inadequate compacting of sands at the time of pavement construction.

Conclusions

The GPR system can determine layer thickness and locate voids, and the Bore Hole Camera can verify the GPR results and provide ground truth data to calibrate the GPR data. Together, these two systems add accuracy and functionality to more common data acquisition techniques. The GPR/Douroscope system is not intended to replace surface and FWD inspection, but rather to complement them and provide the pavement engineer with as accurate a data stream as possible. This procedure increases the value and prediction accuracy of the PMS, thus providing industrial and public users of pavement infrastructure with maximum quality at a minimum cost.

MINNESOTA DEPARTMENT OF TRANSPORTATION EXPERIENCE IN THE APPLICATION OF ADVANCED TRAFFIC MANAGEMENT SYSTEMS

by: Richard A. Stehr[1]

Introduction

The Minnesota Department of Transportation (Mn/DOT) began deploying Advanced Traffic Management Systems (ATMS) in the early 70's. Interstate 35W just south of the Minneapolis Central Business District was experiencing severe reoccurring congestion. A system of ramp meters, TV surveillance cameras, and vehicle detectors were installed and a Traffic Management Center (TMC) was built to monitor and control the system.

Until 1988 the system expanded slowly. In the fall of 1988 a major program expansion was declared with the goal of achieving total freeway system coverage for metropolitan freeways by 1995. New motorist information services and several demonstration projects were launched. The program has achieved much in spite of problems encountered.

----

[1] Director, Office of Traffic Management, Metropolitan District, 395 John Ireland Boulevard, Room 120, St. Paul, Minnesota 55155

Program Achievements

The program is on track to achieve its goal of system-wide coverage of all 260 miles of freeway by 1995.

	Actual 1988	Actual 1991	Planned 1993	Planned 1995
Ramps Metered	66	213	391	535
% of Total	(10%)	(32%)	(59%)	(80%)
Surveillance Cameras	21	59	109	260
% of Total Miles	(8%)	(23%)	(42%)	(100%)

Surveillance includes installation of a fibre-optic communication system and vehicle detection equipment.

Demonstrations of other ATMS features were also built and are being evaluated.

Demonstrations

Call Boxes - Push button	90 boxes	22 miles
Accident Investigation Sites	20 sites	14 miles
Cameral Surveillance on Tall Buildings	5 sites	7 cameras

An area-wide highway advisory radio service was established and supplemented with dynamic signing to support the service. This service broadcasts live traffic reports from the TMC every ten minutes during peak travel periods and broadcasts continuously during major incidents.

The highway advisory broadcasts are supplemented by 27 freeway signs strategically located on monitored sections of freeway. Flashing lights on these signs can be activated individually or in groups by a radio signal from the Traffic Management Center. These lights are activated when an incident occurs downstream from the sign. Continuous broadcasting on 88.5 FM provides information to the motorist when the lights are flashing.

Problems Encountered

The public reacted negatively to the initial operation of new ramp meters. Ramp meters were installed and operated on flashing yellow for several weeks. Suddenly one morning metering would begin on a group of 6 to 8 ramps. Metering rates were initially based on the best engineering guess at the rate required to maintain smooth mainline operation and adjusted with experience. On the first day of metering, ramp queues were long and would often block local intersections, buses would miss connections. The result was that Mn/DOT would receive 25 or 30 letters of protest and a dozen angry phone calls from motorists or transit riders.

The negative public reaction was reduced by easing the transition to full metering of ramps. Signs were placed a week in advance of beginning signal operation which announced which day of the following week the signals would begin operations. Initial metering rates were set as rapid as possible and then gradually reduced to the desired rate over the next week or two. Complaints dropped to one or two for each new group of meters turned on.

Call boxes received negative publicity also. The State Patrol was not a full participant in the decision to install push button call boxes and would have preferred voice call boxes, yet they were expected to use and operate the system. The newspapers picked-up on the negative reaction of the patrol and also the high initial cost of the system (over $10,000 per box installed when the whole system was averaged over the 90 locations). A draft evaluation indicates the system may not be cost effective as a congestion reduction measure even though it may be desirable as a safety measure or motorist service.

The area-wide highway advisory radio service also aroused some controversy. Initial inquiries among commercial stations failed to generate any interest in providing frequent traffic reports. They also indicated that no radio station would allow continuous program interruption for major traffic incidents. Once Mn/DOT did contract with a non-profit radio station for the service, private stations and a private traffic reporting service protested. It was felt that the public sector was inappropriately competing with the private sector, and that the private sector could do a much better, more comprehensive job. In response, a request for proposals was issued to contract with private providers for many additional motorist information services. A strong legislative effort by an unsuccessful proposer to stop the contract succeeded. Individual elements of the plan to expand motorist services, such as cable TV broadcasts and aerial surveillance, then had to be pursued separately, but they were delayed a year or more by the controversy.

Future Plans

Evaluations of call boxes, accident investigation sites, and cameras on tall buildings will lead to decisions whether or not to continue their current operation or broadly expand their use.

Other means of detection will continue to be researched, demonstrated, and, if the demonstrations are successful, broadly deployed. One such device, called Autoscope, has gone through nearly five years of research and development. This video image detection system processes video images to identify vehicle presence in lanes on a TV screen. Three of these detectors have been undergoing field testing for nearly a year. Fifteen additional Autoscope detectors will be deployed this summer along I-394. Several others will be deployed at signalized intersections.

Other ATMS systems planned for field tests this summer include infra-red detectors, radio data communication devices, portable TV surveillance cameras with satellite communication, and a cable TV traffic broadcast service. The search continues for less expensive or more flexible systems and equipment.

Research is continuing on traffic modeling and simulation. KRONOS, a very promising freeway simulation program developed by the University of Minnesota will be field tested on three corridors to evaluate alternatives such as HOV lanes, phased construction, and interim improvements. It is hoped this tool will eventually become the real time simulation tool needed for advanced traffic management systems.

The fibre-optic communication system being installed is intended to provide the public infrastructure "backbone" on which future IVHS systems can be deployed. It has tremendous capacity for two-way communication of data and video images, and should make future roadside monitoring devices or vehicle to roadside communication devices easy to install. This backbone system should also be the key to linking local street signal systems and surveillance systems to a central control center or centers.

Once the database from surveillance is complete and reliable a broad array of technologies is currently available to disseminate that information to motorists. Audible FM broadcasts is currently the prime method of distributing this information, with cable TV and phone information services soon to follow. It is expected data transmission through FM radio or TV FM frequencies or cellular phone will follow shortly. Each of this will be fairly simple to implement centrally once the data is available and their deployment will primarily depend on how quickly devices can be installed in the vehicles.

This is an exciting time for traffic operations. Minnesota is anxious and well positioned to take advantage of the rapidly developing systems. We have the experience, we have the system in place, and we have the commitment. We look forward to sharing our knowledge and experience and having others sharing their success and failures too so we can all move ahead more quickly.

Academic Research Efforts in the GuideStar IVHS Program

Robert C. Johns[1]

Minnesota Foundations for IVHS

Transportation research at the University of Minnesota has dramatically increased in the last four years, coinciding with the establishment of the University's Center for Transportation Studies. The Center attracts and administers research funds, and manages transportation education and outreach programs. Four broad research emphases have evolved in the Center's research program: Transportation Safety and Traffic Flow, Transportation and the Economy, The Transportation Infrastructure, and Transportation and the Environment [1].

The first emphasis, Transportation Safety and Traffic Flow, has been built around the core group of faculty in the Civil and Mineral Engineering Department who focus on transportation engineering. As at certain other universities, this group began looking at advanced technologies for transportation in the late 1970s and early 1980s, well before the title Intelligent Vehicle/Highway Systems was coined. Research efforts addressed modeling and simulation of traffic flow dynamics, advanced traffic control strategies, and vehicle detection. Many of these early efforts have continued and expanded, and now make up important elements of the GuideStar IVHS research program.

While transportation research in advanced technology was increasing at the University, the commitment to implementing advanced technology for traffic management on Twin Cities freeways continued to grow at the Minnesota Department of Transportation (Mn/DOT). Currently, 59 miles of the Twin Cities freeway system are managed using a sophisticated monitoring and ramp metering system. By 1995, 260 miles of freeway and 40 miles of expressway will

---

[1]Associate Director, Center for Transportation Studies, University of Minnesota, 110 CME Bldg., 500 Pillsbury Dr. S.E., Minneapolis, MN, 55455.

be implemented, resulting in approximately 525 ramp metering signals in the Twin Cities.

The GuideStar Program

When the Center for Transportation Studies was established in 1987, it became clear that a closer relationship between the University and Mn/DOT would be valuable, given the work already under way in both organizations. The Center began facilitating discussions with Mn/DOT in 1988 to look for ways to implement University technology.

At this same time, national attention to advanced technology in transportation was increasing rapidly. Mn/DOT and University officials realized that Minnesota was among the leaders in both research and implementation of advanced technology. In 1989, they decided to develop a formal partnership for the development of advanced vehicle/highway systems technologies. It was concluded that the technical components of this joint program would emphasize advanced traffic management systems (ATMS) and advanced traveler information systems (ATIS), and that the program name would be GuideStar.

The Development of GuideStar Research Efforts

Early GuideStar meetings resulted in considerable discussion about the roles of the University and Mn/DOT. To help clarify roles, a matrix was produced that showed development steps in four column headings: basic research, applied research/development, field testing, and deployment. The matrix listed University and Mn/DOT activities, and proved useful for making plans that horizontally integrate academic research efforts with development and implementation activities.

Research efforts in the matrix also needed to be expanded and integrated vertically. The Center for Transportation Studies attracted additional faculty to the IVHS area through requests for proposals. Nine faculty members, representing civil and mineral engineering, computer science, human factors, and aerospace engineering, worked together to develop a research framework that outlines how the various research components relate to each other.

GuideStar Research Components

There are seven major components of the academic research efforts for GuideStar. The summaries below describe them individually and their relationships to each other.

<u>Real-Time Traffic Surveillance/Detection</u>. The University has made a major breakthrough in developing a new

technology that uses image processing for wide area multiple detection and automatic traffic surveillance. This technology uses cameras and image processing to collect real-time data about traffic flows, providing true wide area detection, more data extraction than loops, and increased flexibility and efficiency in placement and maintenance [2]. A private company has developed this technology into a product called AUTOSCOPE^{TM}. Data collected by AUTOSCOPE will be used in ATMS strategies and as input for ATIS. Mn/DOT and FHWA are funding a laboratory on I-394 in Minneapolis that will ultimately have 38 AUTOSCOPE cameras approximately every 1000 feet per direction on a 3.5-mile segment. Current research is focusing on the development of automatic incident detection and on intersection control, both using image processing.

Modeling and Simulation of Traffic Flow. Traffic flow models help traffic managers understand the impacts of control strategies, incidents, construction, and other system changes. The University has worked for several years on the development of a simulation model called KRONOS [3]. Mn/DOT is testing this model for operation in its Traffic Management Center. Research is continuing to improve KRONOS using AUTOSCOPE data for calibration and to improve the model's computer efficiency using parallel processing and other techniques.

Origin-Destination Estimation and Demand Prediction. Even with the installation of systems such as AUTOSCOPE, there will always be the need to estimate and predict traffic patterns. For example, ATMS and ATIS require the prediction of travelers' origins and destinations on a real-time basis -- to support real-time control and improve traffic flow models such as KRONOS, and ultimately to provide real-time routing information to drivers. The University is developing mathematical techniques for estimation and prediction that will be integrated with the development of traffic modeling and navigation tools [4,5].

Demand Responsive Management/Control. In an effective ATMS, the improved data, modeling, estimation, and communications should lead to management and control decisions that improve the real-time performance of the transportation system. Current efforts are focusing on improving Mn/DOT's library-based system to automatically set the timing of ramp meters during the day [6]. Research is also under way to improve the coordination of intersection control. Ultimately, control systems will be developed to link freeways with arterials, so that corridor and network performance can be optimized based on user-defined objectives. Research will also improve other decision-making and management tools, such as incident management and guidance/information systems.

In-Vehicle Navigation. ATIS innovations have been introduced that give drivers navigation information through maps or head-up displays. GuideStar research in this area is based on the important links between ATMS and ATIS. Ultimately, an effective ATIS needs to use much of the same real-time information used by ATMS. University human factors and computer science researchers have developed a driver simulation laboratory that is being used for research efforts [7]. A Honda Accord connected to an advanced graphics system allows a person to "drive" on simulated Minneapolis freeways. Research will test alternative ways of providing navigation information to the driver and their impacts on driving behavior, and will also develop joint ATIS/ATMS databases.

In-Vehicle Collision Avoidance. An important goal of IVHS is to improve transportation safety. Improving system management through ATMS should reduce accidents, but research is also needed to understand driver behavior. With new technologies and management strategies being implemented, it is critical to know their impacts on the driver. Research using the driver simulation laboratory described above has begun to look at these issues, with a special focus on driving of the elderly [7]. Research is also developing tools that can evaluate road segments and their safety impacts for various driving populations [8].

Database Design and Information Systems Management. For IVHS technologies to be successful, large amounts of data will need to be stored, manipulated, and updated on a real-time basis. Many of the ATMS and ATIS research efforts described above will use common databases. GuideStar research efforts in this area will increase as the tools are linked together, developing information systems for traffic managers, drivers, and transit riders.

Resources for GuideStar Research

Funding sources for GuideStar academic research have included Mn/DOT, federal agencies, the Minnesota Supercomputer Institute, and the Center for Transportation Studies. Funding since 1987 is shown below:

Research Components	Funds Awarded Since 1987
Real-Time Traffic Surveillance	$2,400,000
Traffic Modeling/Simulation	441,000
O-D Estimation & Demand Prediction	270,000
Demand Responsive Management/Control	607,000
In-Vehicle Navigation	75,000
In-Vehicle Collision Avoidance	93,000
Database Design & Info. Sys. Mgt.	45,000
Total	$3,931,000

In addition, FHWA and Mn/DOT are funding the construction of the 3.5-mile laboratory on I-394, which is costing approximately $3.0 million. In the future, the Center for Transportation Studies will also work to attract private sector funds for research.

Conclusions

1. Interdisciplinary teams of faculty are able to contribute considerably to IVHS research, developing an integrated research program.
2. New roles are required of faculty: 1) to work as a team, across departmental lines; and 2) to work more closely with transportation agencies.
3. A coordinating center at a university provides essential linkages among faculty and to transportation agencies.
4. The commitment of the state DOT to IVHS is essential for joint programs to be successful. Transportation agencies must realize the value of basic research.

References

1. Johns, R.C. (1991). Towards a Balanced Transportation Research Program. *Trans. Res. Record* (in press).
2. Michalopoulos, P.G. (1990). Vehicle Detection Through Image Processing: The AUTOSCOPE System. *IEEE Trans. on Vehicular Tech.*, Vol. 40, No. 1, pp. 21-29.
3. Michalopoulos, P.G., Kwon, E., and Khang, J.G. (1991). Enhancements and Field Testing of a Dynamic Simulation Program. *Trans. Res. Record* (in press).
4. Davis, G.A., and Nihon, N.L. (1991). A Stochastic Process Approach to the Estimation of Origin-Destination Parameters from Time-Series of Traffic Counts. *Trans. Res. Record* (in press).
5. Stephanedes, Y.J., and Kwon, E. (1991). *On-line Demand Diversion Prediction for Integrated Control of Freeway Corridors*. UMSI91/49, Minnesota Supercomputer Inst., Minneapolis.
6. Stephanedes, Y.J., Kwon, E., and Khang, J.G. (1991). *Demand Responsive Ramp-Metering Control to Improve Traffic Management in Freeway Corridors*. UM-CTS-5048-91, Univ. of Minnesota, Minneapolis.
7. Hancock, P.A. (1991). *Design and Development of a Fixed-Base Driving Simulation for IVHS Human Factors Experimentation*. HFRL Technical Report, 91-03, Univ. of Minnesota, Minneapolis.
8. Davis, G.A. (1990). *A Statistical Method for Identifying Areas of High Crash Risk to Older Drivers*. Interim Report to the Minnesota Department of Transportation, St. Paul, Minn.

ADVANCED TRAVELER INFORMATION SYSTEMS

THE VISION BEYOND

H. MILTON HEYWOOD, P.E.[1]

ABSTRACT - This tutorial describes the author's vision of a mature Advanced Traveler Information System (ATIS), the individual modules, including public transportation, of the system, and the integration of the modules that will form the nationwide ATIS system.

TRAVTEK - The TravTek device can serve as an autonomous, self-contained, navigation/information device and includes a map database, guidance computer and a business directory database. Vehicle location is calculated via dead reckoning and map-matching techniques. Advancements in global positioning system (GPS) technology and quantity production may make GPS a practical vehicle location alternative. Optimal routes, in the autonomous mode, are based on historical link times stored in the in-vehicle device. The autonomous in-vehicle device will be useful in any area or corridor where map and link time databases are available.

The usefulness of the device is enhanced when dynamic traffic information and link time data is provided via the infrastructure. The device then displays traffic information that impacts the driver's route, and derives and displays the best route based on the dynamic link times. The development of the TravTek system is a major ATIS milestone and has been an exciting experience for the partners involved in its design and implementation. While TravTek has many features of a mature system, in reality, it is only a prototype system involving 100 vehicles.

PROBES AND BEACONS - The evaluation of TravTek and other projects will provide information needed to determine the value of "probe" data from vehicles (equipped vehicles provide link travel time data to the traffic management center). The traffic management center then combines and sorts ("fuses") this data with other data and information to derive the link travel times that are furnished to the

---

[1]Technical Program Advisor, Advanced Systems Division, Federal Highway Administration, 400 7th St., S.W., Washingt D.C. 20590

vehicles.

Transmission of data from vehicles to the traffic management center via conventional broadcast data radio schemes is probably not practical for a mature ATIS. One likely alternative to wide-area data radio communications appears to be a "beacon" based communications system. Beacons are roadside devices that provide communications between vehicles and the infrastructure instead of the wide-area data radio communications used in TravTek. These beacons are then interconnected with the traffic management centers via cable communications, or conceivably in rural areas, via satellite. Additionally, beacons can transmit location codes to the vehicle thus frequently calibrating the vehicles' locations.

Typically, beacon spacing may be 1.6 km or less in urban areas and from three to fifteen kilometers in rural areas.

PREDICTIVE/INTERACTIVE CONTROL - Further elements of the ATIS concept are predictive traffic models and interactive traffic control. A highly sophisticated model will, when an incident occurs, predict the new traffic patterns and change signal timing to reflect these new patterns. Outputs from the model will also be used to provide predicted link times to in-vehicle devices.

The congestion reduction benefits of ATIS, very much, relate to the potential of this predictive/interactive control. Traffic will be distributed, with ATIS, throughout the traffic network and, when major incidents occur, even to other modes. Thus a balanced flow will be achieved throughout the transportation network.

IN-VEHICLE WARNING SYSTEMS - Work is underway to provide roadway hazard warnings to in-vehicle devices via permanent or portable devices deployed along the roadway (such as ramps, accidents, etc.). Warnings, such as those for traffic stoppages and flooded roadways, will also be transmitted from the traffic management centers. Data transmitted from the roadside devices and from the centers to the in-vehicle devices will be used to generate voice and/or graphics displays. The technical issues to provide these displays is relatively straight forward; however, there are obvious human factor issues that must also be addressed.

REDUCED FUNCTIONS IN-VEHICLE DEVICES - There will not, of course, be a 100 percent market penetration of the full scale in-vehicle ATIS devices that have been described above. Reduced functions, lower cost systems will be available that will provide traffic information to motorists but will not include the navigation and business directory functions provided by the full scale in-vehicle devices. Drivers will select the areas and corridors for which they wish to receive traffic information. The reduced functions systems will use the same traffic information data as that transmitted to the full scale in-vehicle devices.

COMMERCIAL FLEET MANAGEMENT - Commercial fleet management is more an element of Commercial Vehicle Operations (CVO) than of ATIS. However in-vehicle navigation systems must address the special needs of commercial vehicles. The link times and other traffic information will be transmitted to dispatch centers that then will use their own software and communications system for management of pick-up and deliveries and just-in-time deliveries.

PUBLIC TRANSPORTATION - Transit fleet management systems typically include: an on-board location system and/or beacons; monitoring of on-board status relative to the vehicle, passenger loading, etc.; alarms generated by the drivers; and radio communications between the transit vehicle and a control center. Benefits include on time performance; security of driver and passenger; improved system planning data; and improved coordination at major transfer points of bus/bus and bus/rail.

The costs to deploy transit fleet management systems will be significantly reduced by integrating these systems with the infrastructure deployed for ATIS systems. The availability of real time traffic data and information at the dispatch center will also contribute to optimal transit fleet management especially in the case of major incidents or adverse conditions. Linkage with ATIS will allow transit passengers to be provided with the arrival time of the next bus at localities such as major bus stops, office building lobbies, and residences.

ATIS will also include friendly subsystems that afford matching of paratransit vehicles and passengers thus providing improved passenger convenience and reduced transit costs in low density areas and during off-peak travel times. The operation, as envisioned, would be similar to the operation of pick-up and deliveries in courier management service systems - perhaps it should be termed passenger express.

PRE-TRIP INFORMATION - Information about incidents and delays is extremely useful to travelers before their trip begins. They, at that time, have more options, such as delaying their trip, selecting a different mode, or cancelling the trip. Thus the same link time, incident, alternative routing, and traffic information provided to in-vehicle devices will also be provided to various computer terminals and processed to provide information, tailored to meet the needs of individual travelers.

ADVANCED TRAVELER INFORMATION SYSTEMS 471

Pre-trip travelers information will also be available at kiosks. These are actually computer terminals with simplified, friendly interfaces and displays that will be located in such places as motel lobbies, shopping centers, and rental car stations. They will provide traveler information, best street or transit route to destination, and otherwise emulate many of the functions of the in-vehicle navigation devices.

CLASSES OF SYSTEMS - It can be assumed that an ATIS systems infrastructure will not be deployed in some areas for a number of years. Thus using the autonomous system model, it can be assumed that three classes of systems will co-exist in the U.S. These classifications approximate certain of those provide in Reference [1].

Class 0 - Autonomous in-vehicle subsystem only.

Class 1 - Beacons not provided, vehicles not used as probes. Traffic information and link times obtained from various sources and derived (where available) from signal system and freeway management system sensors. This data is then broadcast by data radio to the vehicles.

Class 4 - Full scale system with beacons, probe vehicles, and predictive/interactive control.

Class 1 and Class 4 systems will also provide for communications of traffic information to reduced functions systems. All three classes of systems will be useful for travelers and generate societal benefits; however, the utility and societal benefits will be greater with the higher class systems.

AN ALTERNATIVE APPROACH - The Siemens Ali-Scout system being used in the LISB test in Berlin offers an alternative model to the autonomous TravTek type device.

The detailed map data base, in the Ali-Scout system, is resident in the traffic management center and the processing for the best routes is accomplished at the traffic management centers. This provides economies of cost and space in the in-vehicle system in comparison with autonomous in-vehicle systems.

The disadvantage is that the device will only be useful in those areas where the beacons and the communications between the beacons and the traffic management centers have been installed to provide Class 4 systems. Communications with the vehicles, in the Ali-Scout system is via roadside beacons using an infrared medium between the vehicles and beacons.

RURAL ATIS - Urban ATIS systems have primarily been addressed to this point. However rural corridors are also a significant element of ATIS. The same in-vehicle full scale and reduced functions subsystems that are used in urban areas will also be useful in rural areas. It is likely that many rural corridors will be operated as Class 0 or Class 1 systems with Class 4 systems deployed only in the heavier travelled corridors.

Functions that will be useful to drivers in rural areas include in-vehicle warnings; incident, traffic and road condition information; route guidance for best routes; and in-vehicle information about points of interest. Motorist services information will also be highly useful and can be provided in real time where beacons are deployed. Thus the driver could be directed to the next motel that was available and met his/her price range, credit card and amenities requirements. Two-way communication, Class 4 systems, would also provide a "mayday" feature, permitting emergency calls from disabled motorists.

ATIS will be especially useful for those routes frequently subject to adverse weather conditions such as fog, dust storms, blizzards and flooding. Drivers will be warned sufficiently in advance to take advantage of alternate routes, if available, or to take a delay at a convenient motel, truck stop, town, etc.

Class 4 rural ATIS systems will emulate Class 4 urban ATIS systems and use probe data to derive link times. Road weather systems developed for maintenance management purposes in many states will also be integrated with Class 1 and Class 4 systems. Class 1 and Class 4 systems will require mobilization of existing resources, such as police and maintenance patrols, to provide information on exceptional conditions. This information will be transmitted to traffic management centers that are actually work stations. The centers will in turn estimate link times and transmit the information as data to vehicles equipped with in-vehicle systems.

REFERENCES

1. Road Transport Research Programme, Organization for Economic Co-operation and Development, "Driver Information and Guidance Systems Using In-car Communications", October 1987.

2. Federal Highway Administration, Draft - "A Program for the Advancement of Intelligent Vehicle Highway Systems (IVHS)," April 1989.

3. "Proceedings of a National Workshop on IVHS Sponsored by Mobility 2000," March 19 - 21, 1990.

Flexible Pavement Distress Evaluation Using Image Analysis

Lan Li, Paul Chan, Ashok Rao, and Robert L. Lytton[1]

Abstract

An unique combination of image histogram and projection histogram are employed to classify major distress types on flexible pavement. This technique has taken into account the presence of non-distress objects which include road markings, oil stains, and tire marks on the pavement surfaces. It has successfully separated the non-distress objects from crackings hence increasing the percent accuracy of the evaluation algorithm.

Introduction

Pavement evaluation data are critical to all pavement management and rehabilitation activities. Promptly collected and analyzed pavement distress data can assist a pavement engineer to make plans and decisions. Distress data is usually collected either by drive over or visual inspections. With the advancement of video technology, the current distress survey employs photo-logging where pavement surfaces are filmed on 35mm film, video tape or video disc. This improvement allows the inspection of pavement surfaces to be performed indoors and saves time and money and improves safety over the option of sending personnel out in the field. The next logical step is the automatic analysis of the collected distress data via image processing and pattern recognition techniques.

Pavement distress analysis has presented a challenge to researchers due to the size, shape, and variations of each distress type, as well as the variation of the texture and color of the intact pavement surface. This paper describes an innovative algorithm to identify major flexible pavement surface distress types. It also considers common scenes of road surfaces which include road markings, oil stains and tire marks with attempts to isolate them.

---

[1] Texas A&M University, Texas Transportation Institute, College Station, Texas   77843-3135

Image Classification Algorithm

A. Image Histogram

The most common road surface pictures can be divided into eight categories, namely: (1) sealed cracks, (2) road markings, (3) oil stains, (4) tire marks, (5) intact pavement, (6) longitudinal cracking, (7) transverse cracking, and (8) alligator cracking.

This categorization originates from viewing video tapes recorded by the Automatic Road Analyzer (ARAN). First, an image histogram is formed by tabulating the total number of pixels at each greylevel (Mahler, 1985). This algorithm will then test if the image histogram is bimodal by locating two major peaks. The test will divide the eight categories into two groups as shown in Figure 1. The first group will be the sealed cracks, road markings, oil stains and tire marks. The second group includes the intact pavement, longitudinal cracking, transverse cracking and alligator cracking which have histograms that are unimodal. The main difference between these two groups is that the first group generally has a large object of different greyscale while the second group has a relatively small object compared to the background pavement. The presence of a relatively large object will change the image histogram to a bimodal in the first group. The second step involves the isolation of the sealed cracks for quantification. This process can be achieved by finding the relative position of the smaller peak with respect to the larger peak and the amplitude of the valley between peaks. The smaller peak represents the sealed crack while the larger peak represents the background pavement (See Figure 2).

B. Projection Histogram

For the second group that have a unimodal image histogram which includes the intact pavement, longitudinal, transverse and alligator cracking (See Figure 3). A moving window technique along with projection histogram is used to process the image (Toshihiko, 1989). The window size (32x32) is chosen based on the relative size of the object (crack) to the window area. Within each window, projection histogram in four major directions is tabulated (0°, 45°, 90°, 135°). A projection in the vertical direction will sum all the greylevel of each pixel within the same column for all columns.

$$H(i) = \sum_j f(i,j) \qquad (1)$$

Figure 4 illustrates the four projection histograms of the window at the end of a longitudinal crack. The projection histogram will indicate the orientation and position of the crack segment within the window. A parameter called Shape Factor is used to describe the relative size of the peak as well as how "narrow" this peak is in the projection histogram.

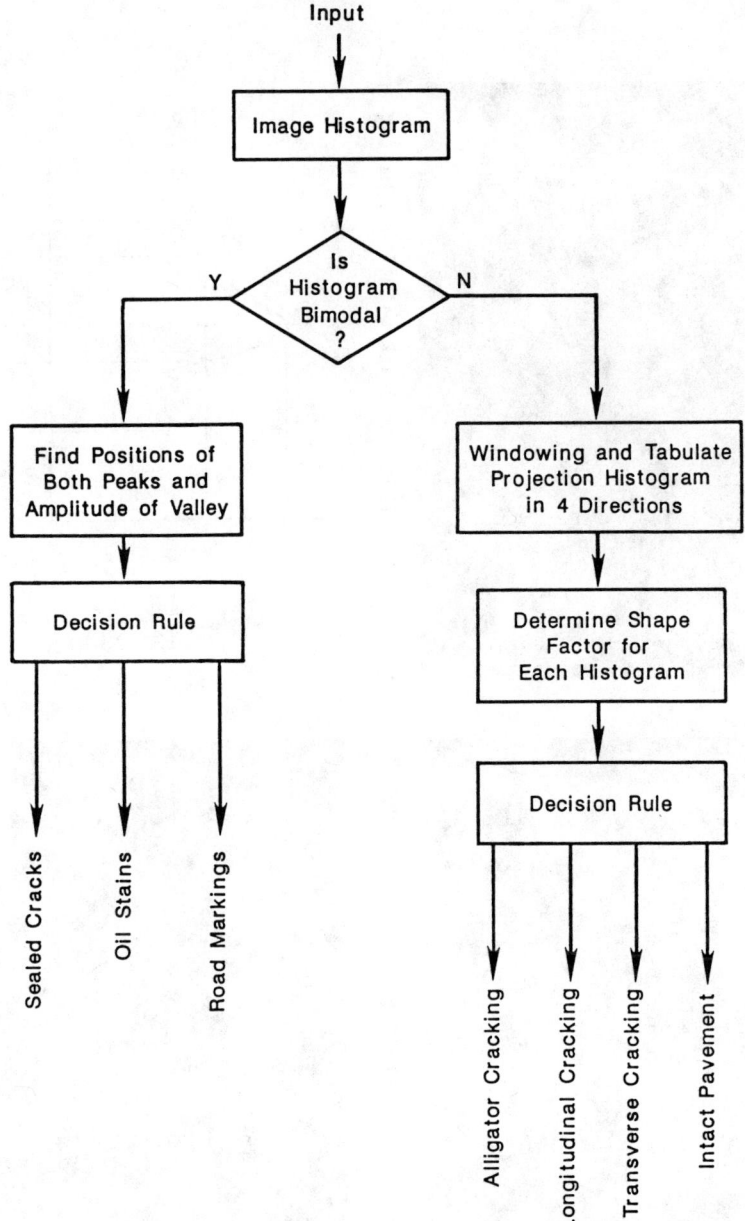

Fig. 1. Block Diagram for the classification Algorithm of Flexible Pavement Distress.

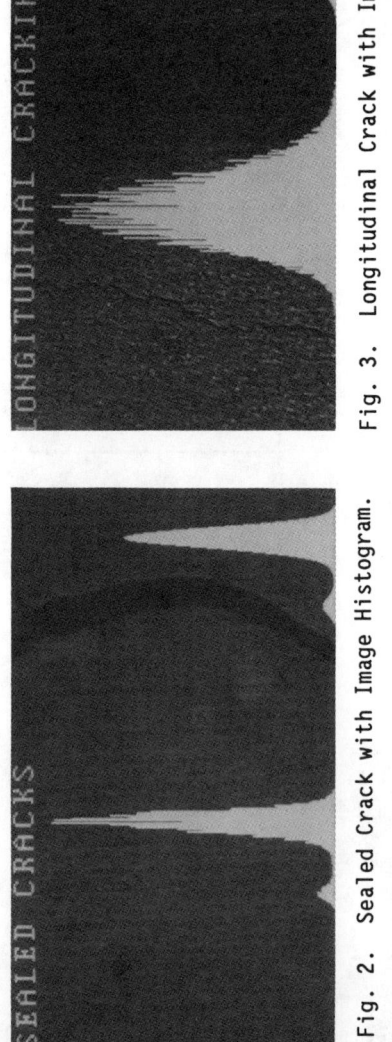

Fig. 2. Sealed Crack with Image Histogram.

Fig. 3. Longitudinal Crack with Image Histogram.

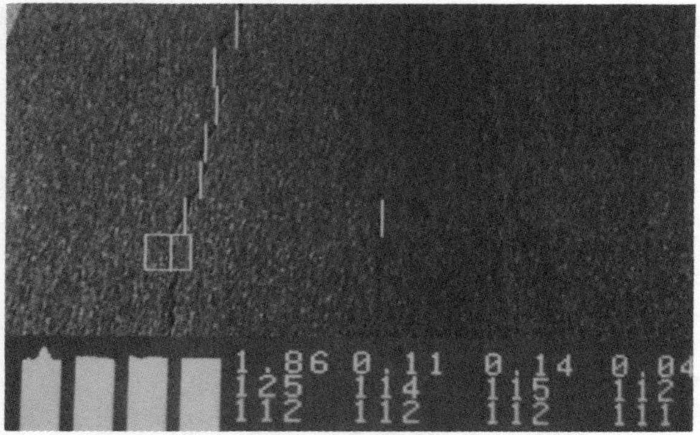

Fig. 4. Long. Crack with Projection Histogram

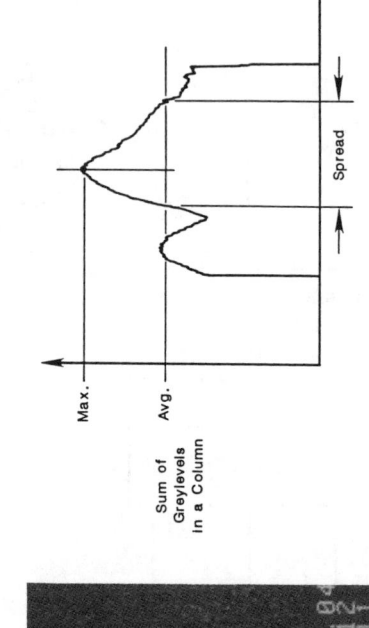

Fig. 5. Shape Factor of Projection Histogram.

$$\text{Shape Factor} = \frac{\text{Max} - \text{Avg}}{\text{Spread}} \qquad (2)$$

The maximum Shape Factor of the four projection histograms is chosen and compared against a set threshold level. This threshold is to ensure the peak in the projection histogram is caused by an edge segment rather than noise effect. A classification decision is then made according to the number of "edge" windows in each projection angle.

Results and Discussion

A small roadway section of about 200 ft (180 m) which contains the major distress types is selected to be processed with this technique. The result from this pilot test indicates an accuracy of 85%. The error includes all images that are misclassified due to intersecting cracks, shadows and paver joints. The next step will be a survey of multiple 1 ~ 2 mile (1.6 ~ 3.2 km) sections to further field test the accuracy and reliability of this algorithm.

References

Mahler, D. S. (1985), "Final Design of Automated Pavement Crack Measurement Instrumentation from a Survey Vehicle." FHWA/RD-85/077, Federal Highway Administration, Washington, D. C.

Toshihiko, F., et al. (1989), "Automatic Pavement Distress Survey System". Proc., 1st AATT Conference, San Diego, California, pp. 33-38.

SUBJECT INDEX
Page number refers to first page of paper.

Aerial surveys, 248
Algorithms, 71, 252, 378, 408
Approximation methods, 121
Architecture, 419, 428
Arterial highways, 248, 393
Artificial intelligence, 126, 147, 188, 213, 223, 252, 297, 307, 312, 363
Australia, 81
Automation, 11, 16, 56, 61, 66, 76, 106, 136, 147, 198, 218, 233, 238, 307, 383, 388, 414, 423, 428, 433, 438
Automobiles, 223, 363, 463

Barriers, 223
Buses, 263

California, 96, 322
Cameras, 106, 111, 131, 142, 198, 453
Capacity, 46
China, People's Republic of, 11, 252
Classification, 11, 147
Communication systems, 162, 167, 273, 278, 322, 332, 363, 414
Comparative studies, 81
Computer aided drafting (CAD), 228, 293
Computer applications, 131, 252, 283, 414, 419
Computer graphics, 293
Computer hardware, 268, 368, 423, 428
Computer models, 393
Computer programming, 423
Computer software, 111, 203, 243, 268, 368, 453
Computerized control systems, 419
Computerized design, 228, 293
Computerized simulation, 203, 258
Construction, 193
Control systems, 51, 86, 91, 268, 288
Control theory, 177
Coordination, 81
Correlation techniques, 1
Cost savings, 162
Costs, 193
Cracking, 56, 61
Cracks, 66, 76

Data collection, 6, 26, 46, 198, 228, 312
Data collection systems, 1
Data processing, 21
Data processing techniques, 16
Databases, 188
Decision making, 307
Design, 26, 182, 248, 268, 302
Design criteria, 76
Digital systems, 332
Distributed processing, 86, 388
Dynamic analysis, 358
Dynamic models, 182, 403, 409
Dynamics, 36

Efficiency, 327
Electronic equipment, 116
Emergency services, 157
Equipment, 21
Europe, 273
Evaluation, 238, 473
Expert systems, 41, 51, 136, 258

Fares, 414
Fibers, 278
Field tests, 91, 131
Flexible pavements, 473
Florida, 16
France, 142, 383
Fuzzy sets, 41, 433

Gates, 414
Global positioning, 468
Guideways, 193

Hazardous materials, 152
Highway accident potential, 157, 373, 378
Highway construction, 56
Highway maintenance, 56, 61, 66, 71, 76
Highway management, 126, 157, 172
Highway transportation, 46, 96, 116, 147, 177, 183, 188, 213, 223, 278, 288, 297, 302, 307, 312, 317, 322, 343, 353, 363, 383, 408, 433, 438, 458, 463
Highways, 11, 16, 218, 243, 448
Histograms, 473

Identification, 16, 136, 147

Image analysis, 96, 101, 106, 111, 233, 238, 353, 383, 388, 448, 453, 473
Implementation, 223
Information management, 188
Information systems, 26, 31, 182, 312, 327, 343, 368, 409, 443, 468
Infrared detectors, 317
Infrared scanning, 11
Infrastructure, 126
Innovation, 21
Inspection, 228, 238, 453
Instrumentation, 6
Interactive graphics, 293, 368
Interactive systems, 26
Interconnected systems, 81
Interferometry, 71
Intersections, 258
Intrusion, 142
Inventories, 448

Japan, 288

Kalman filter, 398
Knowledge-based systems, 208, 213, 388

Licenses, 147
Linear programming, 172
Location, 152

Machinery, 76
Magnetic levitation trains, 193
Maintenance, 193, 208
Management systems, 233
Measurement, 448
Minnesota, 458
Model tests, 46
Model verification, 46
Modeling, 36, 398, 443
Models, 162, 218, 238, 373
Monitoring, 1, 111, 136, 162

Navigation, 358, 463, 468
Netherlands, 343
Network design, 121
Network reliability, 121
North Dakota, 81

Operation, 41, 353
Optical properties, 278
Optimization, 51, 243, 419
Optimization models, 208

Passenger vehicles, 263
Pavement deterioration, 56, 233, 473

Pavement management systems, 238, 448, 453
Pavement markings, 56
Pavement overlays, 71
Pavements, 61, 66, 76, 463
Planning, 167, 208
Portugal, 31
Predictions, 398, 403
Priorities, 263
Problem solving, 41
Prototypes, 383
Public participation, 126
Public transportation, 468

Queueing, 106

Radar, 453
Rail transportation, 198, 203, 208, 283, 414, 423, 428
Railroad trains, 203
Railroads, 419
Rails, 198, 208
Ramps, 172, 177
Reconstruction, 71
Reliability analysis, 121
Renovation, 208
Research, 41, 116
Risk analysis, 307
Risk management, 152
Roads, 121, 448
Robotics, 56, 61
Route preferences, 468
Routing, 152, 167, 182, 183, 327, 332, 337, 343, 358, 393, 438

Satellite communications, 167
Scheduling, 167, 203
Sealing, 66, 76
Sensors, 66, 142, 317, 443
Signal processing, 101, 438, 443
Simulation, 36, 131, 157, 182, 433
Simulation models, 46, 86, 91, 183, 409
Specifications, 423
Speed, 1
Speed control, 51
Stochastic models, 403
Stochastic processes, 408
Subway tunnels, 142
Subways, 423
Supervision, 203
Surface defects, 198, 453
Surveys, 193

Technology assessment, 302, 322
Testing, 6, 36

Three-dimensional analysis, 142
Tolls, 11, 353
Traffic accident analysis, 297
Traffic accidents, 378
Traffic capacity, 213, 218, 243
Traffic characteristics, 252
Traffic congestion, 6, 11, 36, 51, 106, 116, 157, 172, 177, 218, 283, 393, 408, 409, 458
Traffic control, 1, 6, 21, 46, 51, 101, 121, 126, 136, 147, 152, 183, 188, 203, 223, 248, 252, 263, 268, 273, 288, 327, 337, 343, 353, 363, 373, 378, 393, 403, 408, 433, 463
Traffic control devices, 26, 31, 86, 91, 96, 116, 172, 213, 302, 332, 438
Traffic flow, 36, 157, 172, 177, 183, 203, 248, 368, 373, 398, 403, 408, 409
Traffic management, 1, 36, 101, 116, 121, 167, 172, 182, 228, 258, 268, 273, 278, 327, 332, 337, 358, 368, 373, 398, 458
Traffic safety, 142, 152, 223, 297, 302, 307, 312, 423, 428, 463
Traffic signal controllers, 86, 243, 258, 337
Traffic signals, 81, 111, 248, 263, 283, 317, 428
Traffic speed, 111
Traffic surveillance, 106, 111, 116, 131, 136, 142, 152, 162, 297, 363, 378, 383, 388, 408, 458, 463
Transportation, 41
Transportation systems, 293
Travel patterns, 468
Travel time, 183
Trucks, 162
Tunnels, 228

Urban areas, 6
Urban transportation, 6, 31, 101, 136, 177, 182, 183, 218, 228, 233, 248, 263, 273, 283, 317, 388, 443, 468

Vehicles, 11, 16
Vehicular traffic, 6, 21, 26, 91, 101, 111, 126, 131, 147, 152, 162, 167, 188, 213, 223, 228, 273, 288, 297, 302, 307, 312, 317, 322, 327, 337, 343, 358, 373, 383, 403, 409, 433, 438, 468
Videotape, 96, 101, 353
Visual aids, 198, 233, 378

AUTHOR INDEX
Page number refers to first page of paper.

Alfelor, Roemer, 198
Assmann, Susan F., 248
Aswegan, James, 46
Aultman-Hall, Lisa, 373

Ball, William L., 16
Barceló, J., 91
Barfield, Woodrow, 26
Bell, M. G. H., 106
Benedito, S., 91
Berbineau, M., 273
Beskos, Dimitrios E., 36
Bhouri, Neila, 177
Blosseville, J. M., 383
Blosseville, Jean Marc, 353
Bordenave, Henri, 419
Boulmakoul, A., 388
Bruyelle, J. -L., 131
Bullock, Darcy, 66, 268

Caird, J., 312
Cassidy, Michael J., 116
Catling, Ian, 358
Chan, Paul, 473
Chang, Kai-Kuo, 172
Chassiakos, Athanasios P., 378
Chatusripitak, Sompol, 438
Chatziioanou, Alypios E., 96
Chaudhry, Bharat B., 263
Clark, M. A. G., 317
Coussement, Marc, 111
Coyle, M., 312
Cremer, M., 21

Dauchy, P., 423
Daviet, Bruno, 353
Davis, Gary A., 408
Deparis, J. -P., 142
Djemame, N., 383
Donskoy, Boris, 6
Downey, Allen B., 233
Duvieubourg, L., 142

Egea, P., 91
Elahi, S. Manzur, 258

Fehon, Kevin, 283
Fehon, Kevin J., 81
Frank, Andrew, 433

Garner, Margaret, 26
Gartner, Nathan H., 248
Goble, Brian, 26

Goto, Koichi, 414
Goul, K. Michael, 258
Grau, R., 91
Grove, Darrin, 66
Gupta, Ashok K., 293
Guralnick, Sidney A., 71

Haas, Carl, 66
Hakala, Hannu, 443
Hall, Fred L., 373
Hallowell, Susan F., 162
Hancock, P., 312
Hancock, P. A., 188
Harris, Richard, 358
Hasegawa, Toshiharu, 288
Haselkorn, Mark, 26
Hautamaki, Jerry L., 368
He, Guoguang, 252
Heddebaut, M., 273
Heller, Mahlon, 438
Hendrickson, Chris, 66, 268
Herman, Robert, 409
Heywood, H. Milton, 468
Hicks, Gene, 46
Hitchcock, Anthony, 297
Hobeika, A. G., 393
Hockaday, Stephen L. M., 96
Hodge, A., 317
Hou, Dennis L., 248

Iida, Yasunori, 121
Inagaki, M., 453

Jayakrishnan, R., 182
Johns, Robert C., 463
Johnson, S., 312
Johnston, Robert A., 218
Jovanis, Paul P., 307

Kaikkonen, Jari, 443
Kasahara, A., 453
Kaseko, Mohamed S., 238
Katakura, Masahiko, 363
Kawakami, Shogo, 403
Keller, John C., 307
Kenneally, Kent, 66
Khasnabis, Snehamay, 263
Kikuchi, Shinya, 41
Kirschke, Kenneth R., 76
Klijnhout, Job J., 343
Kos, Leon, 203
Koursi, M. El, 428
Koutsopoulos, Haris N., 233

AUTHOR INDEX

Kretschmer, Werner, 203
Kuah, Geok K., 11
Kühne, Reinhart D., 51
Kwon, Eil, 398

Lasaga, Fernando, 248
Lee, Hosin, 448
Lemaire, Frans, 111
Lenoir, F., 383
Leong, Siew, 86
Li, Lan, 473
Lipičnik, Martin, 136
List, George, 86
Liu, Bao, 252
Lu, Baichuan, 252
Liu, Sijiu, 433
Lyall, Bradley, 373
Lyrintzis, Anastasios S., 36
Lytton, Robert L., 473

Macedo, António Lemonde de, 31
Mahmassani, Hani S., 182, 409
Martland, Carl D., 208
Mason, John M., Jr., 223
Matsubara, Hiroshi, 414
Matsuo, Takeshi, 288
Mattila, Pentti, 443
McCarley, Carl A., 96
McGregor, T., 453
McNeil, Sue, 66, 198
McQueen, Bob, 358
Meir, S., 312
Michalopoulos, Panos G., 36, 101
Mierzejewski, Edward A., 16
Miki, Shigeo, 414
Mishalani, Rabi G., 208, 233
Morin, Jean Marc, 353
Morlok, Edward K., 162
Motazed, Behnam, 198
Motyka, V., 383
Motyka, Vincent, 353

Nagai, Noboru, 414
Nathanail, Teti, 157
Nodder, Ronald, 322

O'Hara, Katharine S., 368
Ozello, P., 423

Page, Dorriah L., 218
Pan, Wen-Min, 11
Papageorgiou, Markos, 177
Patten, Michael L., 223
Paulose, Saaju, 86
Peeta, Srinivas, 409
Peters, Daniel, 66

Pfannerstill, Elmar, 1
Phelan, R. Scott, 193
Pierce-Spring, Rosalind, 228
Pierrelée, J. C., 388
Pietrzyk, Michael C., 16
Pollak, Alex J., 228
Postaire, J. -G., 131, 142
Powell, Joseph P., 448
Prasad, Ravi, 293
Price, Mark, 6

Radwan, A. Essam, 258
Rao, Ashok, 473
Ravani, B., 61
Rebolj, Danijel, 136
Rilett, L., 183
Ritchie, Stephen G., 238
Rothery, Richard, 409
Rourke, A., 106
Rowe, S. Edwin, 126

Saito, Mitsuru, 363
Schrijver, P. R., 167
Sellam, S., 388
Shekhar, S., 188, 312
Shi, Yong, 373
Shladover, Steven E., 213
Short, Thomas, 198
Sinha, Kumares C., 116
Smith, Christian, 71
Sol, H. G., 167
Sparmann, Jürg M., 327
Spasovic, Lazar N., 162
Spyridakis, Jan, 26
Staley, Clint, 322
Stehr, Richard A., 458
Stephanedes, Yorgos J., 172, 378
Stewart, J. Allen, 243
Stuparu, A., 428
Su, Shi-Lin, 11
Suen, Eric S., 71
Sullivan, Edward C., 96
Sussman, Joseph M., 193
Szelag, M., 273

Tada, H., 453
Taff, Sam, 322
Tollazzi, Tomaž, 136
Tomita, H., 453
Trahms, Robert, 438
Tritter, D. Bowen, 278
Turner, John D., 337
Turnquist, Mark A., 152

Van Aerde, M., 183
Van Aerde, Michel, 243

Velinsky, Steven A., 76

Wakabayashi, Hiroshi, 121
Walton, C. Michael, 147
Wang, An-Sheng, 11
Weissman, Michael A., 448
West, T. H., 61
West, Thomas, 56
Wichman, Shannon, 66
Winter, Walt, 322
Wright, James L., 46

Xu, Zhimin, 403

Yang, A., 312
Yang, T. A., 188
Yi, Ping, 36, 46
Yoshino, Hideaki, 414
Yoshino, Tsuyoshi, 288

Zechnall, Wolf, 332
Zhang, Wei-Bin, 302
Zhang, Y., 393
Zhou, Tong, 56
Zografos, Kostas G., 157